State Versus Ethnic Claims:
African Policy Dilemmas

Also of Interest

Women and Work in Africa, edited by Edna Bay

**Alternative Futures for Africa,* edited by Timothy M. Shaw

Nigeria in Search of a Stable Civil-Military System, J. 'Bayo Adekson

Interdependence in a World of Unequals: African-Arab-OECD Cooperation for Development, edited by Dunstan M. Wai

Women in Rural Development: A Survey of the Roles of Women in Ghana, Lesotho, Kenya, Nigeria, Bolivia, Paraguay and Peru, Donald R. Mickelwait, Mary Ann Riegelman, and Charles F. Sweet

An African Experiment in Nation Building: The Bilingual Cameroon Republic Since Reunification, edited by Ndiva Kofele-Kale

African Upheavals Since Independence, Grace Stuart Ibingira

Analyzing Political Change in Africa: Applications for a Multidimensional Framework, edited by James R. Scarritt

Ethnicity in Modern Africa, edited by Brian M. du Toit

***Africa's International Relations: The Diplomacy of Dependency and Change,* Ali A. Mazrui

An Anatomy of Ghanaian Politics: Managing Political Recession, 1969-1982, Naomi Chazan

U.S. Economic Power and Influence in Namibia, 1700-1982, Allan D. Cooper

The Economics of Political Instability: The Nigerian-Biafran War, E. Wayne Nafziger

Senegal: An African Nation Between Islam and the West, Sheldon Gellar

Tanzania: An African Experiment, Rodger Yeager

*Available in hardcover and paperback.
**Available in paperback only.

Contents

Foreword ix

Part I
Introduction

1. Managing Competing State and Ethnic Claims
Donald Rothchild and Victor A. Olorunsola 1

2. Francophone Nations and English-Speaking States: Imperial Ethnicity and African Political Formations
Ali A. Mazrui 25

Part II
State Claims and Their Implications for Ethnic Relations

3. The State and Society in Africa: Ethnic Stratification and Restratification in Historical and Comparative Perspective
René Lemarchand 44

4. Problems and Prospects of State Coherence
Goran Hyden 67

5. Ethnicity Versus the State: The Dual Claims of State Coherence and Ethnic Self-Determination
John Stone 85

6. The State and Ethnicity: Integrative Formulas in Africa
Henry Bienen 100

7. The Manipulation of Ethnicity: South Africa in Comparative Perspective
Heribert Adam 127

8. A Response to Heribert Adam, and a Rebuttal
Pierre L. van den Berghe and Heribert Adam 148

Part III
The Claims of Ethnic Groups

9. Modernization, Ethnic Competition, and the Rationality of Politics in Contemporary Africa
Robert H. Bates 152

10. Collective Demands for Improved Distributions
Donald Rothchild 172

11. Comparative Claims to Political Sovereignty: Biafra, Katanga, Eritrea
Crawford Young 199

Part IV
Policies to Manage Competing Claims

12. African Public Policies on Ethnic Autonomy and State Control
Donald Rothchild and Victor A. Olorunsola 233

13. The State, Public Policy and the Mediation of Ethnic Conflict in Africa
Edmond J. Keller 251

Part V
The Dialectic of Autonomy and Control

14. Federalism and Politics of Compromise
J. Isawa Elaigwu and Victor A. Olorunsola 281

15. Geoethnicity and the Margin of Autonomy in the Sudan
Dunstan M. Wai 304

16. The Ogaadeen Question and Changes in Somali Identity
David Laitin 331

Index 350
List of Contributors 354

State Versus Ethnic Claims: African Policy Dilemmas

edited by Donald Rothchild and
Victor A. Olorunsola

Westview Press / Boulder, Colorado

Westview Special Studies on Africa

Published in 1983 in the United States of America by
Westview Press, Inc.
5500 Central Avenue
Boulder, Colorado 80301
Frederick A. Praeger, President and Publisher

Library of Congress Catalog Card Number 82-51129
ISBN 0-86531-503-5 (hc)
0-86531-504-3 (pb)

Composition for this book was provided by the editors
Printed and bound in the United States of America

Westview Special Studies on Africa

State Versus Ethnic Claims: African Policy Dilemmas

edited by Donald Rothchild and Victor A. Olorunsola

How can today's independent African states preserve and enhance their coherence while accommodating the legitimate demands advanced by ethnic groups for autonomy and particular interests? This problem is critically important not only in Africa, but also in efforts to establish mutually beneficial state-ethnic relations the world over. To analyze these issues, leading African and non-African experts met in Bellagio, Italy, to address four central issues: the state and its implications for ethnicity; the claims of ethnic groups; policies to manage competing claims; and the dialectic of autonomy and control. This book, in which policy analysis is applied to the search for creative state-ethnic relationships, represents the results of their efforts.

Dr. Donald Rothchild is professor of political science, University of California at Davis. He has been visiting Fulbright Lecturer in Political Science at Makerere University, Kampala, Uganda; senior lecturer in government at the University of Nairobi, Kenya; and visiting Ford Professor of Political Science at the University of Zambia. He is author of *Politics of Integration: An East African Documentary* (1968) and coauthor of *Scarcity, Choice and Public Policy in Middle Africa* (1978).

After a long academic career that has included faculty posts at various universities, Dr. Victor Olorunsola is now chairman and professor in the Department of Political Science at Iowa State University. He has edited and contributed to *The Politics of Cultural Sub-Nationalism in Africa* (1972) and written *Societal Reconstruction in Two African States* (1977).

To
Alice L. Rothchild
and
Katherine Bradley

Foreword

The idea for this book can be traced to an informal brainstorming session among four very good friends -- Ali Mazrui, Victor Olorunsola, Donald Rothchild and Dunstan Wai. In a real sense, then, as editors we owe a lot to Dunstan and Ali for their intellectual stimulation and for encouraging us to pursue a follow-up to The Politics of Cultural Sub-Nationalism in Africa.

African states have had to deal with the conflicts and tensions generated by the juxtaposition, or, if you like, the interplay, between the state's claims for control over society on the one hand and the desires of various geopolitical units for autonomy and the attainment of their particular objectives on the other. Conflict and cooperation are necessary and unavoidable in societal interactions. It is equally evident that in the quest for stability, conflict and cooperation must be judiciously balanced if the interactions between states and subnational groupings are to be mutually rewarding. How can this be done? What kind of policy options are available? This volume addresses four dimensions of this complex issue we feel are critical: (1) the state and its implications for ethnicity; (2) the claims of ethnic groups; (3) the policies to manage competing claims; and (4) the dialectic of autonomy and control. Should this volume contribute in any way to African policy-makers' search for realizable ends in this area, the authors will be intensely gratified.

The Rockefeller Foundation granted us the use of the Bellagio Conference Center; the Ford Foundation, an anonymous donor, and the University of California at Davis provided financial assistance for the conference. Iowa State University contributed support services. Clearly, without the generosity of these institutions and agencies, the Bellagio Conference would not have taken place. In particular, we wish to thank the director of the Bellagio center, Roberto Celli, and his staff; Dr. Bill Carmichael of the Ford Foundation; and Dr. John Stremlau and Susan E. Garfield of the Rockefeller Foundation. The Bellagio Conference itself was exceedingly stimulating and productive, not only because of the intellectual caliber of the participants, but also because of the collegial atmosphere and attitude engendered.

Verla Schoelkoph, Lois Persinger, Virginia Lee, Cheryl Lyttle, Vickie Zinner, Brenda Peterson, Donna Hughes-Oldenburg and Michael Foley have been most supportive, efficient and tolerant in their handling of administrative and editorial matters. Robin Campbell did exemplary work in editing and typing the final manuscript. Of course, none of the individuals or organizations mentioned should be held responsible for any of the opinions expressed in this volume.

We cannot adequately express our gratitude to Carol Olorunsola and Edith Rothchild for being silent but exceedingly active partners in this venture. They have our boundless gratitude.

Victor A. Olorunsola
Donald Rothchild

State Versus Ethnic Claims: African Policy Dilemmas

1
Managing Competing State and Ethnic Claims

Donald Rothchild
Victor A. Olorunsola

In their search for creative relationships between groupings, all societies require a balance between conflict and stability. Where conflict between the state and subnational grouping is openly expressed and becomes unbearably intense, interaction proves difficult and possibly destructive -- leading, in extreme cases, to a breakdown in the political system itself. Where conflict is held in check by an overbearing, hegemonial actor, interaction is largely stifled and the cosmetics of normality mask deep cleavages in the society. For "constructive conflict" (i.e., growth-producing interactions) to prevail, intercommunity relations must be regulated by recognized rules of exchange. Whenever these conditions are met, a balance may be achieved between the dual claims of state coherence (i.e., regularized patterns of interaction) and ethnic autonomy.

In brief, how can today's independent African states preserve and enhance their coherence while at the same time accommodating the legitimate demands advanced on the part of ethnic groups[1] for autonomy and particular interests? Can the state's "support" demands be reconciled with the subnational group's change demands, especially under conditions of evident scarcity and institutional fragility? Two categories of self-realization, those of the state and the subnational group, are locked in unavoidable encounter, and new ways must be found to adjust these legitimate and competing claims if destructive clashes are to be avoided. As Walter Lippmann observed, "To traverse the world men must have maps of the world."[2]

This is the central problem: as various contending African actors -- states and subnational groupings -- assert valid but conflicting principles, how can public policies be realized which promote a sense of mutual interests? In probing the dimensions of political choice open to these decision elites, social scientists may be in a position to offer useful insights on the management of conflict by setting out the terms and costs of interaction. Not only does this involve an analysis of the likely qualitative costs and benefits of society's adoption of specific policies and strategies on conflict regulation, it also includes a delineation of the process to be pursued in the implementation of the constructive (i.e., morally self-realizing) goals of the African leaders themselves. Although the responsibility for choice clearly rests with African decision-makers, their search for realizable ends can be facilitated by the policy analyst's examination of the issues at hand.

In an effort to deal with the many subtle permutations of this central problem, the participants in this volume have focused on four central issues: the state and its implications for ethnicity; the claims of ethnic groups; the policies to manage competing claims; and the dialectic of autonomy and control. After first describing the elements of national identity and statehood in Anglophone and Francophone Africa (see Ali Mazrui's essay immediately following), the contributors to this volume have juxtaposed the competing claims of the formal political system and of ethnic groups in African societies. Once the contradictions across objectives have been made evident, it will then be possible, most tentatively, to begin to probe the policy process as it relates to African state-ethnic relations. It is our hope that this cooperative effort will come to be seen as a conscious beginning, stimulating systematic comparative policy studies on a much wider range of appropriate choices for coping with the apparently contradictory claims of state and ethnic groups in contemporary Africa.

State Claims and Their Implications for Ethnic Relations

Before discussing the claims of the state on society, it is necessary to take note of the divergent approaches adopted by contemporary analysts concerning the nature of the modern state. Leaving aside the view of the state as the processor of rival demands, we see two main lines of thought regarding the character of the African state. The first tends to look upon the _state as manager_. As the political organization of society (i.e., the framework of institutional relationships, body of legal relations, customs, and conventions), the state is utilized by the dominant political class[3] to deal with such critical tasks affecting ethnicity as the mediation of intergroup competition, the enforcement of basic rules on interaction between groups, the recruitment of elites into the political and economic system, and the mobilization and distribution of resources. Although competition among ethnic groups, subregions, socioeconomic classes, and other interests is assumed to be ever-present, whether in open or hidden form, those adopting a managerial view of the state see these conflicts as subsumed within the state's agreed upon regulatory processes. Acceptance of common rules by the community at large is critically important here. As A.D. Lindsay explains the managerial position on this issue: "The social divisions in society are obvious enough. The fact of class conflict cannot be denied. _If these divisions get sufficiently intense_, the state cannot fulfill its proper function. It either breaks down or its force is perverted from the task of enforcing common rules to that of keeping a government or governing class in power."[4] The consent of the citizens to common rules, then, becomes the basis for active intervention by state agencies to regulate the interactions with and among ethnic groups.

In the African situation, discontinuities between the center and periphery are marked. A professional, corporatist elite at the center, lacking a functional relationship to the prevailing modes of peasant production in most African states, appears to be only partially integrated into the rural economy. (See the chapter in this volume by Goran Hyden.) Consequently, this incomplete linkage requires a process of interelite bargaining to enlarge the core elite's base of political and economic support. Such a bargaining process entails a substantial expenditure of opportunity costs by the dominant political class as the energies of public officials are diverted from mobilizing resources for the expansion of the modern economic sector to fostering basic political and economic integration. Nevertheless, such an effort to promote an identification with the political system and its norms and values indicates the difficult role that an autonomous state must play as it pursues its managerial responsibilities in an African context.

The second line of thought tends to view the state as controller. A dominant section of the population, whether resting upon variables of ethnicity, class, subregion, religion, ideology, or others, employs the power of state institutions to coerce compliance on the part of the community at large. Lenin, writing on class relations in the bourgeois state, articulates this theme of group dominance most compellingly: "The state," he asserts, "is a special organisation of force; it is the organisation of violence for the suppression of some class."[5]

Subsequent neo-Marxist writers, expanding the functions as well as the networks of oppressors, have tended to look upon the dominant class as part of an exploitative "system,"[6] or, alternatively, as a (nonneutral) arena in which various class interests may conflict.[7] Alan Wolfe describes state violence as a "response to a situation," most specifically that of visible dissent.[8] Ralph Miliband distinguishes between the state elite, which wields power in such state institutions as the government, administration, military and police, judiciary, parliament, and local and regional governments, and "the economically dominant class." He is careful, however, to note that it is the latter class which "is involved in a relationship with the state."[9] And Hamza Alavi discusses a "bureaucratic-military oligarchy" which has an independence supported by an autonomous material base and which performs a mediating role between the rival demands of three propertied social classes.[10] In recent years, the analysis has become more complex, yet through it all a connecting thread is apparent: claims to neutrality on the part of the state and its agencies are dismissed and the search for a trans-class consensus is regarded as inappropriate. Rather, the state is depicted as built on group advantage and power, and its instrumentalities are portrayed as reflecting the economic interests of these controlling elements.

Politics, for the neo-Marxists, inevitably assumes something of a zero-sum relationship between rival social classes, while in the case of some non-Marxist analysts, this relationship also appears between competing racial or ethnic sections. Thus in the case of South Africa, state dominance, based on what Heribert Adam characterizes as "conformity pressure among a cohesive ethnic group (Afrikaner)," acts to manipulate the allocation of resources to favor the special interests of the prevailing racial caste. State control of countries in black Africa, maintained by a dominant political class at the center of the country, may also reflect ethnic interests in the political recruitment and resource allocation processes. Although the specific configurations of class/ethnic power distribution differ widely from country to country, there is a general tendency for the new men of power at the center to use state instrumentalities to regulate the activities of the society in terms of their distinctive class interests. As we will see in the discussion which follows, these state institutions may not prove sufficiently effective to allow them full scope for their organizing activities, but the intent to utilize the state as an instrument of sectional direction nonetheless remains.

Obviously, aspects of the state as manager and as controller are both evident in the actual experiences of the African countries. Relatively few observers on the African scene seriously contend that the state is no more than a neutral structure, persevering merely to process demand inputs emanating from the environment. Marxist and non-Marxist analysts seem implicitly to agree on the active role played by the state in establishing a national identity, creating an acceptable authority system, mobilizing and distributing resources, and dealing with the challenge of structural dependency. The two concepts of a class state and a relatively autonomous state are by no means mutually exclusive. Miliband, in an important reconstruction of Marx's thought, reconciles them in the following manner: "The state is indeed a class state, the state of the 'ruling class.' But it enjoys a high degree of autonomy and independence in the manner of its operation as a class state, and indeed must have that high degree of autonomy and independence if it is to act as a class state."[11] Neo-Marxists and non-Marxists can find common ground with respect to the active and relatively autonomous role of the state vis-a-vis the various classes and class segments making up society. Where they tend to disagree is on the extent to which the class consciousness of the dominant political elite necessarily distorts public policy. Are government leaders and their instrumentalities no more than protagonists of class interests, or can they act for a countrywide constituency as they go about instituting common rules and managing the economic and political development of their lands? And there is another dimension to the debate over the state's active role. If the African state is a political superstructure with limited relationship to rural production, as Hyden contends,[12] then its efforts to increase output or transform the production process may represent little more than movement without significant consequence. Obviously states vary enormously in the extent and nature of their active roles or their impacts upon the rural parts. We must consider the implications of limited state capacity for policy implementation if we are to gain an appreciation of what claims the state can realistically make upon the peoples of the hinterland.

"Stateness," as René Lemarchand stresses, involves a wide spectrum of political roles on the African scene. Thus sometimes the state appears overly effective -- even overbearing. Examples of this discussed in subsequent chapters are South Africa's deliberate promotion and enforcement of racial hierarchies (the chapters by Adam and John Stone), and Sudan's various attempts to influence linguistic and religious practices in its south (Dunstan Wai's analysis in this volume). Other evidence, possibly less dramatic, of African state tendencies toward hegemonial leadership abounds elsewhere. Claude Ake, commenting on the "overpoliticisation of life" in one of Africa's most prominent polyarchies, notes as follows with respect to the character of the Nigerian state: "State power is highly developed and used rather freely and the role of the state in the economy and society of Nigeria is very substantial. The remarkable growth of state power in Nigeria began under colonialism. The colonial state needed the apparatus to be effectively repressive and to ensure monopoly control of the economy. Since independence the power and interventionism of the Nigerian state have increased even more because the state had to spearhead economic development in the absence of a strong indigenous capitalist class. As things stand now, the Nigerian state appears to intervene everywhere and to own virtually everything."[13] For Ake, the dynamic and all-powerful state leads inevitably to "a desperate struggle" to secure control over state instrumentalities.

The nature of the campaign against colonialism contributed to the dominant class's determination to maintain -- and augment -- the strong institutions of governance already in place at the center.[14] Success in the struggle for state freedom required each multiethnic society to overcome its differences and unite, if only temporarily, behind the nationalist party and its inspirational leader. "Tribalism," as Guinea-Bissau's Amilcar Cabral put it, was something to be feared wherever it acted "as a function of opportunistic attitudes," that is, whenever it became the basis for socioeconomic "contradictions" within the nationalist movement, as "detribalized" individuals came to champion the cause of their clients over competing ethnoregional interests.[15] To the extent that the struggle for state freedom succeeded, and in the process overcame the pulls of "opportunistic" leaders as well, a critically important sense of unity was forged, temporarily at least, behind party and leader. Because of this, many an African state came to power with the dominant party firmly in control of the central political institutions.

After assuming power, the dominant political class sought to build upon this unity in an effort to cope with the many challenges of administering the affairs of its country. The members of this political class viewed strong state institutions as indispensable in expanding "the domain of its authority" -- both in terms of extending its functions at the center and its capacity to influence the course of events at the periphery.[16] In postcolonial circumstances, only a central government could be expected to prove an efficient and effective motor agency for economic and political development. To the extent that it might assure survival and integrate society as well as mobilize and distribute resources efficiently and secure freedom from external control, a central government was considered a vital instrument for achieving societal goals at the lowest possible cost. The need to display movement and a sense of purpose was imperative. In Kwame Nkrumah's words: "However poor the country, the new government cannot sit and do nothing. Construction must begin. There must be something to show for independence. And if there is nothing to show, popular discontent may split the country apart."[17] Leadership success, then, was to be measured in terms of achieving system goals. And under the difficult circumstances encountered at independence, the value preferences of the state-party elite worked in combination with environmental conditions to make an enlargement of state functions and capabilities instrumental to dominant class and/or ethnic interests.

An important aspect in this trend toward central political-administrative control was the acceptance, sometimes for a transitory stage only, of the need for hegemonial authority systems. Hegemonial systems are characterized here as elite-dominated, bureaucratically directed political orders displaying, for the most part, limited public accountability. Although some reciprocity is present in the relationship between rulers and ruled, an essential attribute of such systems is their potential for low-cost decision-making.[18] In addition, the capacity of these systems to contain conflict -- in the short term at least -- commends them to some members of the dominant African class. For example, the Sierra Leonean foreign minister, Abdulaih Osman Conteh, when arguing that the one-party state would settle the problem of coups and countercoups, contended that his country's "new constitution will put an end to tribalism, instability and other factional tendencies, and no more will brother rise against brother, or family against family."[19] In various ways, then, hegemony was depicted as a logical response to Africa's postcolonial environment: it represented a mechanism by which the new state-party elite might seek to close the gap between the state's diverse burdens and its limited capabilities. Not the least of these burdens were the challenges of national integration and of creating a national identity. Although no national elite could avoid the critically important tasks of building a network of coherent intergroup relations and facilitating the emergence of a society-wide identity, these challenges differed markedly across societies; some societies clearly were more responsive to state efforts to apply public formulas for the identification of ethnic groups than were others (see Henry Bienen's chapter). In its most extreme form, the challenge of national integration led to geoethnic challenges to the integrity of the state itself, leaving leaders little alternative but to take decisive counteraction. In Ghana, Kenya, Nigeria, the Congo (Zaire), and other countries, leaders have had to move swiftly to resist basic thrusts against central authority or see their primary task of ensuring the survival of their state placed in doubt. Not surprisingly, such far-reaching challenges to public order led a number of leaders to consolidate their power at the center and the periphery in an effort to control any manifestations of internal dissension.

If the African state sometimes appears overactive and imperious, it can, paradoxically, also display an evident "softness" and ineffectiveness. With respect to the latter aspect, the hegemonial option was less a manifestation of the strength of state institutions or of the preferences of the dominant political class than a reflection of the frailty of the postcolonial state in the face of burgeoning pressures. Hegemonial institutions and practices tended to obscure the reality of limited governmental effectiveness, both in managing central affairs and penetrating and regulating activities in the hinterland. "All attempts to recreate elite consensus and establish a one-party system in Dahomey," concluded Samuel Decalo, "though 'successful' on paper, only provided a structural umbrella of artificial unity underneath which the tripartite struggle for ultimate hegemony continued unabated."[20] In Uganda, A. Milton Obote's regime might have adopted "tough tactics" in 1981 to quell the threat posed by three guerrilla movements, but his divided and poorly trained army and police force were hardly in a position to carry out the assignment unassisted.[21] The appearance of concentrated state power in these and other instances (Chad, Zaire, Angola) was most deceiving, then, and a balanced perspective requires discussion of what Gunnar Myrdal describes as the "soft state."

For Myrdal, the state in South Asia lacks the ability to promote rapid development because political and social conditions (in particular, a low level of social discipline) prevent governments from enacting public policies imposing substantial obligations on the citizenry. "When we characterize these countries as 'soft states,'" he writes, "we mean that, throughout the region, national governments require extraordinarily little of their citizens. There are few obligations either to do things in the interest of the community or to avoid actions opposed to that interest. Even those obligations that do exist are enforced inadequately if at all."[22] It is possible to question whether the soft state in Africa arises from low levels of social discipline or from resource constraints and poorly conceived policies. Irrespective of cause, however, the outcome is similar -- a state limited in its control over society and therefore incapable of achieving its ambitious objectives. In Africa, the soft state is marked by fragile institutions; not only are these institutions constrained by the ineffectiveness of linkages and the unavailability of human, material and fiscal resources,[23] but the pressure of domestic and international demands may adversely affect performance. Recognizing the limitations upon its capacity to govern, the soft state responds by seeking to gain greater control over the periphery through various combinations of regulation and accommodation. Its bargaining with powerful ethnoregional actors, then, is mainly a reflection of weakness at the core, not the value preferences of the dominant political elite. The leaders' attempts to build coherent links with local notables may well prove to be one of their most critical tasks, i.e., to ensure the survival and integrity of the state itself. For unless political organizations and practices are congruent with the dispersion of ethnoregional power and diversity, as Bienen points out, central leaders may well have a difficult time regularizing patterns of interaction throughout the society.

The reality of the soft state in many African countries raises questions about the relative benefits and costs which hegemonial authority systems incur. If strong central political-administrative structures facilitate the state's efforts to make effective claims upon the citizenry (through regulations, extractions, production incentives, allocations, and so forth), they may also incur efficiency costs which impact upon state-ethnic relations. Dominant elites, recognizing the limitations on their capacity to govern, have reacted most forcefully to perceived threats from ethnoregional groups. A case in point is the quick response in April and May 1980 by Algeria's seemingly secure and authoritative state-party elite to contain protesting Berber university students in Tizi-Ouzou who were demanding official recognition of their language and culture.[24] Similarly, Kenya's newly installed president, Daniel arap Moi, anxious to consolidate his power as well as to promote national unity, startled his country in July 1980 by securing the party elite's support for a resolution calling for the "wind up" of all tribal associations. Although such organizations as the Gikuyu, Embu and Meru Association, the New Akamba Union, and the Luo Union reacted cautiously at first to the suggestion that they terminate their activities, they did begin to dissolve their branch organizations in earnest by November as Moi persisted in his policy.[25] Other examples of determined state efforts to head off geoethnic challenges are at hand, but these suffice to make our point: governments, conscious of the relative lack of central power and society-wide agreement on norms and values, feel endangered by ethnic claims and may act vigorously at times to reduce these strains emanating from the environment.

Particularly in soft state situations, centralizing thrusts have not been without their complications for processes of political and economic development as well as for those of interethnic conflict management. They accentuate the isolation of the core, increase the difficulty of moderating local nationalist demands for autonomous powers, heighten public expectations on the distribution of centrally controlled resources, and bring about administrative and fiscal inefficiencies. In the Central African Republic, where civil service salaries account for some 85 percent of the state's receipts, 77 percent of the country's public servants are settled comfortably in Bangui.[26] And not only does overcentralization produce pockets of privilege and power at the capital (and in the main provincial centers as well), it also makes more evident than ever a lack of fit between the realms of state and ethnoregional autonomous power. An unbalanced growth of state capabilities at the center may have a destabilizing effect. Authority figures at the periphery, operating on the basis of premodern norms, values and institutions, may find themselves working at cross-purposes with bureaucratic injunctions.[27] In their eyes, central institutions, with their continuity to the colonial past, tend to lack legitimacy. Because contact with these distant and encapsulated centers is potentially troublesome, they may withdraw from contact -- and thereby resist the interventions of dominant class interests from the outside. In some instances, face-to-face relationships can be used to make up for the lack of regularized patterns of interaction, but this represents an expedient means for adapting to a situation essentially lacking in coherency.

Clearly, as fragile central institutions become functionally overloaded (see Stone's chapter on this), the state's ability to extract resources from the periphery and to enforce basic rules on interaction between authority centers is constrained. The state presses its claims ineffectively upon the periphery, and, in worst case situations, appears ineffectively linked to the system of production in the hinterland upon which it depends for support.[28] Overcentralization not only opens the way to a costly process of broken connections, but, in less threatening environments, it also becomes the cause of administrative and fiscal inefficiencies. As Sir Arthur Lewis observes, the African villager is inclined to assume that the central government will finance a variety of services and amenities on its own, thereby creating "insoluble financial problems." He writes: "The clear connection between taxation and quantity of service would both check demand and increase the willingness to pay taxes. When the service is on the central budget, the connection between demand and taxes is tenuous; nobody wants to pay taxes to a distant central government, whose use of them is neither known nor approved. The government is then caught between unlimited demands for public service and very limited willingness to pay for what they cost."[29] Overcentralization contributes to the functionally overburdened state. And in more threatening situations, such as the demand for ethnoregional autonomy or independence from the state, overcentralization may lead to a weakening of linkages between different authority centers. By inhibiting the development of what James S. Coleman refers to as "intervening aggregative structures," statist policies can cause demands "to be thrust upon the polity in an unrestrained, intermittent, unpredictable, and totalistic manner."[30] In Chad, for example, increasing ethnoregional claims, from the southern Sara as well as the northern Toubou, upon François Tombalbaye's unitarist-minded and "insensitive" regime brought little in the way of constructive response.[31] Authoritarian inclination combined with a low level of bureaucratic capacity led to rebellion and the issuance of nonnegotiable demands -- the prelude, in this extreme case, to the general disintegration that came to mark Chad's state system by the late 1970s. The overcentralized yet fragile one-party system thus contributed to the undoing of the state itself. Not only had overcentralization proved inefficient, but it had itself contributed to the process of destructive conflict. The failure of the Tombalbaye regime to identify realistic objectives, because it insisted on all or nothing, helped to bring on a ruinous shortfall in Chadian state-ethnic relations.[32]

If carried too far, then, statist tendencies can actually thwart the achievement of African system goals. Such tendencies can strain state-ethnic relationships and complicate the tasks of facilitating development. As Mazrui shows in the next chapter, the British did more than their French counterparts to encourage African administrative independence and fiscal and budgetary autonomy while discouraging the processes of national integration and the emergence of national identities; it follows therefore that in these Anglophone countries the strains brought on by overcentralization can be expected to prove particularly intense. Thus despite the appeal of hegemonic control to members of the dominant political class intent upon utilizing central institutions to deal with challenges of national integration and resource mobilization and distribution, such state formulas, if applied rigidly, can prove counterproductive in practice.

In light of this, constructive interaction requires new thinking concerning the desirable balance in state-ethnic relations between conflict and stability. Each set of leaders must decide for itself what rules to designate for legitimate state and ethnoregional political and economic powers and activities. Attention to this task of defining the spheres of responsibility and action is indispensable to establishing coherent interactions. As will be discussed in the final section of this introduction and in the rest of the volume, new pressures for a changed role for the state have become evident in certain instances. Where these pressures lead to experiments in the state-ethnic encounter, experiments which take into account the state's inability to control society at this juncture, they may represent less a retreat from principles of state control and supremacy than a realistic accommodation to the configurations of group power in the society at large.

The Claims of Ethnic Groups

At the outset, it is necessary to put the nature of ethnic claims on the state into perspective. The effect of the soft state, with its relatively weak institutions and limited fiscal capacity, is certainly to produce a situation in which ethnoregional leaders may seize the initiative and issue unreasonable demands which cannot be absorbed by the state. Nevertheless, restraint on the part of the majority of such leaders most definitely remains the norm; in most instances communal spokesmen are prepared to work within the framework of the state, making what amount to reasonable (i.e., low- or medium-cost) demands upon the state's dominant political class. Thus even if primary public attention seems at times focused on the dramatic calls for autonomy and secession, a more comprehensive and balanced outlook nonetheless requires recognition that most ethnoregional claims actually seek distributive ends and accept the existence (if not the legitimacy) of the state as a framing fact. It is precisely because these reasonable claims are of low intensity and do not entail a serious threat to state coherency that stable patterns of state-ethnic interaction can normally take place.

As indicated in the chapters by Robert Bates and Donald Rothchild, the more commonplace, low-intensity demands arise largely from the desires of Africa's peoples for the benefits of modernity. Competition among societal groups occurs because the general public exhibits a growing desire for scarce opportunities and goods. However this competition among groups is by no means an equal one. Inequalties, in fact, are often quite marked and result from a variety of factors: natural resources and geography, colonial impact, missionary activities, access to decision-makers, political skills and information, resource allocation patterns, cultural propensities, ideologies, and so forth. The upshot may be to exacerbate conflicts already latent in the society.

Nevertheless, given the prevailing scarcities of contemporary Africa, what seems surprising is less the existence of group competition and conflict than the ability of most of these societies to contain conflict within acceptable limits. A partial explanation for this, put forward most tentatively, may be the tendency of peoples in the relatively disadvantaged areas to make fewer demands upon scarce public resources than their counterparts in the relatively more advantaged parts of a country. As shown by Rothchild's three surveys, the peoples of different subregions do exhibit somewhat diverse tendencies to demand the components of modernity. Such a finding is not without importance for state-ethnic relations. If the relatively disadvantaged lack political clout *and* make only limited demands upon the state, the government finds itself, by default, in a position to put minimalizing (or "satisficing") policies into effect. The soft state can, for the time being at least, endure and display a degree of coherency precisely because the modest demands of the relatively disadvantaged permit somewhat inequitable practices in the distribution of scarce resources.

In contrast, the more intermittent claims made by ethnoregional groups to self-determination and independence threaten the very existence of the state itself and generally cannot be reconciled by minimalizing policies. These high intensity demands, more often than not presented to state leaders in absolute terms, undermine the critical balance between conflict and stability. In addition, given the demonstrable fragility of Africa's state systems, the pressing of nonnegotiable demands by ethnic brokers carries a high potential for destructive conflict.

Claims to ethnoregional self-determination and sovereign independence in Africa, as Crawford Young's chapter on Katangese, Biafran, and Eritrean experiences demonstrates, are territorially rather than culturally defined. In each instance, the "right" to self-determination is based upon a spatial division which has its genesis in the colonial administrative process. Although this territorial foundation of the self-determination demand distinguishes the African experience from the more culturally defined pressures in other parts of the world, many of the ambiguities implicit in a quest for such autonomous relationships apply as much to the experience of Africa as elsewhere.

Self-determination -- a principle which holds that a culturally distinct people is entitled to enter into autonomous relations with other peoples -- is in most circumstances an ambiguous concept. Its implications tend to vary in different situations; it is a force at different times for integration or disintegration, positive change or negative change, intervention or nonintervention. Its impact, in brief, is largely dependent upon the conditions under which it is brought into play.

Three specific problems will suffice to indicate the general confusion surrounding the self-determination concept. First, there is the need to identify the people and unit to be accorded political autonomy. For the extreme protagonist of the self-determination ideal, all legitimate peoples ought rightfully to be granted statehood. But what constitutes a legitimate people and what units are valid candidates for statehood? At the peace negotiations following World War I, Robert Lansing, President Wilson's secretary of state, expressed misgivings on precisely these issues. When the president talks of "self-determination," he asked, "what unit has he in mind? Does he mean a race, a territorial area, or a community? Without a definite unit which is practical, application of this principle is dangerous to peace and stability."[33] Self-determination has been plagued by this unit-of-analysis problem ever since.

In the contemporary African context, a number of questions are pertinent in this respect: is an African people/unit which is "colonized" by another African people, as alleged in the case of the Somalis in the Ogaadeen, any less entitled to self-determination than those previously controlled by European metropoles?[34] Is Eritrea, formerly a separate political unit under Italian colonial rule, in a better position to demand self-determination than the Ogaadeen, which was incorporated into Ethiopia by Emperor Menlik?[35] If self-determination is legitimate when a geoethnic group "can no longer feel safe" in a state as constituted,[36] who is to decide at which point separation becomes necessary and what geoethnic minorities can be included in the newly created state entity? Given the subjective and fluid nature of ethnic identities in the modern African context,[37] what measure is to be used to indicate fitness for long-standing political self-rule? Are foreign influences decisive in the political process leading to separatist demands (i.e., in Katanga)?[38]

Is the demand for self-determination a free expression of majority sentiments, or is it merely an expression of dominant minority interests, as in the case of Rhodesia's UDI? If a referendum is set to determine local opinion on the issue, how is the area to be divided for voting purposes? These and other questions about ethnic self-determination point up the difficulties encountered when setting out guiding principles for testing the appeal made by peoples and units for political autonomy. As Rupert Emerson asserts with respect to the Baganda and Kikuyu peoples, "It is difficult to deny them the possibility of nationhood even though the probabilities are against them. If they were to assert their claim to separate national identity and in particular if they were to achieve separate statehood, could a claim be maintained against them that they lacked some essentials of the nation?"[39] In reality, then, self-determination is an indeterminant and inconsistent concept; no precise yardsticks are readily at hand with respect to the size and composition of groups or the circumstances giving grounds for its use. The result of this general ambiguity may well be to make the concept relevant only to the strong, leaving the remaining peoples with little option but to adapt to the existing multiethnic state as best they can.

A second problem highlighting the confusion surrounding this principle is the difficulty in distinguishing between creative and destructive uses of self-determination. Its potential use for destructive manipulation emerged clearly in the 1930s, when the Nazis wielded it as a powerful propagandistic weapon to weaken the Czechoslovakian state. "I saw clearly," Edward Beneš declared in his memoirs, "that it was via these problems that the first and foremost attack would be launched against our independence and existence and _against our state as a whole_."[40] As employed in the Czechoslovakian context of this period, the critical issue became less one of Sudeten German self-determination than of the Czechoslovak state's existence and its majority people's right to freedom from external control.

In contemporary Africa, the principle of ethnic self-determination reveals both a destructive and a creative side. Ethnic Somalis, who form the largest single group in Kenya's North-Eastern Province (previously the Northern Frontier District), have called since the early 1960s for the subregion's independence from Kenya and its union with the Somali Democratic Republic. At the April 1962 constitutional conference, Britain's Secretary of State Reginald Maudling concluded that an investigation of the Somali demand was necessary and promised that there would be no change in the subregion's status before a new Kenya constitution was brought into operation.[41] A commission on attitudes in the Northern Frontier District found that ethnic Somali opinion "almost unanimously favour[ed] the secession from Kenya of the N.F.D., when Kenya attains independence, with the object of ultimately joining the Somali Republic."[42] In the years that followed, the claim to self-determination took a violent turn, with the Somali nationalists, supported by the Somali government, engaging in guerrilla skirmishes with the Kenya security forces. Although Somali-Kenya tensions eased with the conclusion of the 1967 Arusha Accord, the relative calm in the area has, on occasion, been shattered by renewed skirmishing. In the fall of 1980, for example, so-called _shifta_ ("bandit") forces killed the district officer for Dabaab in Garissa district and stole a government

Landrover, reportedly driving it across the border into Somalia. Soon afterward the shifta struck again, this time killing six people during a raid on a shop in Garissa town. The government reacted quickly to contain the situation, causing Somali leaders to make accusations of atrocities perpetrated under the guise of emergency measures to deal with shifta terrorism.[43] In brief, the cultural benefits of ethnic self-determination as sought by Somali leaders[44] must be balanced against the destructive costs for the neighboring state (in this case Kenya, but a discussion of the impact on Ethiopia and Djibouti would also be appropriate). As Alfred Cobban puts the matter so well, "The history of self-determination is a history of the making of nations and the breaking of states."[45]

Finally, claims to ethnic self-determination give rise to a problem of overlapping economic interests. In a variety of conflict situations -- Eritrea, the Ogaadeen, the Western Sahara -- what appear as straightforward political demands for autonomy are at least in part manifestations of collective interests over the manner in which resources and revenues are distributed. Port facilities and transportation linkages may be indispensable to the economic life of the territory as a whole. In the event secession takes place, how are these infrastructures to be shared? Moreover, it should also be noted that discoveries of natural gas and copper in Eritrea, oil and gas in the Ogaadeen, and oil, iron ore, and phosphate in the Western Sahara have all had the effect of intensifying conflict. The separatist movements have been buoyed up by the enhanced prospects of their future viability,[46] and the sovereign countries currently containing the mineral resources have been given additional grounds for adopting an uncompromising stance on the issue of separate state autonomy. To be sure, the demand for ethnic autonomy preceded the discovery of these minerals, yet an awareness of their existence has acted to raise expectations and to heighten a realization of contradictory interests on both sides.

In conclusion, this section shows that demands are highly varied, ranging from the low-intensity demands for distributional benefits within the political system to the high intensity demands for separate statehood and independence outside the system. It is because the great majority of claims upon the state are reasonable in nature that the soft African state is in a position to satisfy these appeals, at least minimally. However, in the exceptional event that ethnic groups insist upon self-determination expressed as an absolute right to statehood, the effect is to polarize state and ethnic elites, making reconciliation within the multiethnic state structure a most trying undertaking. "If the Ethiopian position and resulting proposals seem intractable," Richard Sherman concludes bleakly, "then the Eritrean position is no less so."[47] The "intractable conflict" is a shorthand phrase for a situation of mutually unacceptable choices. In this event, either the principles of state control or of ethnic self-determination must be adhered to absolutely, or public formulas worked out which will enable decision elites to redesign the process of managing conflict and rearrange choices. While recognizing that the margin for maneuver between principles may be limited, it is necessary to look creatively for ways of altering the terms and costs of interaction. Hence we turn now to a discussion of the nature of policy choice in Africa and then, in the final section, to some examples of the rearrangement of choices in state-ethnic encounters.

Policies to Manage Competing Claims

Policy analysis is relevant to state-ethnic relations in that it enables decision-makers to rank priorities on publicly related matters on the basis of qualitative costs and benefits. The position adopted here is that policy analysis is consistent with a view of the state as manager or controller; or, alternatively, with a Marxist or non-Marxist philosophical perspective. Assuming conditions of human, material and fiscal scarcity, it offers guidelines on "rational" choices with regard to such politico-economic matters as the extraction and allocation of public resources, the regulation of public activities, and the implementation and evaluation of public policies. In Third World settings, the policy analyst finds himself constrained by such factors as the unavailability of critically important data, the highly specialized and somewhat ethnocentric assumptions that underlie much of the policy literature, the transcending reality of the "exogenous" variable, and the differences from First and Second World experiences in the nature of group demands, decisional processes, and regulatory capacities. An approach which assumes sufficient governmental capacity to effect change must be modified under the soft state conditions of Africa to take into account limited governmental capacity. Yet despite the presence of these limiting factors, the overriding need for effective patterns of state-ethnic interaction makes an effort at comparative analysis of the efficiency of public choices a necessary undertaking. To plan for creative relationships is to increase the opportunities for tapping new forces for change; such a systematic approach carries with it the possibility for reducing uncertainty in group encounters. The alternative, a "muddling-through" strategy, may well prove highly costly, leaving open possibilities for a drift into destructive conflict.

Any theoretical overview -- particularly one such as policy analysis, which emphasizes elements of choice in a context marked by economic scarcity -- is inevitably limited in its inclusiveness. Its focus on rational calculation may at times only partially take into account the dimension of emotional intensity attached to ethnic symbols. Policy analysis focuses mainly on the qualitative costs and benefits of choice, and less on such subjective factors as a sense of group belongingness, identity, status, or apprehension. (See David Laitin's chapter on the critical role of identity management in this regard.) To put this another way, policy analysis is more interested in decision-making, outputs, and outcomes than in the factors giving rise to group demands, preferences, or actions. Clearly, however, to the extent these factors impinge upon the policy process, the policy-oriented analyst must remain alert to them. For example, an understanding of a decision to utilize the federal form of government requires an appreciation of mutual ethnic fears existing in a society as well as of the integrative benefits of political union.[48] Group fears may not readily lend themselves to negotiation; nevertheless, a study of the terms and costs of interaction may lead to the selection of policies which have positive effects in reducing the intensity of conflicts. It is precisely because the struggle over scarce public resources is so critical to the ethnic phenomenon that we deem a policy focus, with its concern for the state as a distributor of goods, relevant to an understanding of state-ethnic relations in contemporary Africa.

Thus while remaining ever mindful of the linkages between economic values and the political and social-psychological environment, the policy analyst must, to be effective, assume a special responsibility for advising decision elites on preferred policies to maximize their stated objectives.[49] Policy analysis, then, is both analytical and prescriptive. In line with this outlook, both the analyses in this section (see the chapters by Rothchild and Victor A. Olorunsola and by Edmond J. Keller) are concerned with the alternative mechanisms available to public authorities as they attempt to mediate between the different principles of state capacity for governance and ethnic autonomy. Clearly, African societies vary considerably among themselves as to group relationships, organized institutional roles, and elite preferences; it therefore seems quite likely that policy analysts will arrive at diverse conclusions in different societal contexts regarding the most beneficial negotiating mechanisms applicable.

The two policy approaches put forth in this section are suggestive of the many options open to decision-makers intent upon facilitating coherent state-ethnic relations. Rothchild and Olorunsola, cognizant of the presence of low- as well as high-intensity conflicts involving an ethnic aspect, stress the varied uses of both bargaining and hegemonial models to manage competing claims. Eight strategies for regulating conflict are discussed, ranging from subjection (the strategy most obviously linked to hegemony) to sharing (most prominently associated with bargaining). In delineating among strategies, comprehensiveness, not value judgment, is intended. Nevertheless, the authors recognize that choice strategies significantly affect outcomes; these strategies are therefore characterized in the first table of the Rothchild-Olorunsola chapter according to their long-term potential for moderating (or intensifying) conflict.

The main problem dealt with in the Rothchild-Olorunsola essay is the changing of choices: where an ethnic strategy imposes unacceptable costs upon the state, how can decision-makers alter their structures and relationships so as to manage competing claims better? The authors note the presence of significant variations not only among strategies but within the strategies as well. Thus appreciation of a full range of choice options requires attention to the many lesser alternatives present within each choice strategy for reconciling ethnic autonomy and state control. A focus upon suboptimization may hopefully expand choice, thereby facilitating the management of state-ethnic conflicts. By breaking up general strategies into smaller parts, the policy analyst may help in the process of hammering out agreements on new structures and institutions. These can in turn act as arenas for fostering a mutual interplay of interests. To be sure, such mediatory mechanisms may have little attraction for tyrants determined to apply absolute principles of governance, irrespective of human cost; for the main body of African decision-makers, however, an awareness of just how wide a range of choice avails may facilitate efforts to work out unique public formulas to accommodate the divergent interests at hand.

Another application of the public policy literature to the management of competing state-ethnic claims is offered in Keller's chapter, "The State, Public Policy and the Mediation of Ethnic Conflict in Africa." Adapting Theodore Lowi's well-known typology of public policies to the experiences of contemporary Africa, Keller expands the Lowi functional categories of distributive, regulatory and redistributive policies to include additional categories on reorganizational and symbolic policies.[50] With an eye to making the Lowi framework more relevant to conflict management objectives under conditions of economic scarcity and weak institutionalization in Africa, Keller proposes adding to the initial categories (which relate substantially to individual and collective interest-maximization) these two which are more broadly concerned with issues of state coherence. To this end, his reorganizational policies emphasize state initiatives aimed at ensuring state survival and easing the strains of ethnic disaffection, mainly through restructuring political institutions and relationships. And his overarching category of symbolic policies stresses the psychological as opposed to the material resources at the disposal of the African state. Given the imbalance between public demands and state capabilities in many African countries, policies which effectively symbolize the unity, goals, and aspirations of a society as a whole may go far in compensating for the general unavailability of material goods at the government's disposal. By implication, it is as state elites make the best possible choices on utilizing these five types of policies in their own societal contexts that they display what might be described as comparative efficiency in preserving and enhancing the coherence of their state.

The two applications of policy analysis to the issue of managing competing state-ethnic claims discussed above illustrate a conceptual approach we feel holds a significant potential for new insights. In focusing on conflict management questions in Third World environments, some of the analytical rigors normally associated with a cost-benefit or cost-effectiveness approach, which emphasizes the relationships of public expenditures to political system outputs, must necessarily be modified to take into account the qualitative aspects of the issues at hand. In addition, it must also be recognized that policy analysis tends to give only partial insight into the psychological and emotional aspects of behavior, a serious limitation in matters related to ethnicity where symbols of affection are inextricably intertwined. Keller's category of symbolic policies is suggestive of a wider problem here. Although this cluster of policies represents something of a reservoir of nonmaterial resources which can be used by state elites for managerial purposes, the policies are studied essentially for their possible contribution to conflict regulation and less for insights on the psychological imperatives underlying ethnic identification. (On this, see Laitin's analysis in this volume.) In light of these reservations, then, the role of policy analysis in examining state-ethnic relations in Africa may be described as limited but not without importance. It may inform decision elites and their constituents about the comparative terms and costs of pursuing policy alternatives under conditions of scarcity. With such knowledge in hand, it may be hoped that leaders will design structures and institutions and put policies into effect which will balance conflict and stability in state-ethnic encounters.

The Dialectic of Autonomy and Control

For African decision-makers, the issue of ethnic autonomy can pose a most troublesome example of policy-making in the face of decided risks. In many instances uncertainty exists about the validity of the subnational claim to autonomy, the effectiveness of the linkages between center and subregions, and the viability of the state itself. Modernization has awakened new and inflated expectations, causing the contemporary state to be beset with contradictory claims for state-centered and ethnoregional-centered authority. If, at times, sustained economic development requires a "trans-state" solution, ethnoregional self-realization requires a concession of some authority from within the state. In the end, both of these trends raise doubts about the long-term capacity of the state to remain an effective framing fact.[51] Particularly in Africa, where resources are severely restricted and state institutions all too often marked by softness, state-ethnic relations can fall apart. Thus, unless careful attention is given to the public formulas and regularized patterns of interactions occurring within the state, an imbalance between conflict and stability, with its potential for incoherence, may result.

It must be emphasized at the outset that if a lack of fit is sometimes evident between the claims for state control and ethnic autonomy, this is because each of these competing principles can in certain circumstances involve legitimate, albeit contradictory, objectives. Each principle displays an intrinsic consistency and each is championed by committed proponents. Such characteristics lend an inner dynamism to claims made on behalf of these principles, providing the basis for a possibly protracted, and sometimes destructive, conflict. Either an exclusive nonnegotiable demand for ethnic autonomy outside the state or an overweening centralist thrust within the state may cause conflict to surface between these principles. Clearly, such discontinuities must be narrowed to ensure creative relations between peoples and levels of government. It is the possibility of incoherent relationships emerging that necessitates "the risk and opportunity of human choice."[52]

Striking some kind of balance between the pulls of central control and subregional autonomy requires agreement on negotiating formulas which can address, at least minimally, both sets of claims. Once it is recognized that each principle possesses some measure of legitimacy, the bargaining partners can retreat from absolute positions and begin to formulate policy mechanisms and political structures responsive to the two principles. Should agreement -- explicit or tacit -- be reached, the negotiating mechanisms are likely to survive so long as they prove effective in coping with the political and economic problems that gave rise to them in the first place. Such public policies and formulas remain means, not ends. They are needed to keep the two thrusts of central control and subnational autonomy in balance; once this conflict between objectives alters in a significant manner, new perceptions and formulas are likely to materialize.

The bargaining mechanisms discussed in the final section of this volume -- federalism (see the chapter by Olorunsola and J. Isawa Elaigwu), and regional autonomy (the chapter by Wai) -- represent two alternatives for achieving a balance between control and autonomy. The list here is by no means exhaustive, merely illustrative of the choices which may avail under current African conditions. The formulas themselves are not fixed and static but responsive for the most part to changes in the political process. Moreover, as Laitin's chapter on the Ogaadeen question reveals so forcefully, there are always limits to the effectiveness of bargaining mechanisms. This causes him to look, more basically, at the possibilities for finding new political solutions through the management of ethnic and cultural identities.

The responsiveness of political formulas to changes in the political process is shown, for example, in the case of Nigerian federalism, where major shifts have occurred in the number and size of subregional units, central-subregional powers and responsibilities, political party representation and coalition formation, election and recruitment procedures, and revenue allocation formulas. In part at least, these changes reflect the increasing legitimacy of the new presidential system and the new constituency bases from which political elites now operate. Although some elements, such as the demand for additional subregional units, appear to be constant over the years, it is nonetheless possible to detect a change in emphasis here from protecting the autonomous interests of an ethnoregional people to calculating collective benefits in the distributive process. In this respect, the decision to remain in a sizable subregion or to press for its redivision may prove largely a function of how subnational groups perceive the effect on their interests of public policies on political representation and revenue and resource allocation. With the federal arrangement as a framing fact, subnational spokesmen have every incentive to work within the established system to further what they consider the special concerns of their constituents. So long as they play by the rules, and the state-ethnic conflict is channeled along constructive lines, some form of federal mechanism, adjusted to meet new configurations of power and responsibility, is likely to be seen as useful in balancing the two thrusts of central control and subunit autonomy.

In other instances, such as the Sudan, new pressures for a changed role for the state have become evident (see Wai's chapter in this regard). That changed role represents less a retreat from the principle of state control than a realistic assessment of state power and of how critical state objectives can be achieved most effectively. Altered ways of balancing conflict and stability surfaced in the 1970s as the regime of Jaafar el Nimeiry came to recognize the futility of prolonging a savage conflict with the Southern-based Anya-Nya; it thereupon decided to accept the political risks of negotiation with Southern leaders. Putting the harsh "Southern policy" of General Ibrahim Abboud behind him, Nimeiry changed tack and, in an important declaration in June 1969, granted the Southern provinces (Bahr El Ghazal, Equatoria, and Upper Nile) the status of "Regional Self-Government within a United Socialist Sudan." A series of quiet, behind-the-scenes encounters, leading up to formal peace talks in Addis Ababa, culminated in the signing of the Addis Ababa Agreement in 1972.[53] Deservedly, Sudanese negotiators on both sides have received high praise for the bargaining mechanism they put into effect. And Sudanese officials, describing the agreement as "provid[ing] Africa with what we may call a moral for solving the near universal problem of conflicting diversities which now affect the Continent," have gone on to offer to mediate other state-ethnic disputes in Africa (in particular, that between the Ethiopians and the Eritreans).[54] Nevertheless, implementation of the agreement has left the major actors less than fully satisfied on such issues as the extent of Southern autonomy, central revenue allocations, and representation of the subregions in the central decisional process. When he applied the regional system to all sections of the country in 1981, Nimeiry indicated an interest in breaking up the South into three smaller subregions, an idea that has encountered heavy opposition from many members of the currently entrenched elite in the subregion. Irritated at this resistance, Nimeiry has spoken of a need to avoid "mistakes" originally made in the Addis agreement of 1972 when working out the new act of regional government.[55] Nimeiry's remarks were regarded by observers on the scene in Khartoum as a pointed attack on the Dinka elite's unwillingness to accede to territorial redivision -- and to the enhancement of central control this presumedly might entail. By early 1982, however, Nimeiry pulled back from his plan for redivision of the South. This act of prudence is reported to have reflected intense domestic (as well as some external) pressures upon his government.

Finally, ethnically representative coalitions are by definition bargaining mechanisms. Not only their formation but their maintenance is dependent on a continued willingness on the part of factional spokesmen to reconcile divergent interests. In interethnic relations, the negotiation to form political coalitions remains feasible so long as group representatives are prepared to give up one value (for example, unilateral power) in exchange for another (effective participation jointly with others in the decisional process). The outcome, where sustained, is a mechanism offering great potential for mediating the conflicting demands made by ethnoregional sections for public resources.

Conclusion

These negotiating mechanisms on federalism and regional autonomy (as well as others such as coalition government, proportional representation, and mutual veto) act to facilitate regularized patterns of interaction between state and ethnic group and among the ethnic groups themselves. They are not solutions so much as mechanisms and formulas for encouraging ongoing social exchange. As the policy analyst focuses attention on the social costs and benefits of these and other policy instruments in the various African contexts, his recommendations may prove useful to the various decision elites who will then be in a position to adjust the terms of relationship accordingly. The decisions on strategy and policy instruments must remain in the hands of the state elite. Only they can determine whether hegemony or polyarchy and which of their myriad policies and subpolicies are appropriate to local circumstances. Nevertheless, information on realizable options seems indispensable to efficient choice.

Certainly an important aspect in selecting appropriate policy choices is to take the soft state conditions of many African countries into account. Western-based policy analysis assumes an enormous capacity on the part of state agencies to put policies and programs into effect. The African reality allows no such comfortable assumption. Rather, institutions must be shaped which are workable given the government's limited ability to implement desired policies. Thus too great an emphasis on control, bringing on an unbalanced process of overcentralized government, appears likely to prove counterproductive under conditions of human and material scarcity; in these circumstances, prudence dictates central self-limitation and a preparedness on the part of the state's dominant political class to accommodate ethnoregional interests and power. Such a stance reduces the costs of administration and buttresses the legitimacy of the system -- the latter being the critical element in promoting state coherence. State control may be indispensable to stable relations in certain situations. However, to opt for tough tactics, including force, where central military power is in fact lacking (as was the case in Uganda in 1981), is needlessly to court ineffectiveness. Hence the lack of resources and institutional capacity necessitates a willingness to search out and experiment with political formulas to ensure regularized state-ethnic linkages, thereby to balance conflict and stability. Given the dimensions of scarcity at hand, the challenge to African leadership is a great one, yet, as shown in the chapters which follow, the range of choices for such an effort is wider than sometimes is recognized.

Footnotes

1. The ethnic group -- defined here as a distinct category of society self-consciously united around shared histories, traditions, beliefs, cultures, and values which mobilizes its membership for common political, economic and social purposes -- is in essence a culturally based social organization.
2. Walter Lippmann, *Public Opinion* (New York: Harcourt, Brace and Co., 1922), p. 16.
3. Richard Sklar, "The Nature of Class Domination in Africa," *Journal of Modern African Studies*, 18, 4 (1979): 531-552.
4. A.D. Lindsay, *The Modern Democratic State* (New York: Oxford University Press, 1947), p. 203. (Italics in text.) Also see Alfred Cobban, *The Nation State and National Self-Determination* (New York: Thomas Y. Crowell, 1969), pp. 33-34; R.M. MacIver, *The Web of Government* (New York: Macmillan Co., 1948), pp. 31-33; and H. Krabbe, *The Modern Idea of the State* (New York: D. Appleton and Co., 1930), pp. 208-210.

5. V.I. Lenin, *State and Revolution* (New York: International Publishers, 1932), p. 22.
6. The various functions of the state are outlined in Nicos Poulantzas, *Political Power and Social Classes* (London: Sheed and Ward, 1973), pp. 52-53.
7. On this point, we are indebted to Dr. Naomi Chazan.
8. Alan Wolfe, *The Seamy Side of Democracy* (New York: David McKay, 1973), p. 219; also see his discussion of the "franchise state" in *The Limits of Legitimacy* (New York: Free Press, 1977), pp. 128-145.
9. Ralph Miliband, *The State in Capitalist Society* (New York: Basic Books, 1969), p. 54.
10. Hamza Alavi, "The State in Post-Colonial Societies: Pakistan and Bangladesh," *New Left Review*, 74 (July-August 1972), p. 72.
11. Ralph Miliband, *Marxism and Politics* (London: Oxford University Press, 1977), p. 74.
12. Goran Hyden, *Beyond Ujamaa in Tanzania: Underdevelopment and an Uncaptured Peasantry* (Berkeley and Los Angeles: University of California Press, 1980), pp. 26, 88.
13. Claude Ake, "Off to a good start but dangers await...," *West Africa*, May 25, 1981, p. 1162. Reprint of his presidential address to the annual conference of the Nigerian Political Science Association.
14. On the processes of facilitation and impedance, see Donald Rothchild and Robert L. Curry, Jr., *Scarcity, Choice and Public Policy in Middle Africa* (Berkeley and Los Angeles: University of California Press, 1978), pp. 1, 77.
15. Amilcar Cabral, *Revolution in Guinea* (London: Stage 1, 1969), pp. 84-85. Also see Ahmed Sékou Touré, *Towards Full Re-Africanization* (Paris: Présence Africaine, 1959), p. 11.
16. Aristide R. Zolberg, *Creating Political Order* (Chicago: Rand McNally, 1966), p. 134; and David Apter, *The Political Kingdom in Uganda* (Princeton: Princeton University Press, 1969), p. 433.
17. Kwame Nkrumah, "African Prospect," *Foreign Affairs*, 37, 1 (October 1958): 51.
18. James M. Buchanan and Gordon Tullock, *The Calculus of Consent* (Ann Arbor: University of Michigan Press, 1971), p. 72.
19. *West Africa*, November 17, 1980, p. 2275.
20. Samuel Decalo, *Coups and Army Rule in Africa* (New Haven: Yale University Press, 1976), p. 51. For a similar point regarding contemporary Zaire, see Thomas M. Callaghy, "State and Ethnicity in Zaire: Prefects, State Formation and the Coverover Process." Paper presented to the African Studies Association, Bloomington, Ind., October 23, 1981.
21. Michael Holman, "Sort of everything -- including hope," *Financial Times*, July 31, 1981, pp. 12-13; and Donald Rothchild and John W. Harbeson, "Rehabilitation in Uganda," *Current History*, 80, 464 (March 1981): 115-119; 134-138.
22. Gunnar Myrdal, *Asian Drama -- An Inquiry into the Poverty of Nations*, vol. 2 (New York: Twentieth Century Fund and Pantheon, 1968), pp. 895-896.

23. Thus a recent strategy statement on the Sudan describes the planning agency as lacking "an adequate data base, analytical capacity, and implementation and review procedures.... While there is a cadre of seasoned, well-trained senior administrators in Khartoum, quality declines rapidly as one descends the hierarchy, and both quality and quantity decline outward from the capitol." Agency for International Development, Sudan: Country Development Strategy Statement (Washington: U.S.A.I.D., 1981), p. 25.
24. Even though the minister of education did tour the Kabyle area and seek to enter into a process of "dialogue and explanation," few meaningful concessions to Berber cultural demands were immediately forthcoming. See Le Monde (Paris), April through June 1980.
25. Weekly Review (Nairobi), October through November 1980.
26. West Africa, June 15, 1981, p. 1339.
27. C.S. Whitaker, Jr., "A Dysrhythmic Process of Political Change," World Politics, 19, 2 (January 1967): 208; and his book, The Politics of Tradition: Continuity and Change in Northern Nigeria, 1946-1966 (Princeton: Princeton University Press, 1970), p. 12.
28. Hyden, Beyond Ujamma, p. 16.
29. W. Arthur Lewis, Politics in West Africa (London: George Allen & Unwin, 1965), pp. 54-55.
30. James S. Coleman, "The Development Syndrome: Differentiation-Equality-Capacity," in Leonard Binder et al., Crises and Sequences in Political Development (Princeton: Princeton University Press, 1971), p. 96.
31. Decalo notes that "no non-Sara politician, whether loyal or not to Tombalbaye, and few not of the latter's specific clan, were able to hold office for long." Samuel Decalo, "Chad: The Roots of Centre-Periphery Strife," African Affairs, 79, 317 (October 1980): 499; and his "Regionalism, Political Decay, and Civil Strife in Chad," Journal of Modern African Studies, 18, 1 (March 1980): 39; also see Colin Legum, ed., Africa Contemporary Record 1978-1979 (New York: Africana Publishing Co., 1980), p. B528; and Virginia Thompson and Richard Adloff, Conflict in Chad (Berkeley: Institute of International Studies, 1981), chapter 2.
32. See Manfred Halpern, "Changing Connections to Multiple Worlds," in Helen Kitchen, ed., Africa: From Mystery to Maze (Lexington: Lexington Books, 1976), p. 13.
33. Robert Lansing, The Peace Negotiations (Boston: Houghton Mifflin Co., 1921), p. 97.
34. Commenting on two principles of interstate relations still to be resolved, President Mohammed Siyaad Barre of Somalia stated: "First, self-determination must apply to all peoples in the world. Secondly, Somali peoples under colonial rule, just like any others in the world, deserve the right to freedom." Later he pointedly declared Ethiopia "a colonial power, pure and simple." International Herald Tribune (Paris), March 29-30, 1980, p. 1.
35. This distinction is commonly argued by Eritrean spokesmen but denied by other neighboring observers. See Tom J. Farer, War Clouds on the Horn of Africa (New York: Carnegie Endowment for

International Peace, 1976), p. 137; and Standard (Nairobi), April 10, 1978, p. 4.

36. The United Republic of Tanzania, Tanzania Government's Statement on the Recognition of Biafra (Dar es Salaam: Government Printer, 1970), p. 1.
37. On the relativity of ethnic identity, see Crawford Young, The Politics of Cultural Pluralism (Madison: University of Wisconsin Press, 1976), p. 41 ff.
38. On Union Minière's support for Katanga secession in July 1960, see René Lemarchand, "The Limits of Self-Determination: The Case of the Katanga Secession," American Political Science Review, 56, 2 (June 1962): 415; and Crawford Young, Politics in the Congo (Princeton: Princeton University Press, 1965), p. 503, fn. 63.
39. Rupert Emerson, From Empire to Nation (Cambridge: Harvard University Press, 1960), p. 99.
40. Alfred Cobban, National Self-Determination (Chicago: University of Chicago Press, 1944), p. 41; and Memoirs of Dr. Edward Beneš, trans. Godfrey Lias (Cambridge: Houghton Mifflin, 1954), p. 211. (Italics added.)
41. Colonial Office, Report of the Kenya Constitutional Conference, 1972, Cmnd. 1700 (London: Her Majesty's Stationery Office, 1962), p. 11. Also see the discussion in John Drysdale, The Somali Dispute (New York: Praeger, 1964), chapter 11. On the publicizing of these demands by the Somali government, see I.M. Lewis, "Pan-Africanism and Pan-Somalism," Journal of Modern African Studies, 1, 2 (1963): 156.
42. Kenya, Report of the Northern Frontier District Commission, Cmnd. 1900 (London: Her Majesty's Stationery Office, 1962), p. 18.
43. Weekly Review (Nairobi), November 14, 1980, p. 9, and November 21, 1980, p. 9.
44. As recently as 1978, Siyaad called on the Western powers to recognize and safeguard the rights of the people of Western Somalia (i.e., the Ogaadeen and parts of the North Eastern Province of Kenya). Standard (Nairobi), April 4, 1978, p. 4. Subsequently, however, Siyaad reportedly said that he had no territorial claims on Kenya. Legum, ed., Africa Contemporary Record 1978-1979, p. B388; and Africa Research Bulletin (London), 17, 12 (December 1-31, 1980): 5886A.
45. Cobban, National Self-Determination, p. 6.
46. Patrick Gilkes, "Eritrea Could Stand Alone," African Development, 9, 4 (April 1975): 19; "The Rights of the Sahrawi People," West Africa, March 5, 1979, p. 382.
47. Richard Sherman, Eritrea: The Unfinished Revolution (New York: Praeger, 1980), p. 109.
48. Donald Rothchild, Toward Unity in Africa: A Study of Federalism in British Africa (Washington D.C.: Public Affairs Press, 1960), pp. 189-191.
49. Yehezkel Dror, "Some Diverse Approaches to Policy Analysis: A Partial Reply to Thomas Dye," Policy Studies Journal, 1, 4 (Summer 1973): 258.

50. See Theodore Lowi, "American Business, Public Policy, Case Studies and Political Theory," World Politics, 16, 4 (July 1964): 677-715. For criticism of Lowi's schema as "ambiguous and incomplete," see James Q. Wilson, Political Organizations (New York: Basic Books, 1973), p. 328.
51. The Task Force on Canadian Unity, A Future Together: Observations and Recommendations (Ottawa: Minister of Supply and Services, 1979), p. 17. Also see the discussion in Dov Ronen, The Quest for Self-Determination (New Haven: Yale University Press, 1979), chapter 5.
52. Halpern, "Changing Connections," p. 12.
53. For an excellent description of the peace process, see Dunstan M. Wai, The African-Arab Conflict in the Sudan (New York: Africana Publishing Co., 1981), chapter 8.
54. Foreign Minister Mansour Khalid, as quoted in Tristram Betts, The Southern Sudan: The Ceasefire and After (London: The Africa Publications Trust, 1974), p. 143. On the mediation process, see the interview with Sudanese First Vice President Abdel Majid H. Khalil in Sudanow (Khartoum), 6, 4 (April 1980): 14; and Gregory Jaynes, "Sudan Offering to Mediate in Ethiopia's Eritrea War," New York Times, December 4, 1980, p. 3.
55. Sudan News Agency, Daily Bulletin (Khartoum), issue 3882, September 15, 1981, p. 4.

2
Francophone Nations and English-Speaking States: Imperial Ethnicity and African Political Formations

Ali A. Mazrui

For our purposes in this analysis, statehood implies concentration of sovereignty at the center. Nationhood implies substantial cultural homogeneity. The modern nation-state was therefore a merger of the two trends -- the trend toward sovereign political centralization (statehood) and the trend toward cultural homogeneity (nationhood).[1]

Our first proposition in this essay is that Francophone African countries are, on the whole, nearer to becoming nations than are English-speaking African countries.

Our second proposition is that the same Francophone African countries are further away from becoming true states than are English-speaking countries. In other words, the Francophones are ahead in nationhood and behind in statehood despite the fact that the term "the state" is used more often in the vocabulary of Francophone governments than in the political lexicon of former British Africa.

Our third proposition is that these comparative predicaments are substantially due to the interplay between imperial ethnicity (of which racism is one manifestation) and indigenous African ethnicity (of which "tribalism" is one manifestation).

But before we address ourselves to those formidable theses, let us briefly look at the European origins of the nation-state in order to place our own African concerns in a wider global context.

The Legacy of Westphalia

It may well turn out that Europe's most enduring legacy to Africa is the nation-state. This particular mode of political organization originated in European diplomatic history and statecraft, combined with major sociological changes within European societies. The international system as it exists today is often traced to the Treaty of Westphalia of 1648 which ended the internecine Thirty Years' War in Europe. As we have been reminded repeatedly by historians and students of diplomacy, it was upon the foundations of the Treaty of Westphalia that the entire superstructure of world diplomacy and international relations came to be constructed.

African countries were among the last to be admitted to this community of nations with its own rules of the game, its own conventions and codes. International law itself is a child of European diplomatic history and is now supposed to bind the new members of the club of nation-states.

For our purposes in this paper, it is important to remember that the concept of the nation-state is a compound one, bringing together the concept of nation on one side and the concept of state on the other. City-states have existed from time immemorial. They were followed in some parts of the world with empire-states, ranging from the empires of Mali and Ghana in West Africa to the Austro-Hungarian Empire in Europe. These were states with different levels of sovereignty. In a sense they were precursors of modern-day federalism. To put it in another way, a federation is an intermediate category between an empire-state on one side, and a truly classical nation-state on the other.

A nation-state proper concentrates sovereignty at the center. And certainly from the point of view of external relations both federations and unitary states have operated on the basis of the exclusive sovereignty of the central government in foreign affairs and defense under the Westphalian legacy.

As Europe approached this style of diplomatic organization it was at the same time experiencing greater internal integration within each European society. The new actors in international politics became increasingly homogeneous internally. Some of the old empire-states were beginning to erode and to produce subunits which were nations -- in some cases nations seeking separate statehood.

On the other hand, there were also principalities and princely estates in Europe, some of them little more than city-states. Again the idea that states should coincide with nations was beginning to endanger the survival of small political units in Europe's experience. It fell upon Prince Otto von Bismarck to end the era of small German principalities and to accomplish the first major modern phase of German unification. Some of the wars he helped to initiate (with Denmark in 1864, with Austria in 1866, and with France in 1870) helped him in the goal of creating a new German empire-state. Bismarck governed that empire-state from 1871 to 1890.

Italian principalities underwent a similar experience, with the enthusiasm of the Risorgimento and the brilliantly unifying charisma of Garibaldi. Both Italy and Germany were later to become nation-states proper, though the Federal Republic of Germany has retained a residual federal principle. Both countries are substantially homogeneous culturally and linguistically. Both have become nations as well as states.

Europe's transfer of its own state system to Africa was not accompanied by any prior calculation to make statehood coincide with nationhood. There was little quest for cultural congruence between the unit of the state and the national unit or even for cultural similarities among the different units being ruled by the same colonial government. Very few of the new territorial units which were carved as countries could be described as nations in the classical European sense which inspired many Germans under Bismarck and later under Hitler, and which inspired romantic and nationalistic Italians under Mazzini.

Of the half-dozen or so European countries which have ruled different parts of Africa, Belgium and Great Britain were the most culturally divided within their own societies. But in terms of language Great Britain was almost completely Anglicized apart from small pockets of Welsh speakers. In any case two-thirds of the country was English, while the remaining third consisted of the Welsh, the Scots, and the Irish.

The split in Belgium between the French-speakers and the Dutch-speakers is more dramatic. In a curious way the dualistic ethnicity in Belgium seems to mirror the dualistic divide between the Hutu and the Tutsi in two countries which Belgium once ruled, Rwanda and Burundi. But the cleavage in the two former colonies is of course far deeper, and has been much more tragic since independence, than the cleavage between the Waloons and their compatriots in Belgium.

What this means is that Europe itself has not always achieved a neat coincidence between nationhood and statehood. What we can say is that in virtually every European country there is one cultural group which comprises more than half the population of the whole country, and in many cases, more than two-thirds. Highly fragmented European countries like Yugoslavia are more an exception than the rule. Even Switzerland, though opting for federation, has a clear preponderance of German-speakers.

On the other hand, in Africa ethnically fragmented countries are the rule rather than the exception. And in only a minority of African countries, especially those previously ruled by Great Britain, is there one ethnic group which comprises more than half of the population of the country. These observations may be somewhat overstated, and will be qualified later, but for the time being let us simply confront two of the most pressing political challenges facing Africa -- how to move from nationalism to modern nationhood, and how to close the gap between statehood and nationhood. Nationalism is a body of ideas and values which may be held by people who are not yet nations, but long to be. When Africa finally succeeds in satisfying that longing, the gap between nationhood and statehood will at last have been substantially closed.

But where is Africa now in this struggle to close the gap? This brings us back to our compound proposition or thesis -- that Francophone African countries are ahead of English-speakers in nationhood but behind them in statehood, bearing in mind that we are concerned in this context more with how free an African state today is from external control than with how effective it is in controlling its own people. Let us take a closer look at this compound thesis in relation to both indigenous and imperial forces.

Colonialism and Comparative Nation Building

What then were the aspects of British colonial rule which complicated prospects for nationhood in British colonies while strengthening prospects for statehood? What elements in French approaches to colonial rule helped to lay the foundations of nationhood in the colonies? What aspects of France's interaction with her colonies has slowed the pace of state formation?

Let us begin with the textbook stereotypical differentiation between the colonial policies of the two powers. Much has been written about British indirect rule in Africa; at least as much has been written about French policies of assimilation and integration. Far less has been said about the consequences of these policies in terms of state formation and nation building.

The great architect of the British policy of indirect rule was Lord Lugard, perhaps the greatest British administrator to have served in Africa. The idea was to rule the colonies as far as possible through "native" authorities and institutions. The conviction was that people are best ruled through instrumentalities they understand and on the basis of conventions they can to some extent identify with. The great task of the colonial power was therefore to seek out these local instrumentalities and use them as part of the structure of colonial rule. The strategy could be cheap economically since it would be using local manpower as far as possible; but at least as important was the fact that the strategy economized in cultural disruption. It enabled the colonized societies to adjust more gradually to the imperial presence and therefore helped to reduce the tensions of change and acculturation.

The colonial philosophy of indirect rule was in part a derivative of British domestic political culture, which also distrusted rapid fundamental change and emphasized the virtues of gradualism. The same social forces and ideological considerations which have helped to perpetuate the monarchy and the House of Lords in Britain also helped to preserve the power of emirates of Northern Nigeria and the kabakaship of Buganda until military force was used to change this after independence.

But that very military confrontation in Uganda in 1966 and the military-backed revision of the constitution in 1967 illustrate the complications for nationhood in countries previously ruled by Britain. By helping to preserve strong indigenous institutions in Uganda, Britain contributed to the military crisis between Buganda and the central government under Milton Obote in 1966. By preserving Northern Nigeria especially as an area of cultural authenticity and autonomy, with its own strong local institutions, Britain inadvertently contributed to the forces which culminated in the Nigerian Civil War of 1967 to 1970.

But classical cases of indirect rule were in fact few and far between. Britain did not always find appropriate "native" institutions to use in her own imperial order. Sometimes artificial or synthetic chiefdoms were created with no real indigenous roots. In those cases the subsequent disruption has been less severe than in the case of Nigeria and Uganda. But the very creation of such chiefs and native rulers did help to increase ethnic consciousness within the subgroups -- and therefore reduced the prospects for national consciousness in the years ahead.

In those African countries which had a substantial white population during the colonial period, racial discrimination could have helped to unite different "tribal" groups against the white settlers and the imperial order. In some respects racism in those societies did have a unifying impact on those who were discriminated against.

On the other hand, it was precisely such regimes which organized African administration and settlement in "tribal" categories. The origins of South African "Bantustans" lie in the whole philosophy of "tribal reservations" in white-dominated Africa. Kenya and Zimbabwe were among the classical cases where some of the best land was reserved for whites and some of the rest was portioned out as native reserves. At least the South African regime had the good sense at one time to call its policy of homelands "separate development." The "native reserves" in colonial Kenya and Rhodesia were often cases of separate stagnation.

But what matters from the perspective of our concerns in this essay is once again the simple fact that British approaches to colonial rule, by being culturally relative and ethnically specific, helped to perpetuate and in some cases create the kind of ethnic consciousness which could seriously militate against nation building. Even a shared war of liberation against Ian Smith has failed to eliminate the legacy of cleavage between the Shona and the Ndebele in Zimbabwe. Indeed, that war was fought in a compartmentalized fashion, with Ndebele freedom fighters following the leadership of Joshua Nkomo and the Shona warriors responding to their own ethnic compatriot, Robert Mugabe. Kenya's own war of liberation in the 1950s was even more ethnically specific -- a war waged mainly by the Kikuyu and related smaller "tribes."

With regard to internecine wars among blacks themselves, there have been far worse ethnic convulsions in former British than in former French Africa. The longest civil war in the history of postcolonial Africa was that of the Sudan from 1955 to 1972. The British had ruled the Sudan in ethnic compartments. In some ways the South had the de facto status of a "Bantustan." As soon as the British left, the South wanted to push the logic of separate existence to its natural conclusion -- secession if need be. Civil war broke out which was to last some seventeen years.

Lord Lugard's pride and the apple of his eye in Africa was Nigeria. His own policies of indirect rule helped to set the stage for intercommunal massacres and three devastating years of internal warfare.

Uganda has not had a civil war proper, but ethnic strife and periodic violent eruptions have been endemic since independence, and have become more devastating since 1971.

It is true that former French Africa has witnessed a civil war in Chad which is at least as devastating as anything experienced in Nigeria, the Sudan or Uganda. But the Chad civil war has been less clearly ethnic than either the secessionism of Biafra or the separatism of Southern Sudan. In its final phases the Chad civil war is not separatist at all, but is a struggle to determine who should control Chad in its entirety.

Elsewhere in former French Africa nationhood has fared better than in Chad. A variety of factors go toward explaining this apparent success in nation building among the Francophones.

One factor is simply the comparative size of French-speaking countries. Nigeria alone has more people than the whole of former French Africa added together. On the whole it is easier to lay the foundations of nationhood in smaller and more compact units than in larger ones. This is not always true, as the cases of Rwanda and Burundi would illustrate, but, as a generalization, compactness in size facilitates national integration.

A related factor is relative ethnic homogeneity if the unit is small enough. Former French colonies in Africa are more ethnically homogeneous, on the whole, than former British colonies. Ethnic homogeneity, other things being equal, aids the process of national integration.

A third factor is the experience of being a small country in a larger confederation or community. During the colonial period the French ruled West Africa and Equatorial Africa as two federations. Each constituent country in that federation began to evolve its own national consciousness as against the rivalries of other members of the same federation. A framework of peaceful competition among the French colonies helped to enhance their broader regional awareness and at the same time deepened their narrower national consciousness.

As independence approached, the two French colonial federations in Africa were dissolved and the constituent members became independent separately. Senegal and Soudan attempted to form the Mali federation, but personality factors and national consciousness in each unit had already gone too far. The federal experiment collapsed, and Soudan retained the name Mali for itself as a separate national entity.

The fourth factor which helped lay the foundations of nationhood in the French colonies takes us back to the whole policy of assimilation. This included a high degree of educational centralization. The French system was rigorously applied to the colonies, with very little concession to indigenous traditions. While in the British colonies there were variations not only between one colony and another but sometimes between one cultural area in a colony and another, in the French empire no such variations were entertained. The educational system was designed to produce as rapidly as possible black Frenchmen and Frenchwomen. While schools in the British colonies allowed instruction at lower levels to be in indigenous African languages (the so-called vernaculars), in the French colonies the youngsters were plunged almost at the deep end of the French language.

On one hand, then, the policy of assimilation reduced pride in indigenous local languages and cultures in French colonies. On the other hand, that policy helped to create a more culturally integrated elite at the top than might otherwise have evolved.

And even when that elite gets restive and becomes defensive on behalf of African cultures, the tendency is to think of African culture at the macro level of national and continental heritages rather than at the micro level of ethnic legacies. Negritude is only one illustration of Francophone tendencies to think of culture in national and continental terms rather than at the "tribal" level. The macro approach to culture is itself an aid to nation building.

In some of the French-speaking West African countries there is an additional cultural variable which has contributed to nationhood -- Islam. The populations of countries like Senegal, Guinea and Mali are overwhelmingly Muslim. In contrast, there is no English-speaking African country south of the Sahara in which Islam has played as preponderant a role as it has in Senegal, Guinea and Mali. Muslims are the largest religious group in Nigeria, but their numerical preponderance is not as great as that enjoyed by Muslims in Senegal. There is probably a plurality of Muslims in Sierra Leone, but their political leverage has been well below commensurate levels. This is also true of the plurality of Muslims in Tanzania. Most observers do not even realize that there are more Muslims than Christians in

Tanzania.

But in the French-speaking countries of West Africa especially, Islam has played a role of great relevance for national integration.

These are some of the major factors and social forces which go toward making former French colonies more nationally integrated than former British colonies. The relevant factors range from the consequences of British indirect rule to the relevance of size for national integration, from the implications of French educational policy to the integrative role of Islam. Indeed, because Islam has been less preponderant in former British colonies, it has sometimes militated against national integration. Both Christianity and Islam in Nigeria are potentially divisive from a national point of view. Religion in Uganda and the Sudan has also been as much a hindrance as an aid in nation building. Tanzania has fared better, but the risk that Muslims might become increasingly discontented as they witness a disproportionate share of privilege enjoyed by Christians continues to hang over that country, especially in the years to follow the departure of Mwalimu Julius Nyerere.

Former French Africa has indeed witnessed Chad, where religious differences did play a part in the course of the civil war. But in retrospect it would seem that religion was so secondary that by the time the war reached its climax, the two contending sides in the struggle for Ndjamena were both led by Muslims.

On the whole, religion has been less divisive in former French colonies than it has in former British, though the reasons are more complex than can be fully explored in a short essay of this kind.

However, it should be noted that Islam is more than just a religion: it is also a civilization. Islam as a civilization has had a wider influence in Africa than Islam as a theology or creed. For example, the Swahili civilization in East Africa was originally Muslim, but it contributed to several countries in that region, regardless of religion, a very important medium of communication. The great majority of those who speak Swahili as a native language are still Muslims, but the native speakers are only a fraction of the total number of people who use Swahili centrally in their lives. The great majority of those who speak Swahili as a second language are non-Muslims, and are to be counted in millions. The impact of Islam on East Africa has been far deeper than can be gauged by the relatively small number of Muslims in the region.

Similarly Islam as a civilization has contributed to state formation, especially in West Africa. It is to this broad area of the history of statehood in Africa that we should now turn.

Colonialism and Comparative State Formation

Three civilizations played a part in the origins of indirect rule. These were Islam in Africa, British political culture at home, and British imperial experience in India.[2] We have referred to the principle of gradualism in British political culture and its effect on British approaches to colonial rule. The British raj in India had set the precedent of utilizing some of the maharajahs as "native authorities." These Indian precedents influenced the evolution of the doctrine of indirect rule in Africa. The third cultural force in the origins of indirect rule was Islam within the African continent. That is one reason why Northern Nigeria has remained the ultimate paradigm of indirect rule in action. As Lord Hailey came to put it in his classic An African Survey:

> It was in Northern Nigeria that this procedure of using Native Authorities was given a systematic form by Lord Lugard during the years which followed the declaration of the Protectorate in 1900. The area which was brought under British protection was the scene of the most effectively organized system of indigenous rule to be found south of the Sahara. Most of the old-established Hausa Kingdoms had embraced the Islamic faith and under its influence there had by the early sixteenth century developed a well-organized fiscal system, a definite code of land tenure, a regular scheme of local rule through local District Heads, and a trained judiciary administering the tenets of the Mohammadan Law.[3]

In this respect certain aspects of indirect rule, especially in Nigeria, helped to lay the foundations of statehood while at the same time causing a strain on the process of nation building.

Also as a factor behind Britain's impact in strengthening statehood in Africa was its policy of encouraging fiscal and budgetary self-reliance in each colony as far as possible. This contrasted sharply with the elaborate policy of budgetary subsidies from Paris to the African colonies. The English-speaking dependencies were encouraged to live within their means, and apply to Westminster for support only in terms of special projects on an ad hoc basis. London, unlike Paris, discouraged institutionalized budgetary subsidies for the colonies. By thus fostering fiscal and budgetary self-reliance, London was once again laying the foundations of statehood.

The third important difference concerned the local focus of colonial politics in the British empire as contrasted with the metropolitan focus of politics in the French empire. Some major African leaders in the French empire were deputies in French parliamentary institutions. Someone like Félix Houphouet-Boigny served in a number of cabinets of the government of metropolitan France. Léopold Senghor was also often a major participant in the politics of Paris, and is reputed to have been involved in drafting the constitution of the Fifth Republic of France. Such central involvement in metropolitan politics by "natives" of black Africa would have been quite inconceivable within the British empire. The idea of Jomo Kenyatta or Kwame Nkrumah serving as members of Parliament at Westminster, representing constituencies in Kenya or the Gold Coast, is still mind-boggling. Even more so would have been the scenario of their serving as members of the cabinet of Churchill's government, or the government of Harold Macmillan. Jomo Kenyatta and Kwame Nkrumah, once they returned from exile to engage in politics in their respective countries, were forced to focus on the politics of the colonies themselves, rather than the broad policy disagreements among British voters. This local focus in the politics of British colonies, as contrasted with the metropolitan diversion in the French empire, helped once again to lay the foundations of subsequent statehood.

But statehood is not merely budgets and politicians; it is above all administration and control. At least until the projected Mitterrand reforms are fully implemented, the French themselves domestically have had a relatively centralized system of administration and, by liberal democratic standards, the government continues to exercise a good deal of control over important areas of national life, ranging from education to the control of television. It is this same obsession with centralization which made the French empire somewhat more centralized than the British.

But that very imperial centralization slowed down the process of local colonial state formation. A substantial body of administrative directions and instructions had to come from Paris. A significant proportion of the personnel on the ground in the African colonies had to be French. And those French nationals who served on behalf of France in the colonies were not simply in the civil service of the particular African country, but were often members of the French Metropolitan Civil Service directly, usually recruited by metropolitan criteria or on the basis of metropolitan examinations, and subsequently enjoying the same privileges as their counterparts in France. Indeed, many of these colonial civil servants could be transferred to metropolitan France and vice versa.[4]

This high degree of metropolitan orientation in the administrative structure must have considerably slowed down the evolution of sovereign statehood at the local level in Africa. In most of those countries after independence the French factor in the administrative structure still remained quite significant. Major civil servants in key ministries were, in many cases, still French and ultimately accountable to the government in Paris. African presidents had a disproportionate number of French advisors. What is even more staggering is that one or two African presidents after independence have actually retained their French citizenship. This became dramatized when Emperor Jean-Bedel Bokassa was overthrown in the Central African Empire (later renamed Central African Republic). Bokassa's plane landed in Paris, and he demanded his right of access to France on the basis of his being a French citizen. As it happened, he was later persuaded to seek refuge in the Ivory Coast, where French citizenship among Ivorians is again by no means unknown. In English-speaking Africa, on the other hand, it would be inconceivable for an African head of state to be simultaneously a British citizen. Sovereign statehood in former British Africa is taken more literally than in former French Africa. The administrative and political structure as a whole was indigenized faster in English-speaking Africa than in French. The speed of Africanization was faster during the colonial period under British rule, and accelerated on the eve of independence and soon after.

This contrast also applies in the area of military sovereignty. It is about two decades since the following observation was made concerning the Ivory Coast -- and the observation is basically still true today:

> The most striking anachronism to the radical African nationalists is that M. Houphouet-Boigny has practically abdicated sovereignty in the military field. The Ivory Coast has only a small force for internal defense of the

> country. It has been forced to do so, M. Houphouet-Boigny says, because 'we wish to devote our modest means to economic and social development.'[5]

France has military accords or special secret agreements with most of her African colonies, and has reserved the right to intervene. Indeed, French readiness to intervene militarily in Africa contrasts sharply with considerable British reluctance to do so since the early 1960s.

When in 1981 the side supported by Libya gained the upper hand in the Chadian civil war, many French-speaking African leaders blamed France for not intervening in Chad in time. And France dispatched troops to one or two countries neighboring Chad, with full cooperation from the host countries and apparently with full approval of most of France's former colonies in the region. Any movement of British troops to African countries for the purposes of influencing the outcome of an African civil war would cause a major furor in English-speaking Africa, and perhaps on the world scene.

This is not to say that Britain has played no military role in Africa. There is a subtle understanding between London and Nairobi for joint exercises by the two armies. And the government of Daniel arap Moi has also extended special port facilities to American forces in the Indian Ocean. But Kenya is the only Anglophone African country with such an agreement about joint exercises with the former imperial power, Great Britain. None of the other English-speaking African countries have diluted their military sovereignty in that manner. In English-speaking Africa, Kenya is therefore an exception. In French-speaking Africa, on the other hand, special military accords with France, whether open or secret, are the rule rather than the exception. And the deployment of French troops even in former Belgian Africa (Zaire) is by far a more common occurrence than the deployment of British troops in the former British empire.

If sovereignty therefore includes a relatively independent military capability, or at least a genuinely reciprocal arrangement of collective security, former French Africa is in that respect also less sovereign than former British Africa. Most of the French-speaking African countries have, as we indicated, an understanding with France in military matters -- an understanding which is neither genuinely reciprocal nor designed toward maximizing African military independence. To that extent state formation in French-speaking Africa is again hampered.

Finally there is the general area of economic dependency in former French Africa. This has taken the form not merely of French subsidies to African budgets, but the whole monetary system in French-speaking Africa. There is a shared currency in the former colonies, and that shared C.F.A. franc is backed by the Bank of France. There is far less monetary sovereignty in the former French colonies to the present day than there has been in the former British colonies. This lack of monetary independence of course has its benefits. The shared currency of Francophone African countries is almost freely negotiable in the international money markets, and spares most French-speaking African countries the headaches of foreign exchange programs. Dependence on the Bank of France affords some degree of monetary reassurance. It becomes easy not only to import spare parts for much-needed equipment, but even to import bottles of drinking water from France. Monetary sovereignty as an aspect of modern statehood has almost been entirely abandoned by the majority of French-speaking African countries.

Finally there is the general flow of trade and investment. France continues to be the biggest trading partner of almost every former colony of hers in black Africa. This contrasts sharply with Britain's share of trade with her own former empire. In many cases Britain's share might still be larger than any single alternative partner, but almost never larger than all the other trading partners added together. On the other hand, the French share of the external trade of some of her former colonies is sometimes larger than the share of the rest of the world with those particular ex-colonies.

This French preponderance is even more dramatic in the field of foreign investment in the former colonies. France still manages to be successfully possessive in the way it handles the investment potential of its former colonies. There is a high tendency toward monopolizing French-speaking African countries, and other investors have a hard time competing with France.

This contrasts sharply once again with the tendencies within English-speaking Africa. France has had a much easier time cultivating new economic links with a Nigeria formerly colonized by Britain than Britain has had in cultivating economic links with almost any part of the former French empire. This is partly because Nigerians choose their economic partners with greater freedom than do Ivorians and Senegalese. The economic aspects of statehood are more advanced in Nigeria and most other British colonies than they are in most former French colonies.

But how are all these differences related to the comparative ethnic cultures of Britain and France themselves? In what way is imperial ethnicity part of the background to a situation in which we find Francophone Africa ahead of English-speaking countries in level of nationhood attained, but behind those English-speaking countries in level of statehood already established? It is to this background phenomenon of the imperial ethnicity of the French and the British that we should now address ourselves.

Teutonic and Latin Ethnicity

It goes without saying that the British and French have major cultural differences as two nations. But some of those differences go beyond the two nations and constitute broader variations between the Germanic or Teutonic side of Europe, on one side, and the Latin or Romanic side of Europe on the other. Some French characteristics or attitudes are widely shared by Italians, Spaniards, and Portuguese. Some British racial ideals are shared by many Germans, Dutch and white Anglo-Saxon Protestants in the United States.

When some of those aspects of ethnic culture came into contact with a new imperial role by those Western countries, it affected the nature and orientation of that imperialism. On the whole the British in their empire were more racially arrogant than the French, while the French were more culturally arrogant than the British. British racial arrogance included greater obsession with the physical segregation of races in the colonies, building not only social clubs but also separate restaurants, hotels, schools, lavatories, for whites as against blacks. The obsession was with segregating on the basis of color.

The French, on the other hand, preferred to segregate on the basis of culture. A highly Gallicized black man had a much better chance of acceptance in French social and cultural circles than a highly anglicized black man had in comparable British circles. Indeed, in some French colonies, if a black person became adequately French linguistically and culturally, those were grounds for becoming a French citizen. The logic of the policy of assimilation rested on a presumed supremacy of culture, rather than a presumed supremacy of race. Black people could become French -- the key was the French language and culture. This made a difference in social relations between immigrant French people from metropolitan France and the local highly Westernized elite. On balance relations between white Frenchmen and the highly Gallicized African were socially much easier than relations between white Britons and highly anglicized Africans in a colonial situation.

This British phenomenon can be related to the wider Germanic (Teutonic) cultures and racial attitudes. As Dame Margery Perham put it in her BBC Reith Lectures in 1962:

> The Germanic-speaking Europeans -- the British, the Germans, the Americans, the Dutch -- share a deep bias against intermarriage with the Negro race.... This conscious, or sometimes sub-conscious fear of race mixture accounts both for the [Germanic] white man's innermost ring of defense and also for all his outer ring of political, social and economic ramparts.[6]

Dame Margery does not follow this thought through, but it may not be entirely accidental that the architects of both Nazism in Germany and apartheid in South Africa were Teutons or Germanic speakers.

The idea of racial compartmentalization was also highly institutionalized in the United States, at least until the momentous Supreme Court decision of 1954 (Brown versus the Board of Education). Until then the American Constitution had sanctified the principle of "separate but equal." It was assumed that the practice of segregation was not a violation of the principle of equality. But the Supreme Court in 1954 at long last drew the conclusion that to segregate is to practice inequality.

Until then the Americans had called their racism by the euphemism of "separate but equal." Dutch-speaking white South Africans continue to call their own version of racism by the euphemism of "separate development." The United States has ended segregation legally, but residential segregation continues to be a dominant pattern almost all over the United States. Most residential areas which become 20 percent black tend to become before long 90 percent black. After 20 percent the exodus of whites begins, the prices go down, the poor move in while the rich depart -- but even the poor whites very often keep their distance as the neighborhood undergoes a racial metamorphosis.

As for that most imperial of all poems, Rudyard Kipling's "The White Man's Burden," it was perhaps fitting that it should have been written by someone who was in some sense both British and American, and that it should have been designed to persuade Americans to accept more readily their "burden" as members of the "white world" -- and colonize not only the Philippines but also beyond. The Americans did colonize the Philippines, and their racial attitudes toward the "natives" were not very different from those displayed by their British imperial counterparts in India at the same time.

The French have also had their own version of racism, but it has been less characterized by a horror of physical contact or sexual mixture than has the racism of the Germanic "cultures." The child of a French man with an African woman is less likely to be rejected or "abandoned" by the father than the child of a white Briton or white American in a colonial or slave situation. In 1979 Ghana was ruled briefly by a semi-charismatic military figure called Jerry Rawlings. His father was Scottish, and had maintained hardly any contact with the mother or the child. It is reported that Rawlings discovered his father's address in Britain and tried to contact him. This was before Rawlings' meteoric rise in the politics of Ghana. According to reports, the father refused to see him. And when Rawlings threatened to travel to the father's home, rumor has it that the father moved out to stay with some friends for awhile.[7]

Was the senior Mr. Rawlings trying to avoid his son because he did not want his neighbors in Britain to know that he had once mated with a black woman and produced a child? Or were the reasons unconnected with Teutonic embarrassment about racial mixture? Was the senior Mr. Rawlings the equivalent of the father of Chicken George in Alex Haley's <u>Roots</u>? The white American had raped one of his female slaves, and later treated the ensuing son as a slave. He even sold him subsequently to a British buyer, conceivably in Scotland. It was not unusual for American slave masters to sell their own children if the mother was a slave. This tendency was far less common in the Latin world of slavery, such as Brazil, Angola, Cuba and the French Caribbean Islands. A racially mixed child whose father was a white Latin had better life changes than a racially mixed child whose father was a white Teuton.[8]

In this regard we could take note of four different classifications of children of racial mixture. One system of classification might be called descending miscegenation. Under this system any child of racial mixture descends to the less privileged parent. A particularly striking illustration of this system is racial classification in the United States. If a black person and a white person mate, and produce a child, the child is black regardless of whether it is the father or the mother who is black. The child <u>descends</u> in status.

In contrast, a system of ascending miscegenation decrees that the child of racial mixture could indeed move up to the more privileged parent, depending upon whether it is the father or the mother who was more privileged. Among the Arabs, if the father is Arab the child is Arab, virtually regardless of the nationality of the mother. Among the Jews, on the other hand, if the mother is Jewish the child is Jewish virtually regardless of the nationality or race of the father. In the State of Israel, where Jews are privileged, this amounts to ascending miscegenation.

Divergent miscegenation asserts that a child of racial mixture belongs to a third group apart. A dramatic illustration of this is the situation of the Coloureds of South Africa. They are a population of mixed racial parentage and are classified as a group distinct from both blacks and whites in that society.

As for ambivalent miscegenation, this is a system within which children of racial mixture would go either up or down depending on such social indicators as the social class of the parents, the level of education of the children, or the precise shade of color of the children. In this last case, the question is not whether or not a child can "pass" as white but whether the system can accept a child as white because of the shade of color of its skin *in spite of knowing that the parentage was racially mixed*. Racial classification in places like Brazil is decidedly ambivalent in this sense.[9]

In Teutonic societies there is a strong tendency toward either descending miscegenation or divergent miscegenation. Children of racial mixture are either classified with the less privileged parent or constitute a third group apart. The Anglo-Indians in British India were a case of divergent miscegenation. So, as I have indicated, are the Coloureds of South Africa.

Children of racial mixture in Great Britain, the United States and Australia tend to be classified with the less privileged of their parents.

In Latin cultures divergent miscegenation is also quite recurrent. Concepts like mulattos and mestizos are rooted in assumptions about intermediate categories. But alongside those intermediate categories in Latin countries is the whole ambivalence about racial mixture.

Children do often move up to the privileged status of their fathers. And in the slave days in Latin countries it was rare for white families to sell the offspring of the master on the open market even if the mother was a slave. This was in contrast to the "Chicken George syndrome" in Haley's *Roots*. As we indicated, Chicken George was the offspring of a white master who had raped Haley's great-grandmother. Chicken George was classified as a slave -- and sold by his own father to a Briton. Chicken George, in other words, truly descended in legal status to the level of a slave -- in spite of the fact that his father was indeed the master. Such situations were much rarer in Latin slave systems than they were on Teutonic plantations.

To summarize then, Latin classifications of racial mixture were hardly ever based on descending miscegenation. Children of mixed parentage were not automatically "condemned" to the less privileged status. On the other hand, Teutonic systems of classifying children of mixed parentage have seldom allowed for ascending miscegenation. They have also rarely been ambivalent.

What the Latin and Teutonic systems have sometimes shared is the principle of divergent miscegenation -- allowing for the possibility that children of racial mixture should belong to a distinctive third group apart.

These lineage systems provide additional evidence that Teutonic or Germanic civilizations among white people have been on the whole more racially exclusive historically than Latin or Romance civilizations.

In recent times the Dutch have proved more difficult to designate neatly in terms of racism. Their record within the Netherlands has been impressively tolerant and racially accommodationist. But the record of the Dutch when transplanted to other societies abroad has been much less characterized by civility and tolerance. The Dutch in the Netherlands have been known to defy Hitler's anti-Semitism, accommodate large numbers of emigre Indonesians, permit impressively liberal immigration of colonials from Surinam, and try to be relatively restrained under considerable Malako provocation within their own boundaries.

But the Dutch overseas have often been Teutonic to the core. Their record in Indonesia was often one of considerable racial arrogance. As temporary settlers or planters in places like Ethiopia, they acquired a reputation for considerable racial exclusivity. And of course in the Republic of South Africa, whites of Dutch ancestry have constructed the most elaborately segregationist system ever devised by man. Thus, the most infamous contribution of the Dutch language to the vocabulary of African politics is indeed the term "apartheid." The Afrikaners continue to provide evidence for the proposition that Teutonic whites are racially more exclusive than Latin whites.

But what has all this comparative imperial ethnicity got to do with comparative nation building and comparative state formation in Africa? Let us now turn to the linkages.

Imperial Ethnicity and Colonial Policy

One area of linkage concerns the double proposition that while the British were racially more arrogant than the French, the French were culturally more arrogant than the British. French cultural arrogance assumed that Africa was a *tabula rasa*, a clean slate on which entirely new things could be written. There was a tendency to regard French culture as the only truly valid culture in human history. Bequeathing that culture to Africans was itself a supreme favor. Moreover, the French Revolution of 1789 implied that the past of any society could be abolished with one stroke, and a new future written immediately. Indeed, all drastic revolutions are based on the assumption that the history of a society is in some sense abolishable. Cartesian philosophy, which attempted to go back to the ultimate fundamental of "I think, therefore I am," characterized both the domestic metropolitan French revolutionary traditions and the external imperial French assimilationist tradition. French cultural arrogance helped to create normative integration in her colonies, and therefore helped to lay the foundations of nationhood.

By being less culturally bigoted, the British allowed for greater cultural diversity within the empire. Indirect rule was in part a product of that cultural relativism. And indirect rule helped state formation, but decidedly hampered nation building.

An alliance developed between British cultural relativism and British ethnic exclusivity. The cultural relativism led to greater respect for "tribal" cultures in Africa than was found in French colonial policy. Ethnic exclusivity, on the other hand, led to the creations of tribal "reserves" and a greater emphasis on ethnic segregation and the rudiments of "tribal" homelands. British racism in this case helped to encourage African tribalism. This is reflected in an even more exaggerated form among those other Germanic speakers in Africa, the Dutch-speaking South Africans. Their own racism is encouraging the compartmentalization of black South Africa into separate homelands for each "tribe." Germanic obsession with ethnic distinctiveness has led to a variety of African repercussions, ranging from the excessive protection of Northern Nigeria from Christian influences to excessive protection of Southern Sudan from Islamic influences, and from the logic of "tribal" agricultural reserves to the logic of racial discrimination in colonial Salisbury or colonial Nairobi. All of these trends militated against nation building.

In contrast, French tendencies were toward both micro-integration and macro-integration. Micro-integration among ethnic groups concerns intermingling within the same society or even within a single city or suburb. French social attitudes seemed more responsive to at least some levels of integration within each colony, some level of racial mixture within each society.

Macro-integration, on the other hand, concerns the absorption of whole societies into each other, or the readiness of the metropolitan center to regard distant colonies as part and parcel of the metropolitan heartland.

France's readiness to accept African deputies in institutions of the metropole as members of France's own parliamentary framework was an illustration of macro-integration. The designation of certain parts of Africa and the Caribbean as departments or provinces of metropolitan France was a related illustration of macro-integration. Even the tragic insistence that Algeria was part of France, and that the struggle for Algerian independence was ultimately a secessionist movement, was all part of this orientation to regard at least aspects of the colonial periphery as aspects of the metropolitan heartland.

That other major Latin colonial power in Africa, Portugal, moved in a similar direction of macro-integration. Portuguese colonies in Africa came to be designated as parts of metropolitan Portugal. There was a recurrent insistence that these were not colonies but parts of the metropolitan heartland. The Portuguese also had their equivalent of an assimilationist policy, radically different doctrinally from apartheid. While apartheid in South Africa insists on both micro- and macro-segregation among the different ethnic groups, Portuguese policy preferred racial intermingling. While apartheid in South Africa seemed to insist on a substantial degree of cultural separatism, including the discouragement of European languages in what was left of the rural areas, French and Portuguese policies encouraged the acculturation of the natives, the assimilation of Africans into the "superior" culture of the metropole.

But within Africa, France was in any case a better illustration of the Latin trend than Portugal. There was more effort behind the process of assimilation in the French empire than in the Portuguese. And there was more genuine centralization in France's relations with her colonies than there was in Portugal's relations with hers. Cultural assimilation and political centralization were two of the major trends born out of the Latin trend. Cultural assimilation was a contributory factor toward nation building in the African colonies, while political centralization between the colonies and the metropole hampered state formation in the periphery.

A reinforcing factor between Teutonic and Latin cultures was religion. The majority of Teutonic or German-speaking whites were Protestants, while the majority of white Latins or Romanic Europeans were Catholics. Protestantism encouraged religious decentralization, and sometimes religious apartheid. Under Protestantism national churches emerged -- the Church of England, the Dutch Reformed Church, the Church of Uganda and the like. The trend under Protestantism was not only toward individualism but also toward national compartmentalization. This compartmentalization had across nations reinforced the trend toward fragmentation.

In name the Roman Catholic Church might also appear to be nation-specific, and certainly named after the capital of a country, Rome. There has also been the Italian factor in Vatican leadership, as the great majority of popes have tended to be Italian. But the division within the Catholic Church has not been between multiple nations and ethnic groups, but at best between Italian leadership and a more representative global religious vanguard.

What is more significant from our point of view in this essay is that the Roman Catholic Church, like the French empire, has been substantially centralized. The Church has illustrated religious centralization within a multiracial religious community. The pope has been to the Roman Catholic Church what the French president has been to the French empire, a symbol of relatively centralized authority and a relatively integrated system of values and culture. The centralized authority made local authoritative formations in the periphery more difficult, while the integrated culture bequeathed by the center helped to make nation building in the periphery easier.

In the broad struggle for the incorporation of Africa into the legacy of Westphalia, we must therefore include something much older than Westphalia -- Christianity itself. But Christianity had in fact been torn precisely by the religious wars in Europe prior to Westphalia. The sectarian conflicts created their own tensions and their own cultural subdivisions. When these differences interacted with prior differentiations in the national cultures of the diverse European countries, new forms of moral chemistry emerged. The chemistry of Latin Europe produced cultural bigotry alongside relative readiness to tolerate racial mixture. Among the British the historic chemistry produced racial bigotry alongside readiness to accept cultural relativism. Imperial ethnicity was later to condition state formation and nation building in the African colonies. French colonies began to outstrip British ones in the rudiments of nationhood; British colonies were winning the race for the prize of statehood.

Conclusion

We have attempted in this essay to discern the precise relationship between imperial ethnicity and indigenous state formations in Africa. Imperial ethnicity encompasses ethnic perceptions and priorities within the culture of the body politic of the imperial power or powers. In the history of colonialism in Africa, imperial ethnicity has particularly encompassed racialism itself, and the basic cleavage between perceptions of whiteness and blackness in inter-human relations.

Indigenous ethnicity in Africa carries the heavy burden of "tribalism," the myths of shared ancestry in conflict with each other, the alliances of clans across a political divide.

We have sought to demonstrate in this essay that the total impact of British colonial rule, with special reference to the doctrine of indirect rule, impeded the process of national integration and the emergence of a shared national consciousness. Indirect rule paid respect to ethnic consciousness and cultural relativism, and therefore hampered cultural convergence or cultural synthesis in African society. In contrast, French policies of assimilation, combined with more compact territorial entities, facilitated cultural synthesis and therefore helped to promote national integration.

On the other hand, state formation in English-speaking Africa received a greater boost than in former French Africa. The French discouraged their former colonies from "excessive" administrative independence, or "excessive" military autonomy, or "excessive" fiscal or budgetary self-reliance, or "excessive" flexibility in choice of trading partners or sources of investments, or "excessive" monetary distinctiveness. All the usual attributes of sovereign statehood have been, within most of the former French colonies, significantly diluted. This trend has been in sharp contrast to the forces favoring autonomy and independence from Great Britain which have been evident in English-speaking Africa.

But our concerns in this essay have not been merely in contrasting former French colonies with former British. We have also attempted to draw attention to significant differences between French political culture and British political culture, in the wider context of the legacy of the Teutons and the Latins.

French-speaking Africa is not of course entirely former French, since it includes the substantial exception of Zaire, as well as Rwanda and Burundi. English-speaking Africa is not entirely former British either, since it includes Liberia with its American connection. But for our purposes in this essay the drama of Britain and France themselves is of the essence. We have therefore focused on the legacies which these two major European powers have bequeathed to Africa. Those legacies have yet to reveal the full range of their complicated implications for the African continent. What this essay has attempted to demonstrate is that in the slow struggle for viable nationhood in Africa, France has played a disproportionate role. On the other hand, in the painful process of constructing the instrumentalities of genuinely sovereign and independent statehood within the African continent, the British contribution has been the more striking. The Treaty of Westphalia had had a dichotomous effect on the African people. Nationhood is a level of identity, and France has helped to forge it; statehood is a level of authority, and Britain has helped to establish it.

Footnotes

1. The term "state" is used here in the macro sense -- as when we refer to the United States as a state. But there is also a micro sense of state -- as when we refer to the State of Michigan. Unless otherwise indicated, we shall be using the word in its macro sense in this essay.
2. It is arguable that the British imperial experience in handling French Canadians was also important in shaping indirect rule in Africa.
3. Lord Hailey, An African Survey, rev. ed. (London: Oxford University Press, 1957), pp. 453-454. I have discussed the relationship between Islam and indirect rule in other places as well, including my paper entitled "The Reincarnation of the African State: A Triple Heritage in Transition from Precolonial Times," prepared for a symposium "The Problematique of the State in Black Africa," organized by the Société Africaine de Culture and the Faculté des Sciences juridiques et economiques of the University of Dakar, in collaboration with UNESCO. Consult also Ali A. Mazrui, The African Condition, 1979 BBC Reith Lectures (London: Heinemann Educational Books; New York: Cambridge University Press, 1980) lecture 5, pp. 90-112.
4. For a brief account of the administrative structure consult Virginia Thompson and Richard Adloff, French West Africa (London and Lagos: George Allen and Unwin, 1958), pp. 179-226.
5. Henry Tanner's despatch, New York Times, March 25, 1962.
6. Perham, The Colonial Reckoning (Collins Fontana Library, 1963), pp. 64-65.
7. This information was given to me by a Ghanaian scholar who had got it directly from Jerry Rawlings. My conversation with the Ghanaian scholar took place in November 1980.
8. For the case of Chicken George see Alex Haley, Roots (Garden City, N.Y.: Doubleday, 1976). See also C.R. Boxer, Race Relations in the Portuguese Colonial Empire (Oxford: Clarendon Press, 1963); and Ali A. Mazrui, A World Federation of Cultures: An African Perspective, World Order Models Project (New York: The Free Press, 1976), especially chapter 5.
9. These categories are discussed more extensively in Mazrui, A World Federation of Cultures, chapter 5, pp. 81-116.

3 The State and Society in Africa: Ethnic Stratification and Restratification in Historical and Comparative Perspective

René Lemarchand

> The word [state] should be abandoned entirely: no severe hardship in expression will result; in fact clarity of expression demands this abstinence. -- David Easton

> L'Etat c'est Moi. -- A Kenyan Proverb

Some years ago the late J.P. Nettl made a convincing case for "integrating the concept of the state into the current primacy of social science concerns and analytical methods."[1] His urgings have gone largely unheeded by students of African politics. While considerable effort has been spent on exploring the internal dynamics of class and ethnicity, and asserting the claims of one over the other, the terms of the debate suggest that much of contemporary Africa is reducible to a condition of statelessness.

Neglect of the state as a concept is in part a reflection of the antistatist bias inherent in the structural-functional approaches that have dominated the field of African studies ever since the 1960s, and in part the result of a widespread dissatisfaction among "mainstream" political scientists with the postulates of Marxist analysis. Reinforcing these normative dispositions and reservations is the ubiquitous decline of state institutions in many parts of the continent. What sense does it make, one might ask, to give the state a conceptual existence where its basic empirical referents are nowhere to be found, or else so elusive as to defy analysis? Where are the institutionalized roles claiming a monopoly of political coordination, law-making, law-enforcement and coercion in contemporary Chad, Equatorial Guinea, the Central African Republic, Uganda?

Rather than to dismiss the concept altogether, a more fruitful approach is to recognize with Nettl that "more or less stateness is a useful variable for comparing [African] societies and that the absence or presence of a well-developed concept of the state relates to and identifies important empirical differences in these societies."[2] The "more or less stateness" argument is especially appropriate as a point of entry into issues of social stratification in colonial and postcolonial Africa. Not only does the passage from colonial status to independence signal important discontinuities in the relative "weight" of the state, and hence in its significance as an agent of stratification, but it also suggests a major shift in the relationship between state and society. For if the colonial state (and its corporate satellites) must indeed be viewed as the single most powerful agent of societal transformation in the history of the continent, the postindependence period is one in which the structure of social conflict is the key determinant of stateness, or statelessness.

It is this dialectical movement between state and society, in which different ground rules of social interaction emerge at different points in time, that we shall try to analyze and explain. We begin with a critical examination of the relevant literature on the subject, with special attention given to the views of Marxist and developmental analysts, and then move on to an exploration of the ways in which the dual processes of state formation and social differentiation have been linked in time and space.

The Nature of the State and the State of the Literature

If the rescue operation attempted here is to be more than a conjurer's trick it is essential at the outset to note the limitations of our enterprise. The notion of the state in Africa has never been objectified in the sense that it has in Western Europe. The colonial state has taken root in a soil that proved highly inhospitable to the institutionalization of a Western model. Traditional polities were incorporated by accidents of history into aberrant territorial frameworks from which some tried to escape as soon as they had the opportunity.

The state is an abstraction. It refers not simply to "government" but to the manner in which sovereignty is institutionalized. It is a shorthand formula to designate the formal organizations through which authoritative decisions are made. From this perspective it is tempting to visualize the development of African state institutions as a linear process involving the gradual elimination of traditional authority patterns and their replacement by inherited state institutions. Imperial legacies, to be sure, cannot be dismissed out of hand. The concept of the Napoleonic state, with its absolutist overtones and centralizing prefectoral corollary, served as the ultimate model for defining the political physiognomy of France's African colonies. In British and Belgian Africa, by contrast, the concept of the state was never culturally internalized by either colonizer or colonized to the same extent. Constitutional monarchies, after all, are conspicuously inhospitable to a strongly articulated notion of the state, and the British monarchy more so than any other. Yet no definitive conclusions can be drawn about the long-run consequences of the "Westminster model" as against the "prefectoral model" on African societies. Nowhere has either model taken root. The development of state institutions has been shaped much more by the nature of the crises experienced by African societies, and the types of reconstruction that followed in their wake, than by inherited institutions.

The legacy of the colonial state is perhaps best understood at the level of attitudes and political style. Colin Leys sees the most significant impact of colonial rule in "the amateur tradition of recruitment to the higher civil service," "the political power of conservatively orientated ministries of finance," and "elite- and urban-oriented school systems."[3] Perhaps even more important, however, has been (1) the attitude of ethnic or regional favoritism which in many instances characterized the policies and practices of colonial administrators, and which in effect prepared the ground for the collapse of the formal political institutions built on the eve of independence; and (2) the highly personalized, clientelistically oriented style of behavior displayed by the same colonial civil servants in their dealings with Africans, which in many instances helped to legitimize or reinforce similar tendencies already present in African societies. Political clientelism, of course, is by no means unique to Africa. Nor have the tendencies toward violence and political disintegration been everywhere countered through the organization of clientelistic machines. Patron-client networks are nonetheless a critically important component of African state systems. While there are wide variations in the extent to which they are institutionalized and penetrate formal political institutions, their continuing vitality suggests that they are one of the few functional alternatives available to African politicians to fill the void left by the decomposition of inherited state institutions or adjust these institutions to the requirements of stability and territorial cohesion. In short, the absence or weakness of state structures of the kind that one associates with a bureaucratic, legal-rational model does not exclude alternative modes of political organization, of which political clientelism is clearly one of the most significant.

To repeat: the state is an intellectual construct; its reality and historic specificity stem from the types of power relationships that have developed over time through different patterns of interaction and competition. On this elementary and almost self-evident proposition there is little room for disagreement. It is on the question of the relationships between state and society that the professional chasms that separate social scientists become apparent. At this level ideological consciousness creates its own reality and in the process the battles are joined at a level of discourse in which normative commitments and preconceptions inevitably take precedence over empirical observation.

A common characteristic of current Africanist scholarship is to treat the state as a dependent variable; as to what the independent forces are which conspire to rob the state of its autonomy there are fundamental divergences of opinion and interpretation. Thus implicit in the formulations of "liberal" development theorists is the assumption of basically neutral state institutions; although these institutions may respond differently to different social forces they are never able to exercise significant control over them. Social stratification, from this perspective, is not a function of the state but rather the opposite, a view which comes dangerously close to relegating the state to the status of an epiphenomenon. Again to quote from Nettl, "The existence of neutral and transformatory structures that merely process inputs into outputs and live up to the model posited by certain types of social science would run directly counter to the necessary conceptual component of autonomy."[4] This more or less systematic neglect of the autonomy of the state must be viewed as a major weakness in the formulations of development or modernization theorists; another is the tendency to reduce the social forces operating in the political arena almost exclusively to manifestations of ethnicity.

This is where classical Marxist theories depart most conspicuously from the previous school of analysis. Yet insofar as they posit the idea of the state as the instrument of a ruling class they, too, end up taking for granted what needs to be empirically established, namely that the state lacks an autonomy of its own. There are, to be sure, many variations on this theme, beginning with the composition of the ruling class, or the nature of the class alliances which may form the basis of the state, whether or not they are best defined by reference to relations of production or relations of power, and how far they lend themselves to manipulation by metropolitan interests. This is not the place to enter into a detailed examination of the various theoretical strands in Marxist analysis; suffice it to say that as a genre of literature it suffers from a reductionist bias which not only tends to exaggerate the significance of class as the determinant of stateness (while also paying little attention to differences of meaning between social class, political class, and ruling class) but for this very reason neglects the possibility of "fundamental conflicts of interests between the existing dominant class or set of groups, on the one hand, and state rulers on the other."[5]

This said, special attention needs to be paid to Richard Sklar's recent contribution to the subject of "class domination in Africa"[6] because of the conceptual links which it suggests between ethnic and class analysis and because it breaks decisive new theoretical ground in the analysis of class phenomena. The conceptual distinction between class formation, class consolidation, class identification and class action, together with the recognition that classes in Africa are rooted in relations of power rather than in relations of production, are basic to Sklar's argument. Once the principle of the <u>political</u> basis of class formation is conceded, the primacy of the state as an agent of class formation logically follows, along with the corollary proposition that there is considerable latitude in the kinds of mechanisms through which the state promotes social stratification. As Sklar explicitly notes, "The exertion of political power in the form of state action appears to overtake and outweigh more gradual processes of economic and social change."[7] If so, we can envisage situations where political classes crystallize at the center yet fail to consolidate themselves in any durable fashion beyond the narrow perimeter of the state apparatus; in many instances the disjunction between the political class and its ethnic constituencies in the rural hinterland is really a reflection of the gap between class formation and class consolidation, and both ultimately are reducible to the socioeconomic lag to which Sklar refers. In short, class formation does not automatically imply that class consolidation and identification must proceed at the same speed or with the same measure of efficiency.

Sklar's reconceptualization of "class" in terms of power relations also raises the question of what benefit, if any, can be drawn from the use of class as an analytic or descriptive category, as against "elites," "power holders," or "patrons." This is not just a semantic quibble. Each of these terms suggests alternative ways in which the state articulates with its subsystems. Terminological preferences thus tend to prejudge many of the critical issues involved in the analysis of social stratification. The term "class" suggests a rather more oppressive type of behavior than "elite" or "patron," and it implies the presence of a subordinate "social class" whose interests are by definition in opposition to those of the "ruling class." Although this may provide a fairly accurate picture of the situation which obtains in some settings, it nevertheless rules out the possibility of contextual shifts in the perception which rulers and ruled have of each other. Members of the ruling class may conceivably act as ethnic patrons, rural notables, bosses, or functional intermediaries of one kind or another depending on the social field in which they happen to operate. Unless one posits at the outset that every type of state action, including ethnic manipulation, patronage, and political bargaining, is really a class action in disguise, a proposition which I find difficult to accept, these contextual role redefinitions bring the realization that class solidarities are at best only one of several conceivable forms of social stratification generated by state action.

How collective identities interact with the state either to challenge or reinforce the order of stratification which it prescribes is largely a matter of empirical investigation. This process of interaction is not a random occurrence, however. It takes place within a set of parameters which remain fairly constant, irrespective of the time frame that one wishes to consider. Its key dimensions are (1) the strategies of territorial expansion associated with any given state; (2) the manner and degree to which certain "ground rules of social interaction"[8] are institutionalized in the wake of state expansion; and (3) how far, and how long, such ground rules are perceived as legitimate. It is in the light of these variables that we can perhaps best understand the conditions that lie in the background of ethnic and class formations in contemporary and historic state systems.

Class and Ethnicity in Precolonial Africa

In what is probably the most intriguing attempt by a social scientist to generalize about the political culture of precolonial Africa, Adda Bozeman sees in the atavistic belligerence of African societies the key to an understanding of both the "floating and unstable" nature of African identities and the warlike dispositions of their rulers. To quote:

> Political identities were floating and unstable because they were being shaped and reshaped by belligerent actions from without.... Conquest states, snowball states, multikingdom conglomerates, and other political systems arose in this fashion, yet few could maintain themselves in their newly acquired identity for any length of time.... Since territorial boundaries, ethnic components and centers of power were shifting everywhere, political identities, too, were always fluid, and phenomena that are in constant flux naturally elude precise definition by fiat or theory coming from without. Only one common criterion of self-definition

> emerges from the multifarious record, namely hostility to the "other."[9]

The argument reflects a fairly widespread, if misleading, notion of precolonial Africa as a continent where violence is the rule and stable identities the exception: the discerning reader may indeed recognize in Bozeman's formulation echoes of Professor Mazrui's "warrior tradition" thesis.[10] Without in any way denying the fluid character of African identities, the question arises as to just how ephemeral or elusive such identities really were.

There is no need to appeal to the historical record to establish what Bozeman refuses to acknowledge: that state-building involved more, or perhaps less, than just conquest, and that regardless of the means employed some state systems were able to persist as distinctive entities for centuries. The more important point brought to light by the work of historians is that processes of stratification and restratification were part and parcel of the social dynamics of Africa long before the advent of the colonial state, and that the passage from kinship-centered organizations to wider ethnic loyalties, and at times to something approximating a class-based type of social stratification, was related in some significant ways to strategies of state expansion.

Examples abound of kinship loyalties coalescing into a wider framework of social identifications in response to the expansion of centralized state structures. In very broad terms, one can outline two distinctive processes of restratification. The first, by far the most common, is one in which kinship identities are absorbed, as it were, into the cultural matrix of expanding polities. Speaking of the Lozi of Barotseland, in what is now Zambia, the late Max Gluckman noted in 1951, "They have absorbed many thousands of foreigners of several tribes of different cultures,"[11] a statement that would apply to countless other societies, including the Swazi, Sotho, and Tswana societies. Alternatively, the fusion of kinship ties may give rise to a separate form of identity in response to perceived threats from a dominant group acting as the vehicle of state expansion. The result has been either to juxtapose ethnically distinct communities, or, as Weber put it, "to transform the horizontal and unconnected coexistences of ethnically segregated groups into a vertical social system of super- and subordination."

The variations expressed in this statement are perhaps best illustrated by examples from Chad and Rwanda. As I have elsewhere tried to demonstrate, the emergence of a Sara identity is traceable to the nineteenth century slave raiding activities of highly centralized, predominantly Islamicized states, notably Bornu and Baguirmi, among the segmentary societies of southern Chad.[12] "Saraness" as an ethnic label was at first restricted to small groups of captives (mainly drawn from the Madjingaye subgroup); in colonial times, however, the term gained a much wider currency and came to designate a variety of culturally related yet distinct collectivities south of the Chari River. Although the concept of Saraness underwent considerable expansion under colonial rule, its ethnic connotations are thus traceable to the precolonial history of southern Chad. The point here is that the "vertical social system of super- and subordination" originally created by Muslim slave raiders transmuted itself into a pattern of "horizontal ethnic coexistence" during the colonial period (while in urban areas taking on some of the characteristics of a class system, of which more later). Exactly the reverse happened in Rwanda: the pattern of horizontal Hutu-Tutsi coexistence was eventually converted into a rigid system of vertical stratification.

The historic transformations that accompanied the rise of Hutu identity in nineteenth century Rwanda have been excellently analyzed by Catharine Newberry in her discussion of the processes of political incorporation in Kinyaga. Her conclusions apply to other areas as well. She shows how, prior to the nineteenth century, the inhabitants of Kinyaga lacked any sense of "corporate identity as Hutu"; only with the penetration of the peripheral marches by Tutsi chiefs from central Rwanda under the reign of Rwabugiri (1865-1895) did the fringe communities of Rwanda develop a new sense of social and cultural self-awareness. "As the political arena widened and political activity intensified, these [ethnic] classifications became stratified and rigidified; rather than simply conveying the connotation of cultural difference from Tutsi, Hutu came to be associated with and eventually defined by inferior status."[13]

These examples illuminate the two basic processes of segregative change that have attended the rise of "ethnic self-determination." In Rwanda, ethnic self-determination implied a drastic restructuring of state and society along revolutionary lines; in Chad, on the other hand, the result has been the detachment of Sara-speaking communities from the formal political space once occupied by the Chadian state, in other words, ethnic secession.

The case of Rwanda raises the question of the relevance of class analysis to precolonial African societies. Over the last few years a number of historians and anthropologists have drawn attention to the part played by changing modes of production in structuring African societies along class lines; thus in a recent study of "the tributary mode of production" in Western Uganda, Edward Steinhart argues the provocative thesis that "the state was thrown up by society only when a class society had developed, and within which the dominant class became conscious of itself as a ruling class and asserted itself as such."[14] The problem with this formulation is that it leaves unanswered the question of how classes developed in the first place, and whether the so-called "tributary mode of production" is really appropriate to describe the nature of the processes involved in class formation.

To speak of a "tributary mode of production" in precolonial Africa is to subsume under a single phrase a wide range of exchange relationships, not all of which were associated with a "mode of production," tributary or otherwise. If there can be little doubt that state expansion went hand in hand with the restructuring of peripheral communities, the ultimate shape of the stratification pyramid had a great deal to do with the particular forms of social exchange and types of resources through which new sets of relationships emerged between center and periphery. In some cases the result was the amalgamation of exogenous communities within a fairly egalitarian and flexible system of exchange; elsewhere social exchange led to social oppression and the rise of sharply differentiated status groups approximating the characteristics of a class system.

The process by which cattle were converted into a political resource under the *mafisa* system of Sotho, Tawana and Shona societies in the mid-nineteenth century differed in important ways from, say, the *gult* system under which Emperor Menelik extended the political control of the Shoa state to the southern provinces of Ethiopia. The incorporation of diverse ethnic loyalties under the *mafisa* type of "cattle contract" never developed into the kind of rigid socioeconomic differentiation described by John Markakis in his analysis of class formation in southern Ethiopia.[15] The expropriation and distribution of large tracts of land, coupled with the taxes and tribute exacted by the Ethiopian state, led to a situation where "the difference in status and power between the ruling group and the peasant mass in the south rendered the latter quite defenceless against abuse of authority."[16] The consequent perpendicular split between landlords (*balabbats*) and peasants (*gabbars*) may conceivably be described as a class cleavage; yet one wonders whether the oppressed peasants ever conceived of themselves as a social class until the 1974 revolution. Until then the basic rules of social interaction were primarily defined in terms of patron-client ties or landlord-tenant relations or ethnicity, depending on the context.

The task of analysis is rendered even more complex where different types of social exchange happen to coexist (and sometimes conceal each other) within the same political system. Elaborating upon the insights of Claudine Vidal,[17] Newberry emphasizes the critical differences between *ubuhake* and *uburetwa* as types of depending relationships in nineteenth century Rwanda and their continuing significance during the colonial era. Whereas the former involved a relatively benign form of patron-client tie centered primarily on chiefly authorities, the latter implied a far more oppressive form of dependency; while the former brought protection and opportunities for upward social mobility to the client, the latter brought little else besides exploitation; moreover, the *ubuhake* carried interethnic integrative connotations that were totally absent from the operation of the *uburetwa*, and often served to mask and indeed consolidate the dominant position of Tutsi elements.

Institutionalized mechanisms for the appropriation of human labor and agricultural resources thus differed significantly both within and among African societies. This is not the place for an extensive description of these various types of relationships; the essential point to note is that their effect on African societies is not reducible to any single type of stratification, any more than they can be associated with a single mode of production. What we have is a fairly broad spectrum of stratificatory phenomena falling somewhere between two poles: the pole of a vertical system of patron-client relationships based on mutually rewarding trade-offs, and that of a horizontal, class-oriented type of stratification in which coercion and exploitation are characteristic features of the social pyramid. It is the carry-over of this latter type of stratification into the colonial period which, in Rwanda and Ethiopia, provided the basis for a more fully articulated class system. Furthermore, just as in the past patron-client ties often served to mask the seamier side of the social system, in recent times much the same kinds of clientelistic strategies have been employed to conceal and consolidate class solidarities.

Despite frequent assertions to the contrary, especially by Marxist scholars, relationships among groups were just as forcefully conditioned by the ground rules of social interaction edicted by the state as they were by the operation of precapitalist modes of production. Situations of ethnic hegemony were maintained through the diffusion of widely shared religious and symbolic norms, resulting in the kind of hegemony conceptualized by Gramsci, i.e., "an order in which a certain way of life and thought is dominant...informing with its spirit all taste, morality, customs, religious and political principles, and all social relations, particularly in their intellectual and moral connotations."[18] Political symbols and mythologies express normative statements about the social order; they help legitimize societal interactions in certain ways, and lend a moral sanction to the priorities set from above. As such they play a critically important role in consolidating emergent patterns of stratification, including class and/or ethnic hegemonies. The drastic alteration of pre-existing rules of interaction under the auspices of the colonial state is perhaps the single most important variable in the background of identity change in contemporary African societies. It is at this level that one finds the sharpest discontinuities between the precolonial and colonial eras, and indeed the sharpest contrasts in the types of stratification with which each is identified.

Social Restratification and the Colonial State

Of all the forces that have contributed to the restratification of African societies under colonial rule, "social mobilization" impresses itself upon our mental retina as the most consequential. Stripped of all its Deutschian biases the concept brings into focus some major dimensions of social change: its usefulness for the analysis of the processes involved in the crystallization of ethnic and class identities no longer needs to be established. What is often missing from analyses of ethnicity is the element of choice involved in the restructuring of social identities.

The argument, in short, is that the colonial state has had a major impact on the restructuring of African societies by virtue of the political choices it was able to exercise at certain critical moments of its existence, namely at the inception of colonial rule, and on the eve of its demise. This is not meant to deny the significance of the societal transformations brought about through social mobilization. The point, rather, is that the outcome of the competition among socially mobilized groups has often been determined by the decisions of the colonial state. Not that these decisions have always been consistent: different conjunctures called forth different options. It is precisely this lack of consistency in the political choices of the colonial state which accounts for the drastic shifts that have occurred in the social structure of African societies.

To stress the interventionist disposition of the colonial state, as well as its capacity to decisively alter the shape of African societies, is not meant to minimize the existence of structural constraints on choice. Cases readily come to mind where the accelerating momentum of the nationalist crusade left the colonial authorities with no alternative but to yield to its demands (as happened in Guinea, Mali, Ghana and Tanzania). Again, the range of options available at the inception of colonial rule was by no means unlimited. Yet in a number of instances the terms on which collaborative bargains were made were of crucial significance in determining the modes of European expansion and its subsequent impact on the fabric of African societies.

The Expansion of Ethnic Boundaries and the Redefinition of Internal Dependency Relationships. African societies entered selectively into the consciousness of the European colonizer: some were seen as reliable allies, others as sycophants; some were thought to belong to "higher civilizations" while others were relegated to the rank of contemptible fetishists; some were seen as industrious and alert, others as inherently lazy, dumb and repulsive. Expediency, prejudice, cultural preconceptions and personal idiosyncrasies all played a part in shaping European perceptions. Ultimately the mental images which Europeans had of Africans served as a guiding system of norms for restructuring African societies.

The significance of this phenomenon is nowhere more evident than in the collaborative alliances formed between the emergent colonial state and those societies which, for reasons of convenience, opportunism or partiality, seemed most qualified to act as its mediator. As the cases of northern Cameroun, Rwanda and Buganda demonstrate, the result has been to substantially expand the territorial jurisdiction of particular ethnic groups while at the same time perpetuating or reinforcing the structural inequalities in existence within their respective social orders. Thus, in northern Cameroun, "under German rule all pagan populations, whether they were subdued by the Fulbe in the precolonial period or not, were forcibly put under Fulbe rule,"[19] and just as the Germans used the Fulbe as their most trusted and obedient collaborators in the administrative sphere, so did the French. Much the same phenomenon occurred in Rwanda, where the German *Schutztruppe* enabled the Tutsi monarchy to incorporate into its fold a huge chunk of real estate in northern Rwanda.[20] A similar scenario developed in Buganda in the wake of the 1900 Agreement, when "ten new counties were added to the ten historic ones," thus giving the kingdom a territorial base twice as large as its original size.[21] In each case the enlargement of ethnic boundaries was the most tangible proof of the benefits derived by specific societies from their privileged association with the nascent colonial state.

Of more immediate significance from the standpoint of this discussion are the patterns of restratification that have accompanied the emergence of these privileged state-to-state relationships. There has been, on the one hand, a more or less conscious attempt by the colonial state to maintain or reinforce pre-existing social inequalities among groups, and on the other hand to extend these inequalities to areas where they had yet to crystallize in systematic form. A classic example is of course Rwanda, where the "premise of inequality" was not only institutionalized to an unprecedented extent within the core areas of the kingdom but "exported" to the newly incorporated provinces. Extensive use and abuse of the *uburetwa* system, facilitated by the dismantling of the traditional chiefly hierarchies, decisively sharpened Hutu-Tutsi polarities in the central region while at the same time extending social inequalities to the outlying areas. In northern Cameroun and Guinea the abolition of domestic slavery implied a more flexible system of stratification, yet Fulbe hegemony remained basically unaltered: "Since the legal emancipation of slaves was not accompanied by any change in land ownership, the former slaves remained economically dependent on their masters who controlled the land.... So, generally speaking, the legal emancipation of slaves did not radically change their subordination to their Fulbe masters."[22] Nor did the early efforts of the French colonizer in Mauritania to "destroy all forms of hegemony of one race over the other, of one ethnic group over the other"[23] basically alter the traditional dependence of the _tribus tributaires_ on the _tribus libres_. In Mauritania as elsewhere the selective allocation of modern skills, including educational skills, to dominant groups meant that modernization would operate in similarly selective fashion, thus reinforcing ethnic and status differences.

While our previous examples suggest a possible coincidence of class and ethnic cleavages, the case of Buganda reveals a somewhat different situation. Here the expansion of ethnic boundaries went hand in hand with processes of internal differentiation _within_ Ganda society. As numerous observers have had occasion to note, the emergence of rural populism among Ganda commoners (Bakopi) in the 1920s reflected in part the incipient class cleavages created by the 1900 Agreement, which in effect converted the Bakungu chiefs into a landholding "aristocracy" or "rentier class," and in part the fact that "traditional obligation of labour and tribute owing to chiefs began to take on novel dimensions as the cash economy penetrated,"[24] ultimately leading to economic exploitation. Although Bakopi-Bakungu tensions expressed themselves in traditional terms, there is little question that class antagonisms played a significant role in pitting one segment of Ganda society against another. Significantly, just as the roots of Bakopi grievances are traceable to the direct intervention of the British Protectorate, formalized by the 1900 Agreement, it was also as a result of the initiative of the Protectorate that "the initial alliance of British and Bakungu was reversed,"[25] thus paving the way for a settlement.

The "differential incorporation" of African societies thus involved two different levels of differentiation: within the wider setting of the colonial state certain traditional state systems were able to achieve a privileged position in relation to lesser entities (as in the case of Buganda in relation to Bunyoro or Ankole); within the framework of the incorporated polities, however, a further process of internal differentiation took place which either reinforced pre-existing ethnic cleavages (as in Rwanda) or else brought into existence intra-ethnic divisions of a new type (as in Buganda), in which class differences became increasingly significant. What we have, in fact, are situations where incipient class cleavages associated with various forms of clientage arrangements have tended to coincide with, or cut across, ethnic identities.

Each level of dependency -- state to state, and class to class -- is of course intimately related to the other: the metamorphosis of traditional dependency relationships into something approximating class relationships can only be understood in the light of the massive backing of the colonial state to its mediating elites. While there is no gainsaying the tendency of the colonial state to allow certain traditional forms of dependency -- *uburetwa* in Rwanda, *tangata* in Malawi, *horma* in Mauritania, and so forth -- to be manipulated for the purpose of increasing its extractive capacities, it is equally in point to note that they could not have been converted into mechanisms of social oppression without the manipulators enjoying the full backing of the colonial authorities. Traditional forms of social exchange thus became identified with mechanisms of social oppression; patron-client ties increasingly tended to divide rather than unite, and where patrons and clients happened to share different ethnic identities class tensions only served to intensify ethnic conflict.

Let us also note the critical importance of the structure of the political arena, that is, the character of its ethnic *découpage*, in the reordering of collective identities. Where large-scale cultural aggregates were brought into the fold of a still wider arena -- as in Nigeria, Mauritania, the Cameroun, Guinea -- the tensions between patrons and clients were usually submerged and deflected onto a cultural plane, with ethnoregionalism asserting itself as a dominant force. Ethnic self-awareness became especially salient where changes were envisaged in the scale of the political arena which portended a further loss of cultural autonomy. Buganda's apprehension of federation with Kenya certainly helped reinforce the strength of ethnic loyalties among Ganda; and the fear that they might be incorporated into the jurisdiction of South Africa had a similarly catalytic effect on the ethnic consciousness of the Sotho, Swazi and Tswana. Only where the boundaries of the colonial state tended to coincide with those of the traditional polity (as in Rwanda, Burundi and Zanzibar) did the lines of cleavage inherent in the traditional society gain enough salience to eclipse wider cultural loyalties.

The Interplay of Class and Ethnicity. In most instances the political choices confronting the colonial state were the unintended consequences of the economic and social transformations dictated by the imperatives of economic viability and administrative efficiency. Efforts to harness traditional institutions to the requirements of a market economy spelled the breakdown of traditional dependency relationships and the rise of new social forces. Urban-based ethnicity and class formation have everywhere acted as the prime vectors of new loyalties and collective self-images.

At this level the state derives its instrumental significance from the allocation of "differentiated rights of access" to specific communities. In the previous section the emphasis has been on the part played by traditional modes of stratification, including clientage arrangements, in regulating such rights; here we wish to concentrate on the creation and distribution of new rights of access in situations where modern social roles increasingly tend to displace traditional ones.

Cutting through the vast body of literature on class and ethnicity, we shall limit ourselves to a discussion of the two basic processes by which new forms of stratification have come into being: (1) the convergence of class and ethnicity in urban settings, and (2) the extension of urban-based solidarities to the rural sectors. Both processes are intimately related to state policies.

The question of the origins of ethnicity in urban settings cannot be reduced to any single dimension, yet a reasonable case can be made for the view that the most durable and devastating forms of ethnic conflict have inevitably shown a convergence of ethnicity and class formation. An equally convincing argument is that the strength of urban-rural networks, and in particular their effectiveness in extending urban-based solidarities to the rural areas, is very largely a function of the social changes that have affected the rural communities.

At the risk of grossly oversimplifying the complexities of rural-urban interaction, two basic types of situations suggest themselves. In the first, class differences between ethnic groups transcend the urban arena and tend to replicate themselves in one form or another in the rural sectors, thus allowing the rapid consolidation of ethnic solidarities on a fairly broad scale. A more frequent type of situation is one in which ethnicity has relatively weak class connotations, remains a strictly urban phenomenon, and indeed a highly ephemeral one.

Both types of situations have been superbly analyzed by Crawford Young in his treatment of Ngala and Luba ethnicity in Zaire. In contrast with "Ngalahood," a highly transient phenomenon primarily restricted to the urban arena of Kinshasa, "Lubahood" developed into a far more enduring and all-pervasive frame of reference, reflecting in part the historic processes through which a Lulua-Luba differentiation occurred, and, more importantly, the comparatively privileged position of the Luba as a class both in the urban and rural sectors of the Kasai. Summing up the processes that led to the rise of Luba ethnicity, Young notes that "within a populace culturally and linguistically indistinguishable, came first a distinction, then through unequal rates of social mobility, a difference. The distinction (ethnicity) fused with the difference (class)."[26] In the case of the Ngala, by contrast, the distinction was never followed by a difference. Efforts to spread the concept of Ngalahood beyond the precincts of Kinshasa ended up in dismal failure; in the Kasai, however, the mobilization of Luba solidarities met with instant success.

Perhaps the most arresting example of a coalescence of class and ethnicity is that of the Sara in Chad.[27] As in the case of the Luba, the emergence of "Saraness" is a relatively recent phenomenon, and is intimately related to processes of class formation. Unlike the Luba, however, the Sara initially stood at the bottom of the heap. The historic subjugation of some Sara-speaking groups (essentially the Madjingaye) to Muslim rulers and slave-traders found an element of continuity in the social and economic discrimination that came to characterize the French colonial policies toward the Sara as a group. The roles of _déscendant d'esclave_, _maneuvre_, _prolétaire_ (and eventually "Communist") seemed occupationally consistent in the minds of French administrators, thus contributing to lend highly negative connotations to the status of Sara. Furthermore the fact that they were perceived as comparatively more "backward" than the Islamicized "tribes" of the north, and at the same time more physically fit, made them more vulnerable than other groups to labor demands. Tremendous social dislocations accompanied the massive recruitment of Sara laborers during the construction of the Congo-Ocean railway in the

1920s; the introduction of compulsory cotton cultivation in the 1920s and 1930s led to the rapid expansion of a rural labor force whose wages were barely sufficient to meet their tax obligations; meanwhile thousands of Sara were recruited into France's nascent colonial army. The net effect of all this has been to create a kind of rural proletariat of deracinated elements whose class consciousness in time was made to coincide with their own sense of Saraness. The key to an understanding of ethnicity among Sara lies in the relative ease with which modern-day politicians were able to give political expression to their collective sense of deprivation as an urban and rural underclass.

In what sense, then, has state intervention contributed to bring the social realities of class and ethnicity into focus? The creation of a taxable base among rural Africans, and the expansion of commodity production, were critical aspects of the penetration of the colonial state. The resulting dislocations shifted the focus of articulation of rural societies away from the clan and the family to a wider framework of interaction, in which ethnicity emerged as the principal focus of individual loyalties. Simultaneously, the inability of rural Africans to maintain their social obligations accelerated the flow of rural migrants to the towns, where new types of social differentiation came into being. While the towns served as the crucibles in which class and ethnicity fused together, the rural sectors experienced a pattern of restratification which often tended to fragment the rural communities <u>across</u> ethnic lines. The institutionalization of rich and middle peasants in Kenya created lasting divisions among the Kikuyu. So did the creation of a landlord-Bakungu "rentier" class among the Ganda, of a landowning, peanut-growing Muslim "saintocracy" among the Wolofs, of a coffee "plantocracy" among the Baoule, of a cocoa-farming rural "bourgeoisie" among the Ashanti. In each case new patterns of patron-client relations came into being which contained within themselves the seeds of rural class antagonisms.

To the extent that it acted as a vehicle of differential modernization the state was directly involved in the creation of new "tribes"; yet as new collectivities have come into being, their social identities have often tended to crystallize in class terms, and this as a result of the manner in which they gained entry into the modern sectors of the colonial economy. Insofar as the colonial state was able to define the ethnic parameters through which Africans were admitted into modern social roles, the result was a division of labor along ethnic lines. In a number of instances, and most notably in Chad and Kasai, the allocation by the state of "differentiated rights of access" was based on a normative scale according to which some groups were seen as superior to others, and in time these normative evaluations were projected into the collective images which Africans formed of themselves.

Ethnicity, in a sense, developed in dialectical contradiction with the colonial state: both changed in response to their mutually inflicted crises. Just as the penetration of fiscality, labor recruitment, compulsory crop cultivation, commodity production, and so forth, created the conditions that led to an extremely severe social and economic crisis among rural Africans -- a crisis of survival -- the new demand patterns generated by the surge of ethnicity in the towns confronted the colonial authorities with a crisis of "ethnic conflict management," often resolved by the institutionalization of ethnic discrimination. The deflection of ethnicity against the colonial state in time led to a fundamental questioning of colonial omnipotence, and thus to a crisis of legitimacy. It was at this point, where the crisis of colonial legitimacy tended to coincide with the crisis of political participation, that dramatic changes occurred in the collaborative bargains struck between colonized and colonizer.

The Reassessment of Collaborative Relationsips. Decolonization was but a strategic moment in an open-ended process of political disengagement, yet a crucial one. Although the rules of the game were drastically redefined, different rules applied to different places (and sometimes to different players). Depending on the type of constitutional formula devised by the colonizer, different forms of conflict emerged which in turn decisively influenced the character of African polities.

The important fact from the standpoint of this discussion is that the structure of ethnoregional conflict has conditioned the perceptions and choices of the departing colonizer in different ways. Seen in the wider context of African-metropolitan relations, and East-West relations, decolonization was not so much a repudiation of colonial omnipotence as it was a decisive phase in a process of neocolonial reconstruction. Local conflicts were generally linked to wider issues (most frequently to East-West issues), and thus the deciphering of ethnicity remained very largely a matter of identifying friends and enemies.

A wide variety of ethnic paradigms emerged in the wake of decolonization which in several instances came to reflect the basic choices made by European officials. One possible option was to recognize the dominant position already held by specific "intermediary" groups, and thus identify the character of the newly emergent state with their assumed cultural, economic and social pre-eminence. The Creoles of Sierra Leone, the Arabs of Zanzibar, the Ganwa of Burundi thus came to symbolize both the nature of European cultural preferences and the ethnic foundation of state institutions. Ethnicity in each case served as the principal source of objectification of the state. To speak of "ethnic hegemony" to describe the position of each of these groups would be inaccurate, however. They are better seen as mediating elites whose legitimacy, and by implication that of the state which they claimed to represent, lasted only as long as they could maintain a reasonable equilibrium between two other major ethnic constituencies, the Mende and the Temne in Sierra Leone, the Afro-Shirazi and the Mainlanders in Zanzibar, and the Hutu and Tutsi in Burundi. The "coherence" of their respective balance of power systems did not survive the escalation of ethnic demands, however, and by 1967 all three had moved in the direction of ethnic hegemonies.

A second type of option expressed itself in the form of a sudden switch of ethnic partnerships, soon accompanied by a drastic redistribution of power. The cases of Rwanda and Chad immediately come to mind. In Rwanda the trust authorities intervened massively on the side of the Hutu once the Tutsi were seen as potential allies of the

MNC-Lumumba in Zaire, and hence as Communist sympathizers. The Hutu revolution entered the consciousness of the *tutelle* authorities as a genuinely democratic, pro-Western movement; making Rwanda "safe for democracy" thus became the major preoccupation of the Belgian colonial state on the eve of its demise. By the time the country became independent in 1962 the Tutsi oligarchy had been effectively displaced from the seats of power, giving way to a Hutu-dominated state.[28]

In Chad the switch came earlier (in the mid-fifties), and in response to different concerns. Until the outbreak of the Algerian war, in 1954, it was among the communities of the Muslim north that the French picked their "interlocutors"; unlike the "fetishists" from the south, the Muslims were seen as belonging to a higher order of humanity, and French colonial preferences seemed corroborated by the predilection of southern politicians for the RDA, until 1951 affiliated with the French Communist Party. Félix Houphouet-Boigny's disengagement in 1951 paved the way for a reversal of alliances. The turning point, however, was the outbreak of hostilities in Algeria: by then the north came to be viewed increasingly as a reservoir of potential allies for the Algerian FLN, and the south as the most reliable partner in the evolving pattern of conflict between Sara and non-Sara. A very intimate relationship developed between the colonial state and its ethnic satellite, ultimately leading to a transfer of power into Sara hands.

In Chad and Rwanda the structure of conflict made it relatively easy for the colonial state to define its options; ethnic polarization created its own Manichean logic. Elsewhere, however, the power equations were more complex, offering a wider range of political choices. Zaire and Angola are obvious examples. In each case East-West issues served as the basic analytic prism through which "friends" and "enemies" were identified. And in each case massive external interventions followed the demise of the colonial state. Furthermore, in both Angola and Zaire a remarkable East-West symmetry developed in patterns of external intervention, thus creating a powerful multiplier effect on ethno-political rivalries. As several observers have shown, and John Marcum more conclusively than most,[29] from 1974 to 1975 East-West involvement in Angola greatly intensified communal violence between supporters of the UNITA and the FLNA on the one hand, respectively identified with the Ovimbundu and Kongo peoples, and the Mbundu-supported, urban-centered MPLA on the other. The extraordinarily rapid fragmentation of the Zairian political field in the months immediately following independence must be attributed to a similar projection of East-West rivalries into the arena of ethnoregional politics, with the added complexity of competitive rivalries among several presumptive "patron states," including Belgium and the United States.

Space limitations preclude a more detailed examination of these events; suffice it to note that (1) in a number of cases decolonization involved the exercise of fundamental political options by the departing colonial authorities; (2) these choices, though motivated by considerations that had little to do with local issues, have had a profound and lasting impact on the structure of ethnic conflict; (3) the collaborative bargains struck in the final stages of the decolonization process survived in one form or another the transition to independence and thus continued to shape ethnic solidarity patterns.

The Decline of State Autonomy

State-society relationships in postcolonial Africa are often visualized in transitive terms, with the state determining the shape of the stratification pyramid. Implicit in this formulation is the idea of the postcolonial society inheriting the "over-developed apparatus of state" and using it as a means of control over the "indigenous social classes."[30] The evidence, however, suggests a more complex dialectic: the specificity of the postcolonial state is both a reflection of the structure of the environment in which it was born, and of its relative capacity to regulate and control this environment.

The changes in the distribution of power brought about through colonial or neo-colonial intervention, the types of sociocultural configurations in existence in any given arena, the character and scope of traditional forms of legitimacy have all contributed to set the basic parameters within which the postcolonial state has tried to ensure its own survival. Just as the structure of the environment has conditioned the nature and salience of ethnic demands, the ability of the state to come to terms with these demands must also be viewed in the light of environmental constraints.

Not only the structure of the environment but the timing of ethnic demands as well as the degree to which they happen to "pile up" cumulatively deserves attention. Until May 1966 Uganda might have qualified as "the strongest example of a liberal polity surviving in Africa";[31] from then on, however, the issue of Buganda autonomy, coupled with that of the Lost Counties of Bunyoro, the incipient Bantu-Nilotic rift and allegations of corrupt practices within the army inexorably drove the political system in the direction of ethnic violence and instability. Moreover, the timing of ethnic mobilization holds critical implications from the standpoint of the availability or denial of political resources. What the Hutu of Rwanda were able to achieve with the massive backing of the *tutelle* immediately before independence could hardly be duplicated in independent Burundi. What seemed like a hopeless cause for the FROLINAT rebels prior to Col. Muammar Qaddafi's rise to power became a far more realistic goal after Libya decided to give its full military backing to the insurgents.

Crucial as they are in particular instances in shaping the course of state-society interactions, no definitive conclusions can be drawn from the play of systemic or conjunctural variables. Perhaps the single most important determinant concerns the nature of the strategies employed by the state in coping with ethnic demands.

The range of strategic choices available to the state is aptly summarized by Nelson Kasfir in his discussion of "ethnic political linkages" in postcolonial Uganda: "The center," writes Kasfir, "can respond to ethnic demands by taking the role of arbiter, bargainer, unilateral actor or capitulator."[32] Whether performed separately or in combination with each other, these multiple roles provide a useful analytic framework for identifying the kinds of interactions that have attended the restructuring of postcolonial societies.

The role of neutral arbiter is most effectively performed where "ethnic mediators" use their symbolic legitimacy to maintain a rough balance of power between two major groups of ethnic claimants. A prime example is Sierra Leone; until 1967, as Abner Cohen pointed out, the Creoles were able "to maneuver themselves into the position of 'functional strangers' who maintain the unity, stability, and continuity of the polity," a role made possible in part "because no viable political culture had been evolved by the provincials to replace that of the creoles," and also because of the fact that "the major ethnic groups, the Mende and the Temne, are of almost equal numerical strength."[33] The Ganwa of Burundi played a very similar role in relation to Hutu and Tutsi, at least until the fall of the monarchy in 1965. Yet the vertical system of stratification of Burundi society, along with the very uneven numerical strength of the two major ethnic groups, made it infinitely more difficult for the Crown and its Ganwa supporters to play the role of a neutral arbiter. Although neither system was able to sustain the strains of mobilized ethnicity, in Burundi the rise of Tutsi supremacy has led to a vertically ordered system of ethnic stratification for which there is no equivalent in Sierra Leone. The outcome of the 1967 coup in Sierra Leone produced a situation very different from that ushered in by the coup in Burundi: "When the army took over in 1967 it soon found that it could not rule the country and after a countercoup, handed reins of power to civilians -- the All People's Congress (APC) and the Creole-operated administration."[34]

Interethnic bargaining is a standard strategy for maintaining a measure of stability in an ethnically fragmented environment, the aim being to knit together a coalition of sorts at the center or to split ethnic segments so as to lessen their potential threats to the center. The structure of the reward system is crucial to the success of the operation. Where the rewards are essentially limited to the political elites (as in contemporary Zaire), with few benefits flowing to the rural constituencies, the result is to accelerate the process of class formation at the center while creating the conditions for ethnic revolts on the periphery. Interethnic bargaining is equally compromised where rewards are allocated "downward" to privileged ethnic segments. The case of Kenya offers an appropriate illustration: as a bargaining unit and one of Kenya's strongest political associations, the Gikuyu, Embu and Meru Association (GEMA) could hardly conceal the dominant and highly privileged position of the Kikuyu in relation to the Embu/Meru segments, a situation made obvious by the massive investments of capital resources in the Rift Valley, Western White Highland and Mau escarpment, as well as by the strategic position held by its ethnic (Kikuyu) patrons in the central apparatus of the state.[35] The dissolution of GEMA by President Daniel arap Moi in July 1980 indicates a major refashioning of the arena in which bargaining is now taking place, which might conceivably lead to a refusal on the part of Kikuyu "notables" to negotiate on the terms dictated by the state and hence to a remobilization of Kikuyu ethnicity.

Interethnic bargaining can mold individual and collective identities in a variety of ways, depending on the manner in which the Lasswellian formula ("who gets what, when and how") works itself out. Through the accumulation of economic resources at the center, it can generate marked inequalities of income distribution and thus stimulate processes of class formation among an ethnically diverse group of recipients; it can also discriminate among ethnic segments and encourage the mobilization of ethnicity against the state; or it can spawn networks of reciprocity between strategically located patrons and their ethnic clients and help institutionalize various types of clientelistic relationships. While all three forms of differentiation may conceivably coexist within the same political arena, they tend to interact in different ways.

Especially noteworthy in this respect is the differential impact of clientelism on class and ethnicity. Where patron-client exchanges operate across ethnic boundaries and result in a fairly equitable distribution of material resources to the rural clienteles, ethnic claims are usually absorbed within the more cohesive framework of interethnic clientelism. On the other hand, patron-client ties may also operate as a thinly disguised form of co-optation designed to mask the dominant position of the political class.[36] Short-run stability in this case depends on the willingness and ability of the ethnic clients to prevent the political mobilization of their respective constituencies. The scope of reciprocities is restricted to a narrow circle of political intimates, and manipulation usually takes precedence over allocative equity. The result is a highly fragile political coalition. Consider the case of Liberia prior to the 1980 coup: patronage networks were largely synonymous with jobbery; so far from promoting a genuine integration of "hinterlanders" and Americo-Liberians, the result had been to accentuate the former's sense of collective deprivation in the face of a growing coalescence of class and ethnic privilege in the hands of the Americo-Liberians. In Sierra Leone, by contrast, the post-1967 situation saw an extension of patronage networks to the point where they came to incorporate "an increasing range of people from outside the regional coastal core."[37] Summing up the implications of these contrasting forms of interethnic bargaining, Clapham correctly speculated in 1976 about "the possibility of political change in Liberia of a more radical kind than Sierra Leone seems likely to experience."[38]

The breakdown of interethnic bargaining brings into play the role of the state as "unilateral actor." Unilateral state actions may take the form of policies aimed at institutionalizing ethnic dominance, or, conversely, at inducing cultural integration by eliminating or dissipating threats of ethnic self-determination. In either case the result may be quite different from what had been anticipated. As the case of Zaire shows, neither the redrawing of provincial boundaries nor the proclamation of "authenticity" as the guiding ideology of the state was all that effective in diminishing the salience of ethnicity. The narrowing of administrative boundaries only served to displace the boundaries of ethnic identities to another level of interaction, while "authenticity" never went beyond the stage of vacuous sloganeering. Far more promising are President Shehu Shagari's current efforts to refashion the ethnic map of Nigeria. Along with the creation of new states, the selection of a new site for a federal capital, the setting up of new electoral constituencies and a totally new local government system, a crucially significant aspect of his federal project lies in the implementation of an equity-oriented revenue sharing formula for allocating federal income to the states. The structure of the reward system, together with the sheer volume of financial rewards made available through the advent of oil (representing 80 percent of federally collected revenue), has created a set of conditions eminently favorable to the promotion of an institutionally balanced pluralism.

The consolidation of ethnic hegemony through unilateral state action has produced similarly mixed results, and for much the same reason: the boundaries of ethnicity cannot be frozen once and for all by administrative fiat or cultural manipulation. Chad is a case in point: despite the efforts of the late President François Tombalbaye to create a Sara-dominated state, Sara ethnicity proved far less cohesive than had been anticipated. Recourse to the _yondo_ initiation rites as a means of solidifying intra-Sara solidarities, in 1974, was hardly sufficient to prevent internal divisions.

Ngambaye identity was the first to reach distinctive salience, expressing the relative deprivation of the Logone Occidental populations. Next to emerge as a political force was the growth of a Mbaye consciousness in the army, quickly relayed into the countryside via _anciens combattants_ networks. The last straw came with the revival of the _yondo_ ordeal, which led to a further alienation of Madjingaye elements from the regime. Even more striking, of course, is the extraordinarily rapid fragmentation of the northern insurgency once the possibility of capturing power finally entered their vision. Perhaps better than any other state, the recent history of Chad provides a graphic illustration of the "contextually shifting nature of ethnicity," to borrow Young's felicitous expression; and it also offers a telling example of "capitulation" as yet another possible response to ethnic demands.

As a residual category the "capitulator state" calls attention to those situations where the collapse of state institutions has as its immediate result the promotion of total anarchy and factional warfare (as in Chad and Uganda), the breakup of the pre-existing polity into separate territorial fragments (as in Zaire immediately after independence), or the virtual takeover of state institutions by exogenous actors (as in the Central African Republic after the fall of Emperor Jean-Bedel Bokassa, Equatorial Guinea after the fall of Macias Nguema, and the Comoro Islands during the Bob Denard-Ali Soilih interlude). These are by no means mutually exclusive possibilities. As these examples suggest, capitulation is as much a reflection of the growing inability of the state to cope with regional, ethnic, or factional threats as it is a consequence of the intrusion of external forces into the domestic arenas of African states. In a fundamental sense the decline of state autonomy is also a symptom of its extreme vulnerability to external threats; and to the extent that domestic factions or ethnoregional movements tend to identify their fortunes with those of outside forces, ethnic self-determination is likely to express itself in the form of externally assisted secessionist moves. The detachment of Mayotte from the Comoros under the joint sponsorship of Bob Denard and Ali Soilih, the virtual separation of northern Chad from Saraland under the combined efforts of Libya and the Goukouni faction of the FROLINAT, the secession of the Katanga under the auspices of the CONAKAT and its mixed bag of mercenaries, the simmering conflicts in Eritrea and the Western Sahara, all point to the mutually reinforcing effects of domestic cleavages and foreign intervention. The burden of rebuilding crippled polities may fall on different shoulders, including those of the former colonial state (as in Equatorial Guinea, the Central African Republic, and for a while Zaire); although the resulting patterns of stratification are as yet too fluid to be identified in any precise fashion, it seems reasonable to speculate that they will increasingly reflect a coalition of interests favorable to the former colonial states' international connections. At this point it will not be so much a threat of intervention as a threat of withdrawal that will most likely endanger the stability of their social structures.

The shifting sands of ethnic competition have affected African state structures in radically different ways. In some cases the effect has been to reduce the salience of "stateness" to the point where formal bureaucratic and judicial structures have simply ceased to operate, giving way to more or less fluid networks of local clienteles; elsewhere, however, state structures provide the institutional framework within which communities are slowly developing a new civic consciousness. Here the range of feasible compromise covers a wide spectrum. At one end of the scale Nigeria stands as the prime example of a federal formula through which a reasonably effective compromise of ethnic interests has been achieved, and where distributive justice helps solidify this promise. At the other end of the spectrum Zaire shows the structural weaknesses of a system in which federal arrangements only serve to formalize the differential incorporation of ethnic segments while the informal brokerage system continues to accentuate social inequalities. There are clearly significant variations in the extent to which "state coherence" can affect the structure of interethnic relations. What ultimately matters is whether African states can effectively exercise their right to self-determination and in the process mitigate the incoherence of their ethnic environments.

Footnotes

1. J.P. Nettl, "The State as a Conceptual Variable," _World Politics_, 20, 1 (1968): 559.
2. _Ibid_., pp. 591-592.
3. Colin Leys, "Political Perspectives," in Dudley Seers and Leonard Joy, eds., _Development in a Divided World_ (Baltimore: Penguin Books, 1971), p. 121.
4. Nettl, "State as a Conceptual Variable," p. 569.
5. Theda Skocpol, "State and Revolution: Old Regimes and Revolutionary Crises," _Theory and Society_, 7, 1-2 (1979): 9.
6. Richard Sklar, "The Nature of Class Domination in Africa," _Journal of Modern African Studies_, 17, 4 (1979): 531-552.
7. _Ibid_., p. 536.

8. The concept is borrowed from S.N. Eisenstadt, _Revolution and the Transformation of Societies_ (New York: The Free Press, 1978). See p. 29 ff.
9. Adda Bozeman, _Conflict in Africa: Concepts and Realities_ (Princeton, N.J.: Princeton University Press, 1976), pp. 202-203.
10. Ali Mazrui, "Phallic Symbols in Politics and War: An African Perspective," _Journal of African Studies_, 1, 1 (1974): 40-69; see especially Appendix I, "Political Masculinity and General Amin's Image."
11. Max Gluckman, "The Lozi of Barotseland," in Elizabeth Colson and Max Gluckman, eds., _Seven Tribes of British Central Africa_ (Manchester: Manchester University Press, 1959), p. 86.
12. René Lemarchand, "The Politics of Sara Ethnicity: A Note on the Origins of the Civil War in Chad," _Cahiers d'Etudes Africaines_ (forthcoming).
13. Catharine Newberry, "Ethnicity in Rwanda: The Case of Kinyaga," _Africa_, 48, 1 (1978): 17.
14. Edward Steinhart, "Herders and Farmers: The Tributary Mode of Production in Western Uganda," in Donald Crummey and C.C. Stewart, eds., _Modes of Production in Africa: The Precolonial Era_ (Beverly Hills: Sage Publications, 1981), p. 116.
15. John Makarkis, _Ethiopia: Anatomy of a Traditional Polity_ (Oxford: Clarendon Press, 1974); on the _mafisa_ system, see Neil Parsons, "The Economic History of Khama's Country in Botswana, 1844-1930," in Robin Palmer and Neil Parsons, eds., _The Roots of Rural Poverty in Central and Southern Africa_ (Berkeley and Los Angeles: University of California Press, 1977), pp. 113-143. See also Philip Bonner, "Classes, the Mode of Production and the State in Pre-colonial Swaziland," in Shula Marks and Anthony Atmore, eds., _Economy and Society in Pre-industrial South Africa_ (London: Longman, 1980), pp. 80-101.
16. Makarkis, _Ethiopia_, p. 117.
17. Claudine Vidal, "Le Rwanda des anthropologues ou le fetishisme de la vache," _Cahiers d'Etudes Africaines_, 14, 3 (1969): 384-401; Catharine Newberry, "_Uburetwa_ and _Thangata_: Catalysts to Peasant Political Consciousness in Rwanda and Malawi," _Canadian Journal of African Studies_, 14, 1 (1980): 97-112.
18. John M. Cammet, _Antonio Gramsci and the Origins of Italian Communism_ (Stanford: Stanford University Press, 1969), p. 204.
19. Victor Azarya, _Aristocrats Facing Change: The Fulbe in Guinea, Nigeria and Cameroon_ (Chicago: University of Chicago Press, 1978), p. 66.
20. René Lemarchand, _Rwanda and Burundi_ (London: Pall Mall Press, 1970), p. 56 ff.
21. Crawford Young, "Buganda," in René Lemarchand, ed., _African Kingships in Perspective_ (London: Frank Cass and Co., 1977), p. 207.
22. Azarya, _Aristocrats Facing Change_, p. 75.
23. Rapport du Conseil de Gouvernement, June 20, 1910, cited in F. de Chassey, "Les structures sociales en Mauritanie," in _Introduction à la Mauritanie_ (Paris: CNRS, 1979), p. 238.

24. Young, "Buganda," p. 210.
25. Ibid.
26. Crawford Young, The Politics of Cultural Pluralism (Madison: University of Wisconsin Press, 1976), p. 117.
27. The information in this paragraph is drawn from my article on "The Politics of Sara Ethnicity: A Note on the Origins of the Civil War in Chad."
28. For further information see René Lemarchand, "The Coup in Rwanda," in Robert I. Rotberg and Ali A. Mazrui, eds., Protest and Power in Black Africa (New York: Oxford University Press, 1970), pp. 915-923.
29. John Marcum, The Angolan Revolution, vol. 2 (Cambridge, Mass.: MIT Press, 1978), p. 221 ff.
30. John S. Saul, The State and Revolution in Eastern Africa (New York: Monthly Review Press, 1979), p. 169.
31. G.F. Engholm and Ali A. Mazrui, "Violent Constitutionalism in Uganda," Government and Opposition, 12, 4 (July-October 1967): 585.
32. Nelson Kasfir, The Shrinking Political Arena (Berkeley and Los Angeles: University of California Press, 1976), p. 162.
33. Abner Cohen, The Politics of Elite Culture (Berkeley and Los Angeles: University of California Press, 1981), p. 139.
34. Ibid., p. 143.
35. I am indebted to Denis Martin for drawing my attention to the GEMA and the complexities of its interethnic networks.
36. For a further elaboration of this theme, see Peter Flynn, "Class Clientelism and Coercion: Some Mechanisms of Internal Dependency and Control," Journal of Commonwealth and Comparative Politics, 12, 2 (1974): 133-155.
37. Christopher Clapham, Liberia and Sierra Leone (Cambridge: Cambridge University Press, 1976), p. 124.
38. Ibid., p. 122.

4
Problems and Prospects of State Coherence

Goran Hyden

While favoritism based on ethnic affinity, and tension ensuing from this practice, is commonplace throughout Africa, it is interesting, given the artificial nature of colonial boundaries and the insistence of Africa's postindependence leaders to uphold the legitimacy of these boundaries, that so relatively few cases involving claims for ethnic self-determination have arisen. The latter cases, e.g., Biafra, Eritrea, Ogaadeen and Southern Sudan, have stolen the limelight in the literature; yet equally significant from a long-term development perspective are those instances where state coherence has been sustained. It is argued here that by examining how African governments have been able to prevent ethnic self-determination from becoming an issue provides an important clue not only to the effectiveness of ethnic conflict management but also to the overall problem of development in Africa. Thus, this paper tries to bring together three sets of issues -- (1) ethnicity; (2) the nature of the postcolonial state; and (3) development policy -- all of which have tended to be discussed in isolation from each other.

The Political Economy of Africa

The political economy of Africa has many unique features which make it difficult to subsume under conceptualizations and categorizations derived from the historical experience of other societies. Much of the debate on the state in postcolonial Africa has been premised on the assumption that a capitalist mode of production is predominant: the primary evidence produced in support of this argument has been the dependency of the African countries, as suppliers of raw materials, on the metropolitan, industrialized countries of the West. Consequently, there has been a tendency to look at the political economy of Africa mainly as a replica of capitalist society. To be sure, there have been attempts to modify the categorizations to fit the more backward nature of peripheral capitalism.[1] Nonetheless, the looking-glasses have remained essentially Western. The roots of underdevelopment have been attributed to the presumed pervasiveness of capitalism and its alienating and marginalizing effects.

In this perspective, the African bourgeoisie has been invariably portrayed as being unwilling or incapable of reducing the dependency of its countries on world capitalism. Both Leys[2] and von Freyhold,[3] for instance, pinpoint the structural constraints preventing the African bourgeoisie from challenging international capital in an effective manner. John Saul, on the other hand, sees more scope for action for the African bourgeoisie but doubts its willingness to challenge the metropolitan agencies of capital.[4] He argues that it possesses a revolutionary potential but wonders whether it will be realized except in the conditions of a liberation war. These and other participants in the debate about the postcolonial state in Africa have generally seen more of capitalism present on the continent than really is the case. One can legitimately ask whether the pertinent feature of the African scene isn't instead the relatively limited penetration by capitalism of the African economies. Certainly, there is much more to the African political economy than what an analysis of the capitalist mode of production alone suggests.

It is only more recently that writers in the Marxist tradition have begun to recognize that the prevailing economic and social organization of African societies is determined by forces other than capitalism. The notion is beginning to take root that precapitalist formations are still significant and in a position, as for instance Rey and Dupre have argued,[5] to block the progress of capitalist penetration. One reason for the tardiness with which this argument is gaining acceptance is probably the absence in sub-Saharan Africa of the articulation of any already known precapitalist mode of production i.e., feudalism or the so-called "Asiatic" mode. Because of the difficulty of categorizing social realities in Africa using the terms of such modes of organization and production, most authors seem to have satisfied themselves with the application of the nomenclature of a capitalist society. There is now at last a debate emerging about contemporary precapitalist articulations in Africa. Spurred by the crisis in agricultural production experienced by most African countries in recent years, this debate recognizes that immobility of resources is such as to hold back the kind of cumulative development process with which more advanced countries have become familiar.[6] While there are still arguments about how to conceptualize the problem, greater attention is being paid to the social and political implications of the prevalence of miniscule peasant households, a historical anomaly hitherto hardly recognized at all in the literature on Africa.

The present author has elsewhere tried to tackle this problem,[7] advancing the point that underdevelopment in sub-Saharan Africa cannot be adequately conceived unless recognition is taken of the existence of a peasant mode of production and the economy of affection. A particularly interesting and relevant aspect of this argument concerns the relation between state and peasant producers in Africa. Because peasants possess their own means of production, i.e., land, which significantly also serves as their means of subsistence, those who rule do not relate to these producers in the same fashion as capitalist to worker or feudal landlord to serf. Relations of production, the standard determinant of power in the Marxist paradigm, do not enter the social and political equations to the same extent in Africa as elsewhere. Where feudalism, capitalism or socialism has been effectively established, rulers control structures which allow them to shape in an irrevocable manner the perceptions, needs and preferences of other groups of people in society. Because people have been alienated in large numbers from reaping the immediate fruits of their labor, they are dependent on the prevailing economic system and thus in the hands of those who have the opportunity to shape and control it. In his analysis of the concept of power, Lukes aptly refers to this as the supreme exercise of power.[8]

This situation, however, does not prevail where the economy is dominated, or at least strongly influenced, by large numbers of small and independent peasant households. Under a peasant mode, production units are not only largely independent of each other but also of the state. As Hindess and Hirst have argued in their analysis of precapitalist modes of production, in the case of the peasant mode, appropriations by other classes through such instruments as taxation are simple deductions from an already produced stock of values.[9] The state does not enter their system of production in a functionally axiomatic manner. As a result, peasants don't need the state for their social and material reproduction. On the other hand, it is impossible to conceive of other social classes reproducing themselves without access to state power. History shows that a prime target of this state power has always been the peasantry. It is via the back of the peasantry that lords have risen to power, as Barrington Moore has shown with great brilliance.[10] Africa is no exception, but it differs from other parts of the world in the sense that this process is still only at an incipient stage. That is, the peasants are far from effectively captured by the "system," and thus those who control the state do not necessarily control society. To be sure, the degree to which this phenomenon prevails varies from one country to another. For instance, the peasantry of the Kenyan highlands is closer to being fully captured than, say, the majority of the peasants in Tanzania. Yet what we witness throughout Africa is a struggle by those in power to subordinate the peasantry, not a situation where they already have at their disposal the strings to pull, or instruments to apply, for effective results. That is why in Africa, the implementation record in the rural sector has been so poor.

While it has been frequently argued that this is the result of a colonial hangover in the administrative machinery,[11] that is, administrative practices were introduced by the colonial rulers which have little in common with the development ambitions after independence, this point does not go to the root of the problem. Poor policy implementation stems in a more fundamental sense from the structural conditions of a peasant economy which limit access of the rulers to the producers. Thus, the postcolonial state, as much as the colonial one, constitutes a "foreign" body to the peasantry. It has not sprung spontaneously out of material conditions prevailing in society. Instead, it is an institution without roots in society and suspended, as it were, in mid-air above society.

One implication of this condition is that politics in Africa tends to be overwhelmingly clientelist in nature. Where the state is not effectively plugged into ongoing production processes, articulation of ideology lacks the materialist foundation that we are familiar with from other societies. While there is no reason to doubt the sincerity of the ideological outlook of African politicians, it isn't derived from specific relations of production. As a result, it is not surprising that in spite of similar material conditions, the ideologies of two neighboring African states can be radically different. Each tends to justify no more than the outlook of a certain grouping of clientelist networks. Because of the inevitable fluidity of those networks, ideological emphases can change quite radically over a short period of time without having any significant impact on the material conditions in society.

Another implication is that political representatives in a society where the peasant mode of production still prevails do not speak on behalf of established capitalist or socialist institutions in which they have a strong personal stake, but on behalf of an unspecified number of rural producers with an interest in social welfare facilities in their respective home areas. Under colonial rule such facilities were provided in no larger quantities than strict macro-economic calculations permitted. Social welfare policies reflected a fundamental capitalist outlook and it was more often such biases than racial prejudices -- the latter only reinforcing the former -- that in the end made representatives of the colonial authorities unpopular. After independence this situation has radically changed. People making policies are no longer members of a capitalist establishment. As they try to prove their relevance to society, however, they inevitably end up as spokesmen of parochial constituency demands which go contrary to macro-economic considerations. This phenomenon is particularly manifest at times of election,[12] but is being practiced in less visible forms also in regular policy-making contexts. It can of course be argued that this conflict between micro and macro interests is not at all peculiar to Africa. Yet there is a significant difference between the African situation and that of other parts of the world. Where most social actors are caught in a system, certain parameters are accepted as given because both parties to a conflict stand to lose if they are ignored or transcended. The prospects of arriving at a mutually acceptable solution through negotiations are quite high. In a society like the African, however, where the bulk of the production units are not only quite autonomous from each other but also from the state, there is considerable fragmentation and little sense of an economic system setting definite limits to action. The micro as opposed to the macro perspective tends to prevail, or, as Ekeh has described it, the "civic public realm" (i.e., the state) lacks the legitimacy that is associated with the "primordial public realm" (i.e., the local community, the clan or the tribe.).[13] The latter is a reservoir of moral obligations which one works to preserve. The former, however, is a place from which one seeks to gain, if possible in order to benefit the primordial public realm.

The Economy of Affection

The "primordial public realm" is an acronym for "ethnicity" and it captures an essential aspect of African politics. There is no immediate translation of this into political economy language; yet it is a phenomenon that forms an integral part of Africa's contemporary political economy. It is an articulation of the peasant mode of production and we will refer to it here as the "economy of affection." There is some merit in thinking of it in these terms as there is no doubt that throughout Africa there is not only an alternative economy but also an alternative economic outlook to that associated with capitalism or socialism. Through the growing interest in "informal sector" activities, this alternative economy is beginning to become recognized.[14] Its proper conceptualization, however, is still a matter of debate. The notion of the economy of affection is used here because it captures the essence of its mode of operation. It suggests that familial and other communal ties provide the basis for organized activity. While ethnic affinity may be one such criterion, the economy of affection is broader than that. It incorporates organized activities that may cut across tribal or ethnic boundaries, yet are not an expression of either capitalism or socialism. Its principal difference from the latter two lies in the fact that economic action is not primarily motivated by the search for a surplus but is embedded in a range of social considerations that allow for redistribution of opportunities and benefits in a manner which is impossible where modern capitalism or socialism prevails and formalized state action dominates the process of redistribution. Based on social ties growing out of ethnicity and village or regional connections, this economy distributes social welfare, financial support and government benefits in return for political support. Thus it lacks the regular cost-benefit considerations associated with economic action under either capitalism or socialism.

The economy of affection is not just a backyard phenomenon. Nor is it only an expression of fond emotions. It denotes a network of support, communication and interaction among structurally defined groups connected by blood, kin, community or other affinities (e.g., religion). One of its most interesting aspects is that it serves to link systematically a variety of subsets or discrete units which in other regards may be autonomous. It is not only confined to economic action at the rural household level. It is at the bottom of the rural-urban continuum that has been so extensively documented in the literature.[15] What is even more important, its principles penetrate much of the economic and social policy at government level. Although normally pursued under the guise of some modern ideology, particularly any variety of socialism, social welfare policies in Africa really reflect the principles of the economy of affection more than anything else. Had Africa's leaders really been captured by capitalist or socialist principles they would easily have arrived at the conclusion that many of their policies, e.g., universal primary education, provision of domestic water supply points in the rural areas, or the extension of rural health clinics, are unfeasible at the pace and in the form they pursue them. There simply aren't the levels of capital formation in the countries to sustain such an expansion of the social sector, even in cases where foreign aid is generously provided for these activities.

The Instrumental Nature of Ethnicity

But why have African politicians been willing and able to persist in their pursuit of such an approach to development? Here it may be helpful to link up with the debate about ethnicity in Africa. It has gone a long way since the days of independence when ethnicity was seen primarily as a characteristic of the rural masses. This perspective was largely determined by the struggle for independence which brought together educated and "Westernized" members of the emerging African "elite" for political purposes. They were regarded as the "detribalized" members of society having left "behind" the majority of the people for whom thinking in primordial terms was seen as the natural state of mind.[16] The political leaders were the political heroes of the early days of independence. They were the catalysts of the new nations. A combination of modernization measures, notably education and political mobilization, was seen as necessary to bring about a change in the rural areas and thus reorient the minds of the masses in a nationalist and modern direction.

Traumatic events, such as the Nigerian civil war, have since altered the perspective on this issue. In the course of the 1960s it was realized that modernization does not only detribalize people but it also tends to introduce an element of competition at the elite level. Such competition, as particularly the Nigerian example brought home, can easily get out of control. State coherence was being endangered as competition for educational opportunities and jobs made members of the political elite more aware of their ethnic identities.[17] Some have argued that this trend reflected merely selfish interests on the part of the political elite but others have clearly pointed to the fact that fellow-tribesmen put the elite under pressure in their desire for "modernity." In the language used here, it can be said that the elite succumbed to the pressures of the economy of affection. To satisfy the local demands they had to organize, and in many countries ethnic unions emerged as a political force. The dilemma facing members of the urban elite has been captured in the following words by Abernethy:

> What was the best course of action open to the urban migrant who was acutely concerned lest his ethnic group fall behind others in the struggle?... Certainly, the rural masses had to be informed of the problem. If the masses were not aware of their ethnicity, they would have to learn who they really were through the efforts of "ethnic missionaries" returning to the homeland. These "missionaries" would also have to outline a strategy by which the ethnic group, once fully conscious of its unity and its potential, could compete with its rivals. Clearly the competition required enrolling more children in school, particularly at the secondary level for the graduates of a good local secondary school would be assured of rapid...mobility within modern society.[18]

In a review of the literature, Bates concludes that ethnicity is both a dynamic and rational form of political behavior: it represents a sensible attempt to deal with, organize, and benefit from the modernization of societies.[19] Ethnicity is not a phenomenon associated with deeply held values only; nor is it merely the attribute of rural peoples living in social isolation. It is as much associated with the urban elite and is often an integral part of their political calculations. Authors like Kasfir[20] and Young[21] have added to this the observation that the ethnic identities which people assume are both multiple and intermittent.

According to these contributions, the most prominent feature of ethnicity in Africa is its "instrumental" nature. Group boundaries are quite open and flexible. It contrasts with ethnicity of the more "consummatory," i.e., more uncompromising, kind. The latter tends to develop particularly in situations where groups are exposed to attempts at colonization by others.

The instrumental nature of ethnicity in postcolonial Africa is perhaps best explained by reference to the economy of affection. It is by adopting its principles that political leaders have been effective in regulating ethnic conflict. The persistence of African leaders in pursuing redistributive policies irrespective of their costs to the system, therefore, is explained by the relative effectiveness of these policies in maintaining state coherence. It is the elevation of the informal redistributive principles of the economy of affection to official policy levels that has reduced the dangers of ethnic confrontations and claims for ethnic self-determination in most African countries. The consequence has been the emergence of the "soft" state.

The "Soft" State

The concept was originally coined by Gunnar Myrdal in his Asian Drama.[22] By the softness of the state he means the general absence of discipline, particularly in the conduct of public business. Laws and regulations are often circumvented by officials and there is inconsistency in the application of policies and laws. Furthermore, government servants are often in secret collusion with politicians and other influential people whose real task is to supervise the execution of policies. Corrupt practices are commonplace in order to secure objectives other than those officially stated. Myrdal arrived at his conclusion about the validity of this concept from studies primarily in South Asia, but he suggests that it has applicability also to other parts of the Third World. It is surprising that his concept has not acquired wider currency because it touches on a very common phenomenon in underdeveloped countries. One reason may be that the concept carries connotations of inferiority, but as Myrdal points out, the "softness" is not inherent in the character of the peoples of those societies. On the contrary, it is the result of the specific historical conditions existing in these countries.

The concept may have particular validity for the analysis of African conditions because, as Myrdal maintains, the "soft" state tends to develop especially in societies where people are anxious to escape control by public authorities. As we have noted above, the "civic public realm" lacks the legitimacy enjoyed by the "primordial public realm" and thus there is a tendency to enhance the latter by taking from the former.

This phenomenon is well-known throughout Africa and makes the state take on its "soft" character. In some countries, such redistribution takes place outside the scope of official policy, that is, it contravenes the public intention of the policy. In other countries however, it has been possible to pursue such redistribution, as suggested above, under the guise of official policy. In either case, public governance has been carefully calculated so as to allow each ethnic group of significance access to benefits which ensure their loyalty to the regime. The softness of the state emerges because leaders do not control the rural masses through ownership of existing means of production. As a result, they have been able to obtain their loyalty only through purchase, that is, they have been forced to give away benefits to the masses without necessarily getting anything in return to strengthen the system they rule. While this may sound like a paradox, this is precisely the position in which many African countries find themselves today.

The costs of managing ethnic relations in such a way as to avoid confrontation, therefore, have been extremely high to these countries, and it is often only generous foreign aid that has enabled the governments to pursue these policies as far as they have. In fact, large flows of aid funds have often concealed the fact that these policies have borne little relation to existing financial and managerial capacity to sustain them over a longer period of time. Yet it can be argued, of course, that the costs would be even higher if these governments were unable to satisfy demands of various ethnic groups under the umbrella of the economy of affection. A look at the "deviant" cases in Africa, that is, those where claims for ethnic self-determination have emerged, suggests that this is a fair proposition.

Cases Involving Ethnic Self-Determination

Take for example the case of Ethiopia, where an indigenous class society developed over many generations of indigenous rule and where the state consequently came to reflect the interests of a landed aristocracy. The economic surplus generated under this system facilitated the development of a strong state machinery which was used to subordinate other neighboring peoples and extend the territorial boundaries until Ethiopia took on the character of a genuine empire. Although the economy of affection survived, its significance declined as the feudal system grew more powerful. After World War II the latter was considerably modified by the growing contact of Ethiopia with the rest of the world. As foreign capital and aid began to flow into the country in large quantities, and a growing number of Ethiopians were educated in other parts of the world, the validity of the existing system was increasingly questioned. The attempt to overthrow Emperor Haile Selassie in 1960, which coincided with the general process of decolonization in Africa, was particularly important as it sparked off rebellions among the Muslim populations in Eritrea and Ogaadeen. The former wanted to create their own state, while the latter wanted to join their ethnic relatives in a "Greater Somalia." These claims for secession were intensified after the downfall of the emperor in 1975. Both the Eritreans and the Somalis saw this as an opportunity finally to gain the autonomy for which they had been campaigning for such a long time. It is significant that although there were members of the Dergue who were prepared to negotiate a settlement with these peoples, the inclination in the military leadership was to couch the conflict in class terms. There was an implicit recognition of the Ethiopian revolution as being the genuine product of class contradictions and thus giving prominence to the "separatist" demands was a way of diverting attention from the real issues. Following this stand against both "class enemies" and people seeking self-determination, there was little room for compromise. The sense of Amharic hegemony was perpetuated in the minds of the ethnic minorities.[23]

The Anya-Nya war in Southern Sudan had much in common with the fighting in Eritrea and Ogaadeen. Although there was no imperial regime in Khartoum, the policy toward the southern part of the country was that of a colonial power. Efforts at proselytization in the name of Islam took on an ethnic or racial dimension as it was carried out by Arabs among Africans. Thus, the latter had little chance of interacting with the former on equal terms. The door remained closed until the Addis Ababa Agreement of 1972 when for the first time there was a genuine attempt by the North to allow the South a greater say in the running of the country. The essence of the agreement was to grant considerable regional autonomy to the South. This arrangement has now survived for ten years as a result of a careful balancing act by President Jaafar el Nimeiry. Tensions have remained manageable partly because the North and the South are in essence still two separate economies. The tenacity of the present constitutional arrangement will be put to a special test as the two parts of the country grow more dependent on each other, particularly as a result of the exploitation of oil. One of the costs of the political compromise reached at Addis Ababa was the softening of the state by allowing for ethnic calculations in a manner which has been ruled out in Ethiopia. Policy-making as well as development management have been affected by this mode of governance.[24]

The cases of the Southern Sudan, Eritrea and Ogaadeen are interesting because ethnic self-determination has developed in opposition to indigenous colonization. Such claims have proved particularly difficult to contain in Ethiopia where class contradictions have been most strongly articulated. In the Sudan, by contrast, these appear to have been easier to disregard in the interest of some form of ethnic accommodation. In that respect, Sudan resembles Nigeria and Zaire where the struggles by Biafra and Katanga for autonomy were fanned by the economy of affection. Consequently, the rebellious provinces could be more easily brought back into the original fold. As these and most other African cases of ethnic conflict demonstrate, the state is much more elastic than is the case under modern capitalism or socialism. The economy of affection allows for a greater scope of conflict resolution, yet because it implies the acceptance of the "soft" state it also includes the accommodation of political management costs that often eat heavily into available funds for development. This, however, has proven acceptable to African leaders, by and large, because the philosophy of the economy of affection assumes much more explicitly than modern capitalism or socialism does, that the ultimate objective of development is man. Thus, as indicated above, redistributive policies in Africa tend to be pursued irrespective of their macro-economic implications.

The question inevitably arises, however, if such an approach is compatible with long-term development ambitions. The "soft" state may have some validity as a mechanism for ethnic conflict management, but it is proving to be an obstacle to the development of the national economy. Being a remnant of precapitalism, it is inconsistent with the modern development ambitions embraced by the African leadership. Thus, it is relevant to ask how a "soft" state can be made firmer.

The Significance of the Market

Since the "soft" state is an articulation of the economy of affection, which in turn stems from the peculiar features of the peasant mode of production, it is clear that any effort to cope with the issue will entail a diminution of the significance of peasant households in the overall economy of African countries. Transcending the inherent limits of the peasant mode has been difficult to accomplish by political and administrative means alone. The efforts by governments of different ideological persuasions since 1960 have not been very impressive. Existing forms of production have remained relatively intact and the relative contribution by agriculture to development has tended to decline in most countries. Although political means have been used to create alternative organizational forms of production, these have proved difficult to sustain.[25]

The tendency to blame all the negative sides of change in Africa on capitalism is quite understandable, given the association between colonization and the growth of capitalism in Europe. Nonetheless, it is also a dangerous inclination as it tends to blind people to the inevitable complexity of the development process. Because the economy of African countries has been dominated by multinational companies and by non-African minorities (Asians, Lebanese and Southern Europeans), African governments have rightly felt a great urge to rectify this situation. In doing so, however, they have shown little appreciation of the market as either a policy instrument or as an agent of change. The strategy has been to resort to bureaucratic measures to control economic and financial transactions. Consequently, the market forces have been undermined. Business practices have become increasingly corrupted; labor productivity has declined; economic growth has been stifled; and social progress has proved difficult to sustain. Somehow, it appears as if many African governments, in their reaction to the inherited economic imbalances in their countries, have thrown out the baby with the bath water.

By placing all their eggs foremost in the bureaucratic basket, however, they have ended up in a perilous situation. As both Marx and Weber acknowledged in their analysis of the emergence of the modern state, at the root of this process was the growing strength of the market forces. They eroded precapitalist formations and practices, and thus paved the way for the types of behavior and organization that facilitate gains in productivity. The market forces also laid the foundation for the development of new groups in society based on economic, as opposed to religious or ethnic, criteria. The "clean sweep" of the bourgeois revolution through the use of the market forces, Marx and Engels argued, was a precondition for the development of class contradictions which in turn could produce a socialist revolution. While it is true that socialist revolutions have subsequently arisen in societies characterized by considerable hangovers of precapitalist formations, none of these socialist regimes have been able to sustain progress without resorting to some use of market forces. It is their reluctance to allow them greater scope, for fear of subverting the philosophical underpinnings of the present systems, which more than anything else hinders economic and social progress in these countries.

Against the background of the turnabout in the attitude toward the market in Eastern Europe in the last twenty years and the even more far-reaching changes in the same direction in post-Mao China, it is at first a little surprising that so many Third World governments still regard the market forces with such great suspicion. Yet, the less a country has been penetrated by capital, the more likely that the regime will fall back on bureaucratic measures to run society. As an organizing principle, bureaucracy "fits" precapitalist society much better than the market. It is more naturally adopted by leaders and supported by the masses in such societies. Yet because bureaucratic principles easily take root in precapitalist society, it does not follow that progress is easier. In fact, it can be argued that resorting to dependence on bureaucracy is a step backward. This may be particularly so in Africa where small-scale peasant production still prevails. Certainly, Marx warned about the consequences of trying to develop society on such a foundation:

> Proprietorship of land parcels by its very nature excludes the development of social productive forces of labour, social forms of labour, social concentration of capital, large-scale cattle raising and the progressive application of science.... An infinite fragmentation of means of production, and isolation of the producers themselves. Monstrous waste of human energy. Progressive deterioration of conditions of production and increasing prices of means of production -- an inevitable law of proprietorship of parcels![26]

Although smallholder peasants have proved more progressive than Marx anticipated, his summary description of the constraints of development under a peasant mode of production has its validity in contemporary Africa. Whereas effective market forces accelerate not only the development of the social forms of labor but also the socialization of labor, peasant society tends to perpetuate the producers' isolation and separation. To be sure, many African regimes have tried to overcome this hurdle by mobilization of the masses through a strong political organization. While the results of this approach proved impressive in a situation of struggle against the colonial state, it has been difficult to apply with the same effectiveness under postindependence conditions. As the market forces have been replaced by political and administrative measures to change society, the capacity to transcend the limits of peasant society has declined. Simultaneously, the bureaucracy necessary to carry out policies has grown in size and complexity. The inevitable outcome has been a growing gap between public revenue and public expenditures, between stated goals and actual policy results. Particularly harmful to development is the growing alienation from the state among peasant producers which follows in societies where bureaucratization is allowed to precede the effective introduction of the market. The personal contacts between leaders and the masses which arise out of the clientelist nature of politics are not a sufficient substitute to mitigate the negative effects of this alienation process.

The market, by contrast, has proved effective in generating a greater appreciation among peasant producers of the role of the state in development. By becoming more dependent on the economic system at large, the peasants tend to become more aware of the positive contributions by government. They also tend to become more willing to make a contribution to the development of society at large. Thus, labor transcends its "individuality" and grows social in character particularly in those areas where market forces have penetrated society. It is no coincidence that in East Africa, for instance, both self-help activities and modern cooperatives have been more prominent in those places where the market has been allowed the greatest scope of penetration.[27]

The market, which in the late 1960s was blamed by many for fostering tribalism and undermining the postcolonial state, is, contrary to that argument, potentially a major factor in promoting greater state coherence. By generating social groups with a stake in the economic system and an appreciation of what the state can offer, it is progressive. By developing class-based social identities, the market has the potential of cutting across ethnic boundaries. For instance, the relative success of the market economy has helped to bridge ethnic cleavages in Kenya. An increasing number of people have developed a stake in the system and are therefore unwilling to rock the boat by fanning ethnic conflicts. At the same time, it is likely that such conflicts could become more prominent if the market forces are undermined or if they cease to produce the same results as in the past.[28] Another interesting example is Zimbabwe where the existence of a well-functioning market system appears to be a powerful factor in making leaders think twice before introducing far-reaching socialization measures or policies that add fuel to ethnic tensions. It can even be argued that one reason behind ethnic fragmentation in Ethiopia has been the absence, both before and after the revolution, of an effective capitalist market. The potential power of the market as an agent of change is manifested by the fact that, although present there only in a marginal sense, it facilitated the development of a unified opposition to the imperial regime. The revolution, however, undermined the market and thus limited the chances of overcoming the ethnic differences within the country. In Nigeria, the relative success of the new federal regime may be seen as a function of the economic boom that the country has enjoyed in recent years. Economic developments have dragged an increasing number of people into the market economy and there appears to be greater willingness now than in the 1960s to accept the present constitutional arrangements and defend them against subversion. To be sure, the lessons learnt from the turbulence of the Biafran secession attempt have also been important. Nonetheless, the petro-dollar surplus accumulated at the federal level reinforces the capacity of the state to deal with tendencies toward malintegration. As all these examples show, well-functioning market forces help to enlarge the arena of choice and create greater opportunities for ethnic accommodation.

While there is little doubt that in the countries mentioned above state coherence is still fragile and intensification of ethnic conflict a real possibility, countries with real problems of state coherence are those which, intentionally or otherwise, have allowed the market forces to cease functioning properly. The effect of this process has been to narrow down the choice in a number of policy areas and to impair the relations of state to society. The gap between state and society has widened and in some cases, like Chad and Uganda, state coherence has been possible only thanks to support by foreign troops. In this situation of virtual anarchy the natural tendency is to give priority to one's own interests and needs. Display of civility and magnanimity becomes difficult and relations among people show signs of greater suspicion and tension. While this doesn't automatically mean that people will engage in civil war, the tendency to indulge in practices that undermine the legitimacy of the state is increasing. This applies also to countries like Tanzania and Zambia which, in spite of their comparative civility, have suffered from these practices as a result of the virtual breakdown of the market economy. The inclination to use one's official position to distribute public resources through unofficial channels has grown in recent years in both countries and rendered the task of preventing economic sabotage against the state extremely difficult.[29]

The threat to state coherence in this kind of situation is that public resource allocation is brought out of sight of the public eye. Nobody really knows what is happening with available resources and rumors flourish, often undermining state authority. Because people tend to assume that everybody favors his own group in such a situation, the danger of enhanced ethnic competition and conflict is also real; the opportunities of effectively dealing with it, short of armed confrontation, are reduced.

While the potential contribution by the market forces to the development of society at large and to the evolution of greater state coherence is clear, recognition must be made of its possible negative side effect. Particularly in southern Africa, where the national economies have been seriously distorted in the last hundred years, it would be naive to expect that the "hidden hand" will automatically serve official policy objectives. Also important is the fact that not all groups in society are equally placed or socially inclined to take advantage of the opportunities for advancement provided by the market. Thus, there is a danger of ethnic imbalances that can foster great social costs to society. This has already happened in various African countries.[30] The point is, however, that these issues can be handled without undermining the market economy as a whole. Redistributive policies aimed at facilitating the entry of disadvantaged groups into the market can be designed in the same manner as policies aimed at increasing the access to education of neglected groups. Particular attention may be paid to those who struggle for existence in the informal sector.

Building Horizontal Linkages

While political leaders in Africa are likely to find the market more difficult than the bureaucracy to tame, there is reason for a wider recognition of the historical role played by the market in promoting economic development and state coherence. To be sure, ethnic conflict will not disappear with the growing importance of the market, but ample strategies and policies can be developed to deal with such issues. In this respect, the new Nigerian constitution is an interesting example of what can be done. Its federal solution isn't generally applicable, but each country is in a position to consider what checks and balances can be established to ensure that ethnic conflicts are resolved in public through formal structures rather than being swept under the rug because of their sensitivity. To that extent, class formation is a progressive trend in African society today. The bourgeoisie is an indispensable agent of progress. Rather than confining its members to working through bureaucratic organizations where their historical role is curtailed, its full potential should be developed. This is particularly so in Africa, where its formation under colonial rule was very lopsided.[31] Rectifying this imbalance by allowing the bourgeoisie a greater role in commerce and manufacturing is a particularly significant step to take.

As the bourgeoisie consolidates its position in society it also develops an economic system from which other groups in society find it difficult to escape. Thus, there is a paradox here: in order to strengthen the functional relevance of state to society in Africa, it is necessary for one group to take effective command of the state and elevate itself from the pressures of ethnic groups. While the process of liberation from the economy of affection is going to take time to complete, experience to date suggests that the market forces are more effective in this respect than any other factor. The noble argument that the African bourgeoisie has to commit "class suicide" in order to pave the way for socialism seems in retrospect a sure way of undermining any long-term progress in a socialist direction.[32] The objective conditions for the articulation of class conflict are only beginning to take shape in Africa, and they need to be enhanced before any form of socialism is possible.

Greater attention to the market and the needs of the bourgeoisie does not imply neglect of other groups. The market, for instance, provides opportunities also for the small actors, although in order to realize these they require special assistance. While advice may be provided by government, the task of enhancing the capability of groups of small actors is best pursued by nongovernmental agencies. As soon as groups of small actors are brought into the state realm, they tend to become appendages to factional political leaders. This is the danger, for instance, with ethnic welfare associations. It explains why in Kenya they were banned.

The linkages that need to be created are horizontal rather than vertical. Groups of small actors, be they peasants or workingmen, must be allowed to join hands in popular movements that are created under a unifying banner other than ethnic identity. Nurturing class-based organizations, therefore, is a progressive measure at this stage of development in Africa. Workers' solidarity, however, is only one aspect of the struggle. In the rural areas, for a long time to come, the articulation of class solidarity will be feeble and thus the horizontal linkages will have to be provided by other organizations, e.g., church groups. The latter have the capacity to bridge ethnic and other parochial identities. While politicians may view them with suspicion because they fall outside the framework of the ruling party, they are in the long run a guarantor of greater state coherence. Governments which circumscribe the actions of religious and other such groups and fear them because they have a different view on substantive issues from their own display a particularly shortsighted view of social realities. They are leading their countries down a blind alley from which it is difficult to return. The existence of movements with horizontal linkages and strong mass support provides, as history from other parts of the world suggests, the best base for effective development. Thus, whether the issue is economic development or state coherence, support of nongovernmental organizations is likely to have a great payoff effect. Where such organizations are weak or nonexistent, special efforts must be made to strengthen them. While ethnic tension cannot be ruled out in these organizations, the probability that it will cause harm to society is much smaller than in the state realm because the organizations do not control more than a modest share of societal resources and, furthermore, the ideologies of these organizations often serve as a more effective restraint on "parochial" action than a general political ideology with little immediate relationship to day-to-day issues. By participating in the process of resource allocation, nongovernmental organizations also reduce the danger of intense ethnic conflict that arises in a situation where the state exercises monopoly over that process. By being more actively involved in the development process, nongovernmental organizations may reverse the "soft" state phenomenon. In short, the task of making the state firmer cannot be tackled adequately only by reorganization of the state machinery or by training of public officials. The answer lies outside the state realm, where enhancing the power and capacity of nongovernmental agencies is likely to have particularly significant developmental payoffs.

Conclusions

It has been the argument of this paper that the "soft" state has greatly contributed to effective ethnic conflict management in postindependence Africa but at the same time it has consumed considerable public resources and thus held back the opportunities for economic development. Given the great costs incurred by the "soft" state, it is not likely to be a phenomenon that African countries can long afford.

Making the state firmer and more resolute, however, is not achieved overnight, nor is it without costs. It is likely to be a long and painful process, implying, among other things, greater risk for ethnic tension. As the state "closes the doors" to the economy of affection, some ethnic groups may feel more directly deprived. Their inclination to raise claims for self-determination may grow. Yet, as this paper has tried to show, through a careful and enlightened use of the policy instruments provided by the market economy, it is possible to take preventive steps and thus reduce the dangers of ethnic conflicts running out of control.

The biggest question mark is the pace with which a firmer state will develop over the next decades. Africa badly needs it for development, yet it is related to historical processes, notably the rise of an indigenous bourgeoisie, which in other parts of the world has taken generations to complete. Somehow, therefore, all efforts must be concentrated on the task of accelerating this process, even if this implies steps that differ quite considerably from current views in African governments or donor agencies.

It is understandable that government policies in the first two decades of independence in Africa in large part have been aimed at negating the policies pursued under colonial rule. One of the consequences of this orientation, however, has inevitably been to reduce the range of policies under consideration. The argument here is not in support of a return to colonial policies. There are certain forces, however, which greatly enhanced the capacity of the colonial state but which have since been abandoned and which at this stage could be reintroduced with even greater benefits. Foremost of these are the restoration of the market economy and the empowerment of nongovernmental organizations, both of which can give society a greater social dynamic, yet assist it to contain tendencies toward ethnic fragmentation.

Footnotes

1. Both the bourgeoisie and the peasantry have been divided into subcategories supposedly fitting the African situation. See, for instance, Issa Shivji, _Class Struggles in Tanzania_ (London: Heinemann Educational Books, 1975); and Lionel Cliffe, "Rural Political Economy of Africa" in Peter C.W. Gutkind and Immanuel Wallerstein, eds., _The Political Economy of Contemporary Africa_ (Beverly Hills: Sage Publications, 1976), pp. 112-130.
2. Colin Leys, "The 'Over-Developed' Post-Colonial State: A Re-Evaluation," _Review of African Political Economy_, 5 (January-April 1976): 39-48.
3. Michaela von Freyhold, "The Post-Colonial State and its Tanzanian Version," _Review of African Political Economy_, 8 (January-April 1977): 75-89.
4. John S. Saul, "The State in Post-Colonial Societies -- Tanzania," in Ralph Miliband and John Savill, eds., _The Socialist Register 1974_ (London: The Merlin Press, 1974), pp. 349-372.
5. Pierre-Phillips Rey and G. Dupré, "Reflections on the Pertinence of a Theory of the History of Exchange," _Economy and Society_, 2, 2 (1973):131-163.
6. See, for instance, Akin L. Mabogunje, "The Dilemma of Rural Development in Africa," _Regional Development Dialogue_, 2, 2 (1981).

7. Goran Hyden, Beyond Ujamaa in Tanzania: Underdevelopment and an Uncaptured Peasantry (Berkeley and Los Angeles: University of California Press, 1980).
8. Steven Lukes, Power: A Radical View (London: Macmillan & Co., 1974), p. 24.
9. Barry Hindess and Paul C. Hirst, Pre-Capitalist Modes of Production (London: Routledge & Kegan Paul, 1975), p. 197.
10. Barrington Moore, Jr., The Social Origins of Dictatorship and Democracy: Lord and Peasant in the Making of the Modern World (Boston: Beacon Press, 1966).
11. This is the principal argument, for instance, in the vast literature on "development administration." It is also echoed in the considerable number of public reviews commissioned by individual African governments to look into the shortcomings of public administration.
12. For an analysis of how this phenomenon has been articulated in elections in Kenya and Tanzania, see Goran Hyden and Colin Leys, "Elections and Politics in Single Party Systems: The Case of Kenya and Tanzania," British Journal of Political Science, 2, 1 (October, 1972): 389-420.
13. Peter Ekeh, "Colonialism and the Two Publics in Africa: A Theoretical Statement," Comparative Studies in Society and History, 17, 1 (1975): 91-112.
14. See, for example, Marvin Miracle, Diane Miracle and Laurie Cohen, "Informal Savings Mobilization in Africa," Economic Development and Cultural Change, 28, 4 (July 1980): 701-724.
15. For one of the interesting case studies documented, see Joyce L. Moock, "The Content and Maintenance of Social Ties Between Urban Migrants and Their Home-based Support Groups: The Maragoli Case," African Urban Studies, 3 (Winter 1978-79): 15-21.
16. Rupert Emerson, "Nation-Building in Africa," in Karl W. Deutsch and William J. Foltz, eds., Nation-Building (New York: Atherton Press, 1963), pp. 95-116.
17. See, for example, Abner Cohen, Custom and Politics in Urban Africa: A Study of Hausa Migrants in Yoruba Towns (Berkeley and Los Angeles: University of California Press, 1969); David Abernethy, The Political Dilemma of Popular Education: An African Case (Stanford: Stanford University Press, 1969); J.S. LaFontaine, City Politics: A Study of Leopoldville, 1962-63 (Cambridge: Cambridge University Press, 1970).
18. Abernethy, Political Dilemma, pp. 107-108.
19. Robert H. Bates, "Modernization, Ethnic Competition, and the Rationality of Politics in Contemporary Africa," in this volume.
20. Nelson Kasfir, "Explaining Ethnic Political Participation," World Politics, 31, 3 (1979): 365-388.
21. Crawford Young, The Politics of Cultural Pluralism (Madison: University of Wisconsin Press, 1976).
22. Gunnar Myrdal, Asian Drama: An Inquiry into the Poverty of Nations (New York: Twentieth Century Fund and Pantheon Books, 1968).
23. The literature on the Ethiopian case is large. One useful source for purposes of understanding the claims and counterclaims in the Eritrean and Ogaadeen wars is Tom Farer, War Clouds on the Horn of Africa: The Widening Storm (New York:

Carnegie Endowment for International Peace, 1979).

24. For an overview of the Southern Sudanese issue, see Dunstan M. Wai, "Geoethnicity and the Margin of Autonomy in the Sudan," in this volume.

25. This is documented in a study by the present author, titled _Efficiency Versus Distribution in East African Cooperatives_ (Nairobi: East African Literature Bureau, 1973). See also Peter Dorner, ed., _Cooperative and Commune_ (Madison: University of Wisconsin Press, 1977).

26. Karl Marx, _Capital_, vol. 3 (New York: International Publishers, 1967), p. 807.

27. For an account of this phenomenon, see Frank Holmquist, "Class Structure, Peasant Participation, and Rural Self-Help," in Joel D. Barkan and John J. Okumu, eds., _Politics and Public Policy in Kenya and Tanzania_ (New York: Praeger, 1979), pp. 129-153.

28. The abolition of ethnic associations and provincial parliamentary groups in Kenya took place in 1980 at a time when the country was facing harsher economic conditions and the risk of intensified ethnic competition was becoming more imminent.

29. It is now recognized in both countries that the greatest threat to the country's future is the widespread economic sabotage. Attempts to deal with it by dismissing officials have not yielded any marked change in the situation.

30. The politicization of the issue of access to markets is covered, for example, by Peter D. Lloyd, "The Changing Role of Yoruba Traditional Leaders," _West African Institute of Social and Economic Research Third Annual Conference Proceedings_ (Ibadan, 1956); and by John Lonsdale, "Political Associations in Western Kenya," in Robert I. Rotberg and Ali A. Mazrui, eds., _Protest and Power in Black Africa_ (New York: Oxford University Press, 1970).

31. V. Subramaniam, "The Role of the Middle Class in Developing Countries: A Reassessment in the African Context," _Universities of East Africa Social Science Conference Proceedings_ (Dar es Salaam, 1970).

32. Attributed originally to Amilcar Cabral during the course of struggle against Portuguese colonialism.

5
Ethnicity Versus the State: The Dual Claims of State Coherence and Ethnic Self-Determination

John Stone

The Withering Away of Ethnicity: The European Legacy

The modern state has become so much a part of contemporary political life that its relatively recent historical roots are often forgotten. One explanation for this, as several political and social scientists have been at pains to point out, may simply be that it is a product of faulty historical scholarship or even a complete absence of any such historical dimension in the analysis of ethnic-state problems. Thus Walker Connor comments: "The total lack of anticipation of recent events (the rise of postwar European nationalist movements) is not so much a case of no history as of poor history. Those who perceived the surge of ethnonational demands throughout Western Europe in the late 1960s as without forerunners were overlooking numerous, well advertised portents."[1] A simple list of the new states to emerge in twentieth century Europe -- Norway, Bulgaria, Albania, Finland, Czechoslovakia, Hungary, Poland, Yugoslavia, Ireland and Iceland -- hardly suggests that ethnic nationalism can be dismissed as a primordial relic of a bygone era confined to Africa and the rest of the Third World.

However, this was the dominant interpretation of ethnicity adopted by most political and social scientists throughout the 1950s and 1960s. The abolition of European colonialism did not secure the new states of Africa and Asia from internal cleavages of a profound and dangerous kind.[2] In fact the removal of the colonial power, which had so often promoted ethnic divisions and rivalries as part of the strategy of divide and rule, left behind a crippling legacy of intergroup tensions and communal mistrust. The uneven development of different regions, differential educational policies that had favored particular ethnic and tribal groups, merchant minorities that had been introduced to fill intermediary commercial and bureaucratic roles in the colonial economy, and the selection of the so-called "martial races"[3] to monopolize the military and policing functions of empire combined to place an ethnic curse on so many postcolonial regimes. What is more, the almost universal adoption of the colonial boundaries of the continent, under the sacrosanct principle of "national" sovereignty, produced a remarkably arbitrary ethnic lottery which set in motion a continuing tension between nations and states.

Superficially, there seemed to be little in common between these circumstances and the ethnic scene in Europe, but the possible relevance of the European experience to the African situation has been increasingly recognized. As Brass and van den Berghe observed: "Many of the issues and demands in dispute among those involved for and against these movements [of ethnic and nationalist groups in postindustrial societies] bear a striking resemblance to the issues which have arisen around the demands and conflicts of ethnic groups in developing societies."[4] The misleading duality between the concept of a "plural society," derived from Furnivall's classic interpretation of colonialism, and the "pluralism" used to describe ethnic and racial divisions in industrialized societies, has been superseded by the notion of pluralism as a variable.[5] This has removed one conceptual barrier that prevented closer comparison between African and European ethnic-state relationships. Another indication of this trend has been the resurrection of the term "internal colonialism" as a fashionable mode of analysis for both racial minorities in the United States[6] and regional movements in Europe,[7] which is a further illustration of the intellectual cross-fertilization that has begun to take place between the Third World and the First World.[8] Of course, in as much as social and political scientists have based their theories of societal change on an inadequate interpretation of the scope and nature of ethnicity in European states, this has further impeded the accurate analysis of African conditions by Western and Western-educated scholars.

In this paper I want to re-examine the European experience of ethnic-state conflict in relation to the biases to be found in the political and social science literature on these questions. Then I will consider the recent revival of ethnic nationalism and some of the new theoretical interpretations that have been placed upon it, in an attempt to isolate some of the important factors that may also be relevant in the African context. Finally, I will consider the special case of South Africa which raises additional issues about the dynamic interaction between ethnic groups and state institutions.

A number of systematic biases in the traditions of both sociology and political science have reinforced the tendency to interpret the development of European society from a state rather than an ethnic perspective. Modern "nationalism," a confusing enough term in itself,[9] dates from the time of the French Revolution, but the legitimacy of the state has also been consistently under attack since that time. An inherent tension caused by the ambiguity in the definition of "national" self-determination results from the fact that the true nation-state, despite much nationalist rhetoric, has been the exception rather than the rule, and so the potential conflict between the state and its constituent ethnic groups has been latent even during periods of apparent "national" cohesion and development. This fact has been overlooked by many scholars influenced by the classical writings of political science and sociology which only contain oblique references to the problems and issues that we would now associate with regionalism and ethnic self-assertion.

Most writers within this classical tradition had a conception of the future development of society that shared one common assumption: that forces in society would create larger and more integrated social units. As a result of this there has been a tendency to focus attention on the state, because it represented a larger scale of analysis, rather than on the ethnic group. Although they had very different visions of the political nature of that future society, most classical writers accepted this basically evolutionary approach, a fact that is equally true of Marxist and non-Marxist schools of thought.

Marxism has been particularly troubled by the resilience of ethnic nationalism which has posed problems at both the theoretical and practical levels. Although in theory Marx anticipated the eventual withering away of ethnicity and the state (in that order), the Marxist approach to nationalism and ethnicity tends to be highly complex and full of ambiguities and contradictions. In relation to Ireland, Marx changed his interpretation during his lifetime, increasingly recognizing the role of nationalism in the conflict,[10] while Engels' contrast between the "historic nations" of Western Europe and the "ruins of peoples" to be found in Eastern Europe and on the periphery of Western European states was based on political calculation and not on an appreciation of the depths of nationalist sentiment.[11]

Marxist thought has placed a considerable emphasis on central planning mechanisms to replace the capitalist market structure and has been generally opposed to forms of decentralization which might challenge the monopoly power of the party. At the same time, Marxist theorists attach little formal importance to national institutions and to what has been disparagingly called the "fetish of frontiers."[12] However, in reality, Marxist strategists have often been less than internationalist in their outlook and certainly Third World Marxist rulers have clung no less tenaciously to the arbitrary boundaries of postcolonial states than their non-Marxist contemporaries. Lenin and the other Bolshevik leaders were compelled to work with nationalist groups in order to consolidate the revolution in Russia, and the right to secede was guaranteed as part of the terms of this arrangement.[13] Once in power, however, the Soviet authorities have regarded any expression of nationalist separatism as the counterrevolutionary action of "bourgeois reactionaries" and have suppressed it with ruthless efficiency.[14]

The relative neglect of the problems of centralization in the Marxist literature can be found equally in the non-Marxist socialist tradition.[15] Among more liberally-inclined sociologists and political scientists, Weber's concern with bureaucracy and Tocqueville's discussion of administration under the _ancien régime_ are probably the most relevant contributions to this question. While Weber was basically pessimistic about the chances of controlling the social and political consequences of centralized rationalization, Tocqueville did propose specific measures to counter the threat of centralized dictatorship. These included the familiar constitutional checks and balances, the promotion of intermediary associations between the individual and the state, and, most significantly, administrative decentralization. Although this analysis was formulated in terms of the individual's relationship to the state, it could easily be recast in a framework linking ethnic and national minority groups to a central political authority. Regional and ethnic institutions could be built into constitutional structures and the maximum degree of autonomy granted to local decision-making bodies.

The emphasis on centralization has a critical importance for the study of separatist movements and ethnic-state relations in general. It is interesting that it was a Frenchman who was so sensitive to this dimension of the problem. The domination of Paris over the provinces should not obscure the fact that the French state expanded from a small central core area over the centuries. Provence was inherited in the fifteenth century, Alsace acquired by conquest in the seventeenth, and Lorraine purchased in the middle of the eighteenth century. In addition to these regions, Brittany, the Basque country and Corsica have to be included in any complete analysis of the ethnic basis of French autonomist movements. It is true that the Jacobin tradition, and the "strategy of silence" adopted by most French political scientists, have frequently given the impression that France was one and indivisible when, in fact, it was no more ethnically homogeneous than Britain or Spain.

Ethnicity Reinterpreted: The European Reality

The myth that European society had evolved beyond the point where ethnic mobilization could threaten the integrity of the state was finally exposed by the simultaneous resurgence of ethnic nationalism in several different parts of the continent during the 1960s. It is interesting to consider some of the factors behind this ethnic revival to see what relevance they might have for ethnic-state conflict in Africa. Some scholars have argued that there is a direct connection between political changes in Africa and European societies, claiming that the end of empire not only produced postimperial fragmentation in Africa[16] but also stimulated metropolitan separatism as well. As Andrew Greeley, one of the celebrants of the "new ethnicity" in America, has suggested: "Just as the collapse of the Austro-Hungarian Empire increased tension in central Europe, so the collapse of the old colonial empires has opened a Pandora's box of tribal, linguistic, religious and cultural conflicts. It may be also that the 'turning in on oneself' that follows the relinquishing of imperial power has given rise to the new nationalisms in Western Europe."[17] Direct confirmation of this thesis can be found in the statements made by the leaders of the Scottish Nationalist Party, who have argued that union with England was worthwhile when there was a worldwide empire to exploit, but to be a junior partner of one of the less successful European economies is a totally different proposition. Corsica is another example of the complex implications of decolonization for ethnic relations in the metropolitan society. The end of the Algerian War not only stopped the emigration of islanders to Algeria, but also caused an influx into the island of *pieds noirs*, French settlers who were repatriated after independence. These former settlers were supplied with government credits and grants to compensate them for the loss of their Algerian property, and they used this financial support to buy land and become modern agriculturalists. The resulting competition made many smaller Corsican farms unprofitable, leading to local resentment (well-captured in the wall daubings "I Francesi Fora") and increasing support for the Corsican autonomist movement.[18]

Apart from such direct connections between developments in Europe and Africa, it is also possible to isolate a number of major factors that appear to be closely linked to the rise of ethnonationalism in Europe and that may have some relevance for the African scene.

Centralization. One of the most striking facts about recent European separatist movements is their concentration in the long-established, centralized states of Britain, France and Spain. It is true that there are many other examples of separatist movements to be found in states with varying degrees of centralization both in the capitalist and communist sectors of the continent. Belgium and Yugoslavia are interesting illustrations of such cases. In the former, Flemings and Walloons are struggling to find an acceptable solution to the intricacies of ethnic politics in a situation described by Tindemans as "federalization without federalism";[19] in the latter, the post-Tito government has inherited a conflict with Croatian separatists, not to mention the other less crucial regional groups, who are accused of plotting a violent campaign to establish a new state by breaking the links with Serbia and incorporating Bosnia, Herzegovina and parts of Montenegro.[20] However, Germany and Italy, the newest and least centralized of Western European states, have been among the least influenced by the wave of rediscovered nationalism. In the German case this appears to be in part the result of deliberate policy. After the end of the Second World War, the Allies insisted on drastic decentralization, a measure designed to impede a resurgence of militant German nationalism, so that the Federal Republic of Germany consists of ten states or länder, each having a separate parliament with powers to control the police, education and cultural affairs. In Italy, too, although more as a result of a slow evolution of historical circumstances than as a product of deliberate planning, there has been a considerable measure of local municipal and regional autonomy.[21] Most, though by no means all, independent African states took over the basic institutional framework of the former colonial regime, not by any conscious design but simply because the winds of change in Africa blew faster than the colonial regimes anticipated. This left little time for nationalist movements to turn their attention away from the immediate problems of liberation to debate the merits of centralized or decentralized constitutional and administrative arrangements. In this way many new regimes inherited a centralized system and attempts to move away from this pattern have only occurred after the unity of the state has been severely threatened by ethnic conflicts and civil wars, such as those that took place in the Sudan and Nigeria.

While centralization is clearly important in certain situations, it is not the only factor involved in the promotion of ethnic separatism. If we look again at the European situation, the campaign of the French-speaking Jurassians to establish the first new Swiss canton for 160 years is a case in point. No one could suggest that Switzerland, famous for its decentralized political and administrative structure, suffers from these problems. Thus the centralization of the state may be a significant contributory factor in ethnic mobilization, but it is neither a necessary nor a sufficient cause in itself.

The Functional Efficiency of the State Unit. A second general hypothesis concerns the size and efficiency of the traditional European state. It has been claimed that these units are too small to deal effectively with global problems, such as defense strategy or the control of multinational corporations, and yet too large to handle local community and regional matters like education, welfare, local planning and cultural affairs. The nineteenth-century European state, what the Breton writer Morvan Lebesque has called the "garrison state," was geared to a situation of conventional warfare and under those circumstances was able to inspire a degree of loyalty and a sense of collective identity sufficiently strong to incorporate a wide range of ethnic groups. After the end of the Second World War, and particularly with the growth of the European Economic Community, the traditional basis for such "national" cohesion has been severely strained if not actually undermined. Of course, this argument has to be qualified as some European nationalist groups, like the Scots, have been generally hostile toward the concept of European unity, while others, like the Catalans, have seen tactical advantages in blurring the edges of the traditional European state. In the latter case, most Catalan nationalists have preferred to split power between Madrid and Brussels thereby creating autonomy in the interstices of a looser federation.[22]

Somewhat similar questions can be raised concerning the size and effectiveness of African states. While the emergence of the Organization of African Unity as a supranational political forum represents some measure of political cooperation, attempts to create economic unions and free trade areas have been notably less successful than their European counterparts. It is true that the problems of controlling the activities of multinational corporations are probably more severe in the African than in the European situation as a result of the greater imbalance in economic power. However, the fundamental problem of the African state has been less a usurpation of functions by suprastate organizations than a functional overloading of state institutions. So important have been the rewards of controlling the state, as a means of determining the allocation of economic and social resources, that the fear of political exclusion held by minority ethnic groups has resulted in severe ethnic tensions.[23] Loss of control of the central government does not simply mean a period in the political wilderness, it spells total economic disaster.

The Quest for Community. A third factor associated with the development of European ethnonational movements concerns the problems of communal cohesion and individual identity posed by the processes of industrialization and urbanization. It is, of course, closely related to the earlier factors that we have already discussed. Mayo has emphasized such forces in her search for the roots of European identity, while French analysts like Serant and Crozier have argued that extensive urbanization combined with the rapid industrialization that took place in France during the 1960s produced a mass of frustrated individuals who felt that they had little or no control over their lives.[24] The desire to reinstate cultural or linguistic traditions, particularly where these had been subject to neglect or outright hostility by central governments, is one possible solution to these pressures.

While language and national identity are often related, language alone is not necessarily a viable basis for separatist movements.[25] *Plaid Cymru*, for example, has had to be very careful in balancing the demands of its Welsh-speaking supporters with those of the non-Welsh-speakers of South Wales. As the Archbishop of Wales observed in 1968: "There is a real danger of a kind of *apartheid* based not on race but on language. It is intolerable that those who do not speak Welsh should be regarded as second class citizens, or less genuine lovers of their country than their bi-lingual compatriots."[26] Much depends on the complex relationship between language and the nationalist movement as the very different experiences of Gaelic in Ireland and Afrikaans in South Africa illustrate.[27]

The extent to which linguistic affiliations and ethnic group membership are strengthened as a result of urbanization and industrialization depends partly on the disruptive consequences of these processes, partly on government policies, and partly on the degree to which there are nonethnic institutions capable of offsetting the strains of industrial alienation and urban anomie. This is not to deny the purely instrumental functions of ethnic parties and urban ethnic associations, but merely to stress that these institutions have important symbolic, social and expressive functions as well.

Resource Allocation and Relative Deprivation. A fourth set of factors proposed to account for the emergence of autonomist movements is concerned with regional economic changes. There is a clear link between economic forces and regionalism but the relationship is by no means a simple case of absolute economic deprivation. It is true that many ethnonational movements have developed in areas with depressed economies and income levels well below the average for the state as a whole. But there are others in regions of relative economic prosperity, such as the Basque and Catalan areas of Spain, even though they may define their situation as one of exploitation.[28] It is a sense of *relative* deprivation that is the crucial determinant of social action: the Catalan nationalists argued that the government in Madrid was drawing off local wealth for the benefit of other regions while swamping them with migrants, a subtle ploy aimed at destroying their language and undermining autonomist sentiment at the same time. Although the Scottish case is complicated, the sense of deprivation derived from the belief that North Sea oil revenues, that rightly belonged to Scotland, were being used to bolster the English economy was a contributory factor in the rapid growth of nationalism during the 1970s. In Africa, the separatist movements in Biafra, Katanga and Cabinda not only present difficulties for a crude economic deprivation thesis but also suggest that the relative deprivation experienced by comparatively prosperous regions, particularly when they do not possess political power commensurate with their wealth, can equally stimulate violent ethnic conflict and civil war.

Ethnicity and Regionalism: Reactive and Competitive Models

In addition to these four basic sets of factors, a number of attempts have been made to present a more coherent account of the origin and dynamics of ethnonational movements. Of the recent models of ethnogenesis, two have provided particularly valuable insights into these questions and may be broadly described as <u>reactive</u> and <u>competitive</u> theories. In certain respects, they are not mutually exclusive explanations of regionalism since the one presents a plausible account of the persistence of ethnic identity over considerable periods of time, while the other tries to provide an explanation for the political resurgence of ethnicity at a specific moment in history. The reactive model, or the internal colonialism model as it is often called, is particularly associated with the work of Michael Hechter[29] and a considerable literature has now been developed by scholars either working along these lines or in opposition. Hechter himself has extended and modified his earlier formulation of these questions in the light of the considerable controversy they have provoked.[30]

In an effort to test the relative merits of the internal colonialism theory of regionalism against the notion that regionalism will tend to disappear as part of the evolutionary process of industrialization -- the diffusion approach -- Hechter looks at British national development from the sixteenth century until the middle of the 1960s. The diffusion perspective can be found in the work of nineteenth century social theorists like Spencer, and has permeated the contemporary writings of structural functionalist sociologists and political scientists concerned with communications theory. It postulates an evolutionary development whereby the core and peripheral areas, in this case London and the South-East as opposed to Scotland, Wales and Ireland, slowly interact to form a single socioeconomic, cultural and political unit. After an initial preindustrial phase, when core and periphery remain mutually isolated, a second stage is reached involving industrialization and significant core-periphery interaction. The diffusion model predicts that such interaction will cause the core social structure to merge into the periphery and while, in the short run, certain "traditionalist" resistance may appear as a reaction to the dislocation of rapid social change, in the long run industrialization flowing from the core regions will engulf the whole state. In the final stage, inequalities in regional wealth will tend to disappear, cultural differences will cease to be socially meaningful, and the political process will produce statewide parties that ensure the representation of all ethnic and regional groups. This model presents a somewhat optimistic picture of what Parsons has called the "inclusion process,"[31] a kind of civic integration of regional areas and ethnic groups. It calls to mind the warnings that Blumer raised in his discussion of the relationship between industrialization and race relations,[32] particularly the assumption that there are any necessary causal links between a process as broad and diffuse as industrialization and one particular social structure or pattern of ethnic relations.

It is possible to compare this approach with the internal colonialism model, which predicts that core-periphery interaction, far from leading to the convergence of social structures, will in fact result in the economic and political exploitation of the former by the latter. The initial, often fortuitous, advantages held by the core areas result in an unequal distribution of resources and power which is subsequently institutionalized into a rigid system of stratification. While ethnic identification persists, the limited measure of industrialization that does occur in the periphery tends to be unbalanced, dependent on and subsidiary to the core. The peripheral economy is more sensitive to the vagaries of international trade, investment decisions take place at the core, and regional disparities in wealth increase. If cultural differences are superimposed on the stratification divisions between core and periphery, the periphery will develop a sense of separate nationhood and will start to seek greater autonomy and independence. Proponents of this view would claim that such a sequence of events resulted in the secession of southern Ireland in 1921, and that the same processes have been at work in Scotland and Wales during recent years. Thus the internal colonialism model presents one picture of the underlying causes of the persistence of regionalism in the United Kingdom, even though it does not explain why separatist movements erupt in relatively advantaged regions or why they occur at one particular moment in time. There may be some reluctance to apply the term "internal colonialism" to Africa in the postcolonial era, but whatever language is used, the processes of regional exploitation are by no means unfamiliar on the African continent.

The question of the timing of ethnonational and regional protest movements has been looked at in more detail by Anthony Mughan in his attempt to develop a theory of ethnic conflict out of the interaction between the process of modernization, a sense of relative deprivation and the regional distribution of power resources.[33] Mughan argues that the politicization of ethnic-regional divisions results from a condition of relative deprivation, which in turn is generated by the uneven distribution of the benefits of modernity among ethnic groups. The relative deprivation hypothesis outlines a necessary but not a sufficient condition for political conflict between groups because it emphasizes the motivational side of conflict without considering the actual possibilities for achieving the desired change. It is important, therefore, to focus on the distribution of power resources and particularly shifts in the balance of power during the modernization process. Mughan illustrates his argument with reference to the Belgian, British, Canadian and, what is particularly significant for this paper, the Nigerian experiences of ethnic conflict.[34]

The reason why modernization and democratization took so long to precipitate regional-ethnic conflict in the first three cases was that they failed to disturb the ethnic status quo consolidated in the early stages of the industrial revolution. Subordinate ethnic groups came to possess more power resources but not at the expense of the dominant ethnic group, so the balance of political power remained unaltered. This general situation persisted until the end of the Second World War. By contrast, ethnic conflict came to dominate the political life of many African states after independence because modernization did not systematically favor one particular ethnic group. The amorphous distribution of power resources in Nigeria invited political conflict as the Ibos sought self-determination against the numerically superior Northerners. Thus "ethnic relations in the former countries were characterized by cumulative status deprivations for a long time after the start of the modernization process, while in Nigeria they were characterized by politically volatile status inconsistencies right from the time of the departure of the colonial power."[35]

This approach adds an important element to our understanding of the development of ethnic-state conflict by stressing not only regional imbalances in resources and a sense of relative deprivation, but also the way in which changes in the balance of power can result in a competitive struggle, leading to ethnic mobilization and demands for national self-determination. Such models, as Mughan shows, can be as easily applied to African as to European states. Indeed, attempts to industrialize in Africa in the space of a short period of time are likely to accentuate these strains both because of the rapidity of social change and also as a result of the method of industrialization through centralized political planning.

Two further general points need to be considered in the analysis of ethnic-state conflict. There are the questions of, firstly, the power of central governments to contain regional and separatist movements, and, secondly, the degree to which "separatist" movements seriously wish to establish a new sovereign state. If we take a relatively short historical perspective and confine our attention to the postwar period, then it would appear that very few political separatist movements have been entirely successful. Only Bangladesh stands out as an important, yet very distinctive, exception.[36] We are left, therefore, with a paradox, noted by Cynthia Enloe: "Separatism, on the one hand, seems a more common and widespread political phenomenon than ever before, but, on the other hand, is perhaps less successful than it has been in the last two centuries."[37] Part of the explanation for this lies in the misleading diagnoses of outside observers who fail to realize that many movements use the separatist slogan as a form of leverage to gain greater autonomy from central governments. There are an enormous number of points along the assimilation-separatism continuum and it is important, although difficult, to decide where a particular movement actually stands. This problem is compounded when groups are mislabeled by strategists at the center, when different factions in the movement make different demands, and when the movement itself varies its aims over time.

The analysis of separatism also requires careful consideration of the policies pursued by central governments: how they manage to exploit divisions in the internal cohesion of popular movements; whether government elites are perceptive enough and sufficiently secure to grant concessions at an early stage of mobilization, so preventing the transition from regionalism to separatism; and whether they can convince an incipient separatist movement that the military and material costs of such action would be unacceptably high.

No one factor or theory can provide a comprehensive explanation for the resurgence of European autonomist movements, or for ethnic-state conflict in Africa. A complex causal chain is involved combining the strains of overcentralization and bureaucracy, the emergence of supranational organizations in certain important spheres, the activities of linguistic and cultural movements seen as an antidote to anomie and alienation, and the role of economic forces producing a sense of relative deprivation.[38] All these factors have been at work behind the current revival of the "stateless nations" of Western Europe and are variables relevant to the rise of demands for ethnic self-determination in the African context.

Ethnicity Promoted by the State: The Paradox of Apartheid

While the experience of ethnic conflict in many independent African states parallels the European situation in certain important respects, South Africa remains a rather exceptional case. This is because it is the one society where the state is deliberately trying to promote ethnic and racial divisions with a policy that would appear, at least superficially, to be one of nation building but state destroying. In reality, the aim is to strengthen state control over all racial and ethnic groups by a variation on the familiar theme of divide and rule. The claim made by certain theorists of apartheid that the policy is simply a strategy to foster the development of true "nation-states" in southern Africa does not bear close scrutiny, a policy of "forced separatism" being the only technique by which a small racial minority can attempt to reconcile the notion of popular sovereignty with minority political domination. As Leo Kuper comments: "The new policy represents the rights of nations to self-determination, but of course, in the South African sense, the self-determination is other-determined."[39] Its appeal for the protection of minority groups against attacks emanating from a unitary system of government -- the Westminster model -- is an ironic reminder that it was this very system that has been used to such effect to entrench the ethnic domination of the Afrikaner.

One of the key issues to arise under this system is the question of definition -- when is an ethnic group not an ethnic group? The strategy of the National Party has been to fragment Africans to the maximum extent by insisting on the ethnic distinctiveness of the Zulus, Xhosas and other African tribal groups, and claiming that as a result they cannot be considered as a single political constituency. When it comes to "whites," however, the definitional criteria shift so that the white group is regarded as a single category despite linguistic divisions, institutional segmentation, religious differences and a diversity of national origins. To maintain this political double standard requires a form of intellectual schizophrenia which has finally collapsed under the pressure of demographic trends, the ineffectiveness of "influx" control and the manifest failure of the fiction of the homelands.[40] Verwoerd's stark vision of separate development has been replaced by the less clearly defined practices of neo-apartheid. However, the white monopoly of political power and the administrative structure of racial domination remain essentially unaltered.

Given this situation, what are the prospects for the resolution of racial conflict in a constructive and peaceful manner? One set of predictions seems to be largely irrelevant because of a refusal to recognize racial and ethnic categories as crucial social forces in South African society. Thus the neo-Conservative and neo-Marxist schools continue to claim, against all the evidence of twentieth-century South African history, that society will be deracialized as a result of the dynamics of the capitalist system -- in the first case, by the operation of colorblind market forces, and in the second, by an alliance between black and white workers intent on a socialist revolution.[41] A more convincing argument emphasizes the role of external factors in forcing concessions out of the white elite. However, the immediate prospects of such pressure having much influence seem remote. The report issued by the Study Commission on United States Policy toward Southern Africa in 1981 concludes that the main military threat to the Republic will be from guerrilla activities rather than from a conventional conflict with South Africa's northern neighbors.[42] Nor does it seem that Western influence is going to be used to demand significant change as the policy of the Reagan administration -- the Crocker doctrine of "constructive engagement" -- begins to take shape.[43] Recent revelations concerning the collusion between the Soviet authorities and De Beers Consolidated Mines Ltd. over fixing the world price of gold, platinum and diamonds suggests that it is not only Western investments that will inhibit attempts to undermine strategic sectors of the South African economy.[44]

As the threat of external pressure does not appear to be a major consideration for the next few years, it remains to be seen whether internal action, such as industrial strikes, riots and sabotage campaigns, can lead to meaningful social change. Adam and Giliomee have suggested that the "pace and scope of racial reforms increase with racial polarization."[45] While this may be true in certain limited fields like sport, there is less evidence to support the claim that intensified urban terrorism by the African National Congress and other liberation groups will result in government efforts to co-opt moderate black leadership or to increase African political representation in anything more than a consultative capacity. The National Party remains committed both to Afrikaner unity and white supremacy, and in the 1981 elections retained 131 out of the total of 165 seats. To get the party to accept even moderate reform requires a leader of imagination and charisma. P.W. Botha, like his predecessor John Vorster, is a pragmatic politician who, in the words of one perceptive commentator, "is prepared to take some risks to modernise the system but who has no intention of becoming an Afrikaner Samson bringing the temple of apartheid crashing down around his ears."[46] Under these circumstances the prospects for a constructive resolution of racial conflict seem remote. By the time that the white minority is ready to make genuine concessions in the direction of power sharing, it seems increasingly unlikely that African nationalist leaders will be prepared to compromise and accept anything less than total liberation.

Footnotes

1. Walker Connor, "Ethnonationalism in the First World: The Present in Historical Perspective," in Milton J. Esman, ed., _Ethnic Conflict in the Western World_ (Ithaca: Cornell University Press, 1977), p. 34.
2. Dunstan Wai, "Sources of Communal Conflicts and Secessionist Politics in Africa," _Ethnic and Racial Studies_, 1 (July 1978): 286-305.

3. J. Bayo Adekson, "Ethnicity and Army Recruitment in Colonial Plural Societies," Ethnic and Racial Studies, 2 (April 1979): 151-165; and Anthony Kirk-Greene, "'Damnosa Hereditas': Ethnic Ranking and the Martial Races Imperative in Africa," Ethnic and Racial Studies, 3 (October 1980): 393-414.
4. Paul Brass and Pierre L. van den Berghe, "Ethnicity and Nationalism in World Perspective," Ethnicity, 3 (September 1976): 197.
5. M.G. Smith, "Some Developments in the Analytical Framework of Pluralism," in Leo Kuper and M.G. Smith, eds., Pluralism in Africa (Berkeley and Los Angeles: University of California Press, 1971); and Malcolm Cross, "On Conflict, Race Relations and the Theory of the Plural Society," Race, 12 (April 1971): 477-494.
6. Robert Blauner, "Internal Colonialism and Ghetto Revolt," Social Problems, 16, 4 (1969): 393-408; and M. Barrera, Race and Class in the Southwest: A Theory of Racial Inequality (Notre Dame: University of Notre Dame Press, 1979).
7. Michael Hechter, Internal Colonialism: The Celtic Fringe in British National Development, 1536-1966 (London: Routledge and Kegan Paul, 1975); and John Stone, "Internal Colonialism in Comparative Perspective," Ethnic and Racial Studies, 2 (July 1979): 255-259.
8. The flow of ideas has been a two-way process. At a Paris conference on regionalism held in 1977, one Occitan scholar, who was also an Africanist, remarked to the author: "I had to go to Africa to discover French tribalism."
9. I am generally following Walker Connor's approach to terminology. See Walker Connor, "A Nation is a Nation, is a State, is an Ethnic Group, is a...," Ethnic and Racial Studies, 1 (October 1978): 377-400.
10. John Stone, "Race Relations and the Sociological Tradition," in John Stone, ed., Race, Ethnicity and Social Change (North Scituate, Mass.: Duxbury Press, 1977); and E.E. Davis and Richard Sinnott, "The Role of Political Institutions in the Evolution and Maintenance of Ethnic Conflict," Ethnic and Racial Studies, 4 (October 1981): 398-413.
11. See Christian Coulon, "French Political Science and Regional Diversity," Ethnic and Racial Studies, 1 (January 1978): 80-99; and Anthony Smith, "Towards a Theory of Ethnic Separatism," Ethnic and Racial Studies, 2 (January 1979): 21-37.
12. S.E. Finer, "The Fetish of Frontiers," New Society, September 4, 1975.
13. See Walker Connor, The National Question in Marxist-Leninist Theory (Princeton: Princeton University Press, forthcoming).
14. D. Kowalewski, "National Rights Protest in the Brezhnev Era," Ethnic and Racial Studies, 4 (April 1981): 175-187.
15. Saint-Simon and Comte, for example, with their three-fold faith in positivism, industrialism and internationalism, blithely ignored the problems of national and ethnic awareness.
16. Ali A. Mazrui, Post-Imperial Fragmentation, Studies in Race and Nations (Denver: University of Denver, 1972).
17. Andrew Greeley, Ethnicity in the United States: A Preliminary Reconnaissance (New York: Wiley, 1974), p. 11.

18. In addition, the unintended consequences of attempts at industrial stimulation from Paris have resulted in mainlanders occupying two-thirds of the top managerial posts. See also Eleonore Kofman, "Differential Modernisation, Social Conflicts and Ethnoregionalism in Corsica," *Ethnic and Racial Studies*, forthcoming.
19. A. Zolberg, "Splitting the Difference: Federalization without Federalism in Belgium," in Esman, ed., *Ethnic Conflict in the Western World*, p. 127.
20. Croatian nationalist leaders include academics, students and disaffected intellectuals who so often play a prominent part in separatist movements. See William Beer, "The Social Class of Ethnic Activists in Contemporary France," in Esman, ed., *Ethnic Conflict in the Western World*, pp. 143-158.
21. The South Tyrol (Alto Adige) is a rather interesting case. See P.J. Katzenstein, "Ethnic Political Conflict in South Tyrol," in Esman, ed., *Ethnic Conflict in the Western World*, pp. 287-323.
22. On this point Finer comments: "It is supremely ironical that at the very moment when the six countries of the E.E.C. have become nine -- a movement of fusion -- this should be countered by demands for the creation of half a dozen new European mini-states to offset it." *New Society*, September 4, 1975.
23. Hans van Amersfoort and Herman van der Wursten, "Democratic Stability and Ethnic Parties," *Ethnic and Racial Studies*, 4 (October 1981): 472-485.
24. Patricia Elton Mayo, *The Roots of Identity* (London: Allen Lane, 1974).
25. The public notices in Brittany -- "Défence de cracher par terre et de parler Breton" -- being a classic example of the linguistic insult that can be turned to good use by nationalist leaders.
26. A. Butt Philip, *The Welsh Question: Nationalism in Welsh Politics, 1945-70* (Cardiff: University of Wales Press, 1975), p. 334.
27. For an exaggeration of this factor, see Edward Sagarin and J. Moneymaker, "Language and Nationalist, Separatist and Secessionist Movements," in Raymond L. Hall, ed., *Ethnic Autonomy -- Comparative Dynamics* (Oxford: Pergamon, 1979), pp. 18-46.
28. Marianne Heiberg, "Insiders-Outsiders: Basque Nationalism," *European Journal of Sociology*, 16 (1975): 169-193.
29. Hechter, *Internal Colonialism*.
30. See Michael Hechter and Margaret Levi, "The Comparative Analysis of Ethnoregional Movements," *Ethnic and Racial Studies*, 2 (July 1979): 260-274.
31. Talcott Parsons, "Full Citizenship for the Negro American?" in Talcott Parsons and Kenneth B. Clark, eds., *The Negro American* (Boston: Beacon, 1967), pp. 709-754.
32. Herbert Blumer, "Industrialization and Race Relations," in Stone, ed., *Race, Ethnicity, and Social Change*, pp. 150-166.
33. Anthony Mughan, "Modernization and Regional Relative Deprivation: Towards a Theory of Ethnic Conflict," in L.J. Sharpe, ed., *Decentralist Trends in Western Democracies* (London: Sage, 1979); see also Anthony Mughan and Ian McAllister, "The Mobilization of the Ethnic Vote: A Thesis with some Scottish and

Welsh Evidence," Ethnic and Racial Studies, 4 (April 1981): 189-204.

34. Ibid.
35. Ibid., p. 12.
36. H. Schuman, "A Note on the Rapid Rise of Mass Bengali Nationalism in East Pakistan," American Journal of Sociology, 78, 2 (1972): 290-298.
37. Cynthia Enloe, "Central Governments' Strategies for Coping with Separatist Movements," in W.H. Morris-Jones, ed., The Politics of Separatism, Institute of Commonwealth Studies, Collected Seminar Papers, no. 19 (London: University of London, 1976), pp. 79-84.
38. The classic formulation of the concept of relative deprivation can be found in the writings of Alexis de Tocqueville. See John Stone and Stephen Mennell, eds., Alexis de Tocqueville on Democracy, Revolution and Society (Chicago: University of Chicago Press, 1980).
39. Leo Kuper, "South Africa, Human Rights and Genocide." Paper delivered to the African Studies Program, Indiana University, April 4, 1980.
40. John Dugard, "Independent Homelands: Failure of a Fiction." Presidential Address, South African Institute of Race Relations, 1979.
41. See Lewis Gann and Peter Duignan, Why South Africa Will Survive (London: Croom Helm, 1981); and Darcy du Toit, Capital and Labour in South Africa: Class Struggles in the 1970s (London: Routledge, 1981).
42. The Star, May 23, 1981.
43. The Times, January 30, 1981.
44. Ibid., April 7, 1981.
45. Heribert Adam and Hermann Giliomee, Ethnic Power Mobilized: Can South Africa Change? (New Haven: Yale University Press, 1979), p. 143.
46. The Observer, May 31, 1981.

6
The State and Ethnicity: Integrative Formulas in Africa

Henry Bienen

Introduction

The processes of group and personal identification are complicated ones. Ethnic affiliation is "self-defined in the sense that members of the group rather than outsiders draw its boundaries."[1] However, many studies of communal identifications in Africa have stressed that individuals and groups receive labels and identifications from others and that over time these may become accepted (or not) by those who are so named or labeled. Studies of rural migration and of urbanization have shown how language or ethnic groups may appear rather quickly.[2] The idea that communal identifications are "primordial" and that they are "givens" has been much revised over the last two decades.

We have come to recognize also that the place of groups within a state system is in part determined by the deliberate ethnic categorizations used by the central institutions of the state.[3] Legal and political structures affect communal and class definitions and competition. Different rules of the game exist for defining groups, legitimizing certain ones and declaring others out of bounds. Ideological formulations vary. Sometimes the rules and norms are explicit and other times they are implicit. It is important who sets those rules and norms and whether they are widely accepted. How congruent the centrally defined and imposed values and rules for group definition are with the values of particular groups in society will be critical for the resolution of conflict and prospects for cooperation.

In Africa, as elsewhere, the administrative and political formulas for encompassing ethnicity are tremendously important. Explicit and implicit quotas for recruitment to civil services and to military and educational institutions are crucial to personal life chances and to group mobility. Individuals will define themselves in and out of groups to get access to valued places.

Beyond administrative mechanisms are broader political formulas that may be even more critical to the evolution of communal developments. The administrative mechanisms may reflect far-reaching ideas about conflict in society. These ideas may have wide acceptance or not. Frequently, there is a struggle over the administrative and constitutional mechanisms which express wide disagreement between communal groups, or between certain elites and non-elites over the political formulas for defining and identifying groups and regulating their conflict and cooperation. South Africa's rulers, for example, define certain groups as belonging to homelands outside the Republic of South Africa. These identifications are rejected by members of the various ethnic groups such as Zulus or Xhosas.

In the very process of rejecting central identifications, ethnic groups define themselves as they seek to react to state actions and formulas. The more centralized the state or political institutions, the more crucial are the formulas for dealing with ethnic cleavage. The reach of state power is greater; this also politicizes groups who for the first time may have to define themselves vis-à-vis central authorities and the social or communal groups that control the center.

The process of group identification involves interactions between the self-determinations of groups, the interactions between different groups, and the actions of, and reactions to the formulas of, those who control the state. Official identifications may be stated in laws or codes, e.g., Nuremberg-type laws or South Africa's system with its racial and national classifications which are expressed in laws defining representation, habitation, and national and homeland status. Also important are more diffuse beliefs about the proper relationship of groups in society. Varieties of African socialisms, for example, have insisted that Africa was free of class divisions and ought to remain free of them. Various statements of traditional relationships between communal groups in Africa have argued that these relationships were essentially conflict-free prior to colonial impacts or at least could be free of conflict in the future. Expression is given to the ideal of absence of communal conflict by the prohibition against reference to communal issues during electoral campaigns in many African countries.

This essay explores various meanings of national integration in Africa. It explores the ways that central institutions -- army, bureaucracy and party -- have tried to provide ideas and formulas for dealing with communal cleavages in society. The focus is on the institutions of the state as they define meanings of integration and are involved in the process of group identification because this focus has been relatively neglected in favor of attention given to communal groups themselves. As already noted, processes of identification are interacting ones between the state and groups within society.

National Integration and Communal Identification

Within any society, there usually abound at any one time various understandings of what national integration means. It would be surprising if groups within culturally, communally, and geographically heterogeneous societies all shared the same ideas about the meaning of national identity, the proper place of groups within society, or the relationship of groups to the state. Moreover, those who control the state may have a vested interest in one understanding of national integration and, if they should lose control, find themselves propounding another idea of group relations. Formulas for national integration relate to desires for centralization and decentralization of authority. Commitment to models of a naturally integrated society depends in part on where one stands in relation to other groups and to state power.

In Africa, the dominant models have been assimilationist types. Within the assimilationist model, there can be assimilation to a particular core cultural identity or to the presumed characteristics and political affiliations of a dominant ethnic group. Assimilationist models can also refer to a particular strand of a culture or political-communal affiliation. That is, all might have to be identified as Moslem, but having been so identified, other identities would be permissible, based both on differentiations within the Moslem community (e.g., belonging to a particular brotherhood) and on ethnic-language affiliation. Assimilation can also be to some new identity. This identity can forged out of old strands or be newly minted.

Tensions exist between accepted models in Africa and the reality of the multiple identifications and commitments that individuals have to ethnic-language, religious, and lineage groups. When integration has been understood as the absence of communal identifications less general than those to the nation-state, that understanding has been frequently rejected. And identification as a member of a group may be all right as long as the identification carries no political baggage with it, but this is rarely the case.

So far, I have been referring more to the form or structure of formulas for national integration than to content. There are also the actual norms of what constitutes proper or successful ethnic behavior and attachment to the correct symbols. Language and religion are two clear examples of this. If a person or a group does not possess the acceptable religion or language, there may be formal or informal bars to occupations, credit, education, etc. Also, religions and languages carry, or affect, depending on one's view, perceptions of reality. Thus the insistence on a certain religion or language as a state religion or language (aside from the practical matter of making it hard for those who do not follow the religion or use the language to compete -- even if they could, in principle, adopt them) means the insistence on certain values and cultural characteristics, a way of defining experience, success and failure.

If a nation becomes, constitutionally, an Islamic republic, certain values are obviously enshrined and given special force. Official memberships in international Islamic organizations have been avoided in Nigeria in order to avoid official status for any one religion. The most contentious issue during constitutional debates in Nigeria was whether or not to have a court of appeal based on Sharia law.

Just as conflict over choosing a particular African language has led to the maintenance in place of European languages, so in Africa there has been a search to suffuse the state with nonthreatening symbols to which all might attach themselves, such as authenticity in Zaire, the idea of Negritude put forward by Léopold Senghor in Senegal, and Swahili as a nonethnic national culture in Tanzania.

Elites are not always the formulators of the symbols and meanings of ethnicity, nor are they the only ones who have ideas about a "properly" integrated society. But those who stand to win or lose the most from official language or nationality policies are those who have a chance of getting public jobs or of taking exams -- that is, elites. Thus whether or not one believes that ethnicity is a mask for class privilege and that elites simply stir up or create ethnicity to further their class interests (and I do not), it is the case that elites interpret meanings of integration and communal identification to wider audiences. It is through the process of defining formulas for national integration, and other political formulas, and establishing them as legitimizing doctrines, that elites attempt to close the gaps between themselves and non-elites.[4]

Not all meanings of national integration have to do with communal relations, of course. Territorial or spatially defined identities are involved. Gaps between classes and also between non-economically defined status groups are crucial too. Relationships between institutional actors and factional groups must be taken into account. Any basis for social conflict can become the core around which meanings of national integration may be developed. Categories for describing human inequalities are variable and various.

Before taking up the role of the state and of its institutions in the process of defining formulas for group identification and national integration, it is necessary to come to grips with understandings of class and ethnicity because the relationships between them are often at the heart of formula-making by the state centers of power.

Class, Ethnicity and Inequality

There is a long-standing debate and a growing literature on the place of classes in Africa. We have many studies arguing about the existence of classes, class consciousness, and the meaning of class in the African context. No one doubts that groups organize for competition and cooperation on the basis of economic interests and motives, nor that occupational distinctions and income differentiations have political consequences. What is at stake in the debate is whether by class politics we mean that people have to be conscious of their position in relation to the means of production and of the social and political consequences that flow from those relations, or whether objectively classes can exist regardless of whether people are conscious of their situation or not. Also, does any hierarchy of inequality constitute a class system? Should conflict over access to public resources be defined as class conflict? That is, are all distinctions between rulers and ruled to be defined as dichotomous and zero-sum and as distinctions between classes?

In part, these issues are definitional. That is, we define class and ethnicity in a particular way when we say that a group is an ethnic one. We make an analytical distinction, a simplification for analytical purposes. In the real world, people have bundles of feelings, attachments, motives, all at once. Individual identities, that is, the way one describes oneself, are a function of many things, including how one was "named" at birth by one's parents, one's ties to relatives, and how one is perceived by outgroups. An individual can accept or reject these identifications. The way one works, lives, moves about, affects self-identification over time. And other groups give identities which sometimes are hard to reject. The state may insist on an ethnic identity for an individual.

Although distinctions between communal and class groups are definitional and analytical, these distinctions have an empirical base. That is, the definitions ought to make sense in terms of what is going on in society. The distinctions ought to be contextually rooted.

Ethnic affiliations may be hard or fluid in different societies, and they can change over time. They are not always "givens" or primordial. The distinction between ascriptive and nonascriptive variables is not always very helpful. These are subjective phenomena, not objective ones. Jean Gallais has observed that in the Western Sudan, traditional ethnic identity involves not only the "givens" of descent, but also techniques of exploitation of nature, history, religion and spatial organization, and that some of these components are based on achievement as well as ascription.[5] Identifications may be derived from language or skin color or place of origin in some societies, but not in others. Political factors are usually important in the process of identification, just as economic ones are.

When we try to isolate analytically the way that communal and ethnic relationships are different from class relationships, it turns out not to be very easy to do. It is said that communal groups encompass the full range of sex and age divisions within society. This is true for class groups too. Communal ties are said to provide a network of groups and institutions extending throughout a life cycle. This can be also true for occupational and class groups who form sports societies, burial societies, health societies. Within what we normally refer to as class and communal groups, internal differentiation of status, power, and lifestyle usually takes place. True, if by class we mean income group, then differentiation by income group can only take place <u>within</u> communal groups by definition while it must take place <u>between</u> class groups by definition. Common cultural identification may occur for class as well as communal groups. Most communal groups, and certainly large ethnic-language groups, do not share cultural symbols across the whole group. Even language communication may be hard for an ethnic group when there are many dialects within a group tracing its origins to a common ancestor. There also are societies where members of one class group cannot be easily understood by members of another class group even when they are all members of the same language group.

We may get left with the idea that an ethnic-communal group is simply one where people claim to share a commonality of identity across many relationships. Some class groups would then be "communal" with this understanding of the term. And indeed, the term "Cockney" implies not just the use of a language dialect but shared food tastes, lifestyles, income, and status.

Normally we refer to class affiliations as those stemming from people's social and economic relations, which are in turn derived from their place in the processes of production. More broadly, we refer to economically based ties as achievement-oriented rather than ascriptive-oriented memberships. Class is thought of by many anthropologists as non-birth-ascribed status. But while it is true that class status is identified by such features as income, education and occupation, it is also true that our idea of class has birth and holistic lifestyle characteristics.[6] We refer to upstart people and classes, parvenus, nouveau riche. How people spend their income, that is, their consumption style, seems related to broader, nonincome characteristics. These associations with class may be more applicable to a society that has many occupational and income differentiations within it as compared to a society where most people are rural and very poor. But we should not think that highly industrialized societies are the only ones in which the notion of class is slippery in application.

We cannot assume that understandings of community, class and inequality will be the same within countries any more than they are the same between countries. Moreover, processes of identification are fluid. Terms to designate individuals and groups may vary as groups have different interactions or individuals appear in particular roles to others. A Senegalese from the Sine Saloum in central Senegal who travels to the Senegal River areas in the north of the country might be seen as a Wolof, or as a member of the Mouride Brotherhood, or as a follower of a particular marabout, or as a Moslem, depending on with whom he interacts and what subjects and issues are involved in the interaction. These identifications are communal but they are also economic and political, and they can be issue-specific, depending on currently outstanding issues between individuals and groups.

Just as the salience of communal issues rises and falls, so specific identifications change over time or coexist at the same time as different groups perceive individuals and other groups in occupational, institutional and political roles. Thus institutions, political groups and economic groups can all be defined in communal terms at times. Military institutions may be understood to be not just dominated by a specific group but the provenance of that group -- the institution which creates and molds their identity. Some discussions of the meaning of Nubians in Uganda suggest that this may have been the case for that community.[7]

The fluidity, heterogeneity, and complexity of processes of identity and group formation suggest that state interventions may be very important for outcomes.

State Interventions and Communal Identifications

All states define rules and regulations for citizenship, but individuals and communities outside the legal boundaries of the nation-state may be part of the national community. Irredentist movements want to bring such groups into the nation-state. The demand for the ingathering of Somali peoples is an example. Also, states may define out of citizenship and membership in the nation certain categories of people. South Africa has been doing this with its shipping of people back to some presumptive homeland. African countries have also deported individuals as foreign nationals. Some of these were clearly identifying themselves as foreign nationals. But at least some people of Nigerian origin had lived in Ghana for a long time and thought of themselves, perhaps, as Ghanaians or had intermarried with Ghanaians before they were sent back to Nigeria.[8]

There are also examples of special reserved rights or privileges which have come about during the transition from colonial or minority government to a new independence constitution. Thus in Zimbabwe, special seats were set aside for whites in parliament. This kind of special treatment or privilege is designed to shore up weak positions for a time and is a sign of a perception on the part of the community demanding such representation or special status that it is weak and vulnerable and perhaps not quite part of the wider nation.

This is but one example of the special relationship that a particular region or community may have with regard to a central government. Arrangements to protect minority rights run from federal systems which "guarantee" spheres of action to subnational units to confederal arrangements to arrangements for major areas of regional autonomy. From independence in 1962 until 1966 Buganda operated inside Uganda with its king, the _kabaka_, and with its own parliament, or _lukiko_. The kabaka was also president of Uganda, and operated as its chief of state, while the prime minister, Milton Obote, was head of government. The tangling of the skein of Baganda nationalism with Uganda's own independence movements and the splits within Baganda by religion, lines of descent, district and class have been described well.[9] Here I only want to note that the administrative arrangements developed by the colonial regime in Uganda to coexist with the Baganda reinforced separatist feelings and helped to strengthen political and cultural subnationalism, and these arrangements had a ripple effect in the Western Kingdoms and in Busoga.

The colonial regime may or may not have had a clear image of the nation it was bringing into being in Uganda. It was compromising with different political pressures as it was carrying out a transition to independent rule. However, the administrative, constitutional and political arrangements colonial regimes established or bargained over in Uganda and elsewhere were critical points of reference or parameters for later struggles. Sometimes those arrangements were scrapped, as when the _majimbo_ or "regional" constitution, which was designed to protect the rights of smaller tribes in Kenya, was replaced after independence. Sometimes, as in Uganda, the struggle over constitutional and political issues was fierce.

Were Kikuyu communal feelings less strong than Bagandan? In part, the Kikuyu were able to dominate the central structures of the state after independence and thus did not have to retreat to a more parochial entity. In part, the fact that the colonial regime had cut the Kikuyu off from central power in the 1950s, but had not provided them with any idea of an autonomous region of their own, and the fact that they were widely spread outside of the Central Province of Kenya, led them to identify their own nationalism with that of Kenya itself.[10] Since the colonial regime did not allow political parties to organize beyond a district base until late in the game of Kenyan nationalism (and this fact was important for the weakness of Kenyan parties at the center and for the continuing importance of the districts in party factionalism), the districts did not provide a base from which to confront any political center with demands for autonomy.[11]

The colonial regime could in Kenya, and in Tanzania too, create vehicles through which political demands could be made. Indeed, where the colonial regimes created federations of "tribes" or brought into being tribes such as the Sukuma or Chagga through administrative arrangements and the appointment of central chiefs in communities which had not necessarily had such figures, they often created new actors in communal competitions; they sometimes did this deliberately. But perhaps there has been a certain fragility to the ethnic group that has been formed so quickly. Another way to put this is to say that where the integration of groups results primarily from administrative requirements and pressures rather than from ties of custom, kinship and shared cultural symbols, there may be a highly expedient aspect to communal membership and definition. Over time, however, members may become quite committed to the new entities. What is highly instrumental for one generation may become primordial for another.

The colonial and the independent African states have intervened in important and frequently deliberate ways to form and reform group identities and relationships. Can we say anything about the ways in which particular state institutions deal with ethnicity and communal identifications in contemporary Africa?

Armed Forces Organizations and Civilian Bureaucracies

The picture is complex when we come to reflect on whether or not armed forces as compared to parties or bureaucracies have been associated with particular models of communal and class conflict resolution, and whether they can be associated with particular formulas for group identification.

Armed forces are hierarchically organized and their command and control systems are relatively centralized compared to parties and even to functionally organized civilian bureaucracies. Frequently, military leaders express a rhetoric that they are above ethnicity, that the armed forces embody the whole society. Armed forces put a high premium on corporate autonomy and the maintenance of their own organizations. Militaries express a fear of factionalism and a determination to socialize individuals to the values of the organization per se rather than to those of subgroups. Structural adaptations to fulfill unique, primary functions of combat give armed forces their special concerns for emphasis on hierarchy and chain of command, or so the literature on the armed forces and society tells us. But militaries do not always successfully socialize individuals, nor do they always maintain corporate cohesion and professionalism.

We know that African armed forces have had to deal with social and ethnic cleavages in their societies and that they have also been internally divided by ethnic, social, personal, age, functional, rank, and service branch cleavages. There have been ethnic formulas for recruitment into various armies, sometimes formally stated, as in Nigeria prior to military rule, sometimes unofficial, as when Ugandan or Kenyan armed forces did not recruit in certain districts. There often have been special tensions inherent in the militaries' claims that they are above factionalism and ethnicity on the one hand, and the fact of cleavage and cliques based on social and class origins on the other hand. That is, it has been the very commitment to a single national identity on the part of armed forces in the face of the reality of their own internal divisions and their frequent exacerbation of communal conflict which has led to a heightened sense of a disjunction between the formulas associated with armed forces and their practices.[12]

When the normative definitions of the militaries and the realities within their ranks conflict, we cannot conclude that the values the institution claims do not matter. Quite the contrary, the failure of armed forces in Africa to deal flexibly with the realities of communal cleavage in society and in their own organizations is partially a consequence of the mismatch between proclaimed ideas about the institution being "above ethnicity" and the realities on the ground. Militaries may be explicitly denying the facts of ethnic diversity and tension while taking account of them in their own internal recruitment and promotions, but not in their formal training and socialization procedures.

The Nigerian military was a relatively professionalized military organization in Africa. But when communal tensions in Nigerian society overwhelmed the boundaries of the armed forces, order broke down, and at least some killings of officers had strong communal motivations. The shock to the military's own self-image and the violation of proclaimed values perhaps accentuated the internal disintegration, and once the army was perceived to be communally divided and motivated, communal violence spread throughout parts of Nigerian society.[13]

Generalizations about military regimes tend to be weak and there has been as much variety within the category of military regime as between military and civilian regimes. Armed forces have been linked to civilian groups and institutions in many different ways across political systems. Moreover, in any particular situation it is hard to separate out a response which is particularly military from other strands, both historical and contemporary, in policy formulation.

Should we attribute the Ethiopian military's decision to prosecute wars in Eritrea and the Ogaadeen to the special features of the organization and composition of the Dergue? To the historical continuities and desires at the center of Ethiopian power to maintain an empire? To the nationality policy of a professed Marxist-Leninist regime which has stressed the unity of oppressed classes?[14] To the recognition that giving way in any one place would make control less tenable in another?

After many contradictions in policy and in policy statements, the Dergue wound up pursuing and intensifying the war against Eritrean nationalists.[15] It even invoked overtones of a Christian crusade against Moslems. These policies were developed while factional infighting was proceeding within the military regime; the doves on Eritrea, headed by General Aman Michael Andom, lost, and Andom's Eritrean positions were probably important in the factional outcomes and in his own murder in November 1974. Concepts held by members of the Dergue on ethnic conflict resolution differed. General Andom, himself an Eritrean, wanted a peaceful resolution of the conflict. He also held positions with regard to the composition and the size of the military council's membership and the treatment to be accorded members of the old regime which went to the heart of the matter of who would hold power. Lt. Col. Mengistu Haile Mariam, who won out in the struggle, was committed to a different policy, which led to continued military conflict. Laitin and Harker attribute Mengistu's (and his allies') intransigence to his ideology of scientific socialism.[16]

It is true that Marxist-Leninist regimes have had a bias for centralized rule. It is also true that Mengistu's perceptions that class enemies were using communal feelings and that outsiders were using Eritrean nationalism to destroy the Ethiopian revolution hardened his positions. But there were strong elements in the Eritrean movement which, after all, had good socialist credentials and were backed by at least some "radical" regimes including, at various times, Iraq, Syria and Cuba. Thus Mengistu's own conception of Ethiopia and his own bias for tough solutions and for dealing with enemies in a ruthless and military fashion have been critical elements. There came together in the politics and person of Mengistu a number of strands which moved the regime toward the fist as a response to Eritrean nationalism. Indeed, there are parallels with Stalin's career and personality. For Mengistu had been regarded as a Galla, not an Amhara (the dominant and core cultural and political group in Ethiopia), just as Stalin was a Georgian, not a Russian. The Ottaways state that Mengistu was regarded as a _baria_, a black-skinned slave from the deep south of Ethiopia, an upstart from the conquered areas, low in social and communal status. However, it is believed that Mengistu's father was low-status Amhara and that his mother was from the south. Perhaps both Mengistu and Stalin had to prove their bona fides on the issue of maintaining the nationalism of "their" societies.[17]

Since Ethiopia was close to territorial fragmentation and social dissolution in the mid-1970s, one could make an argument for the need to reject all compromises. And, as suggested, the issue was fought out, as it had been in the Russian revolutionary movements and then after 1917, in the context of factional struggle and political expediency.

Laitin and Harker suggest that militaries in power have a low tolerance for ambiguity and a need to bring problems to a head and to try to solve them. They cite not only Ethiopia but also the Nigerian military's response to potential secessions. It is true, as they suggest, that while the Nigerian military moved quickly toward reconciliation with the Ibos after the Nigerian civil war and even prosecuted that war, for the most part, with a sense of the need to reconcile afterward, it was impatient in 1967 with negotiations over communal relationships that were to be defined in new constitutional arrangements.[18] It is also true that Nigeria had a history of bargaining over ethnic relationships which was at least as important as its history of overt ethnic conflict. Moreover, the military had within it important components from minority areas, including General Gowon himself, and these elements needed the Ibos back in the Nigerian system in order to balance that system against the power of the Moslem north and Yoruba west. It is difficult to say how much one can attribute the emergence of Gowon the reconciler as compared to Mengistu the conquerer to Nigeria's perceived needs. Does the man fit the needs of the time or does he define those needs with his own views and positions? The process is an interactive one. Widely shared views of what was appropriate in Ethiopia and Nigeria were related to differing historical models and experiences,[19] as well as to the different personalities and factional struggles within the respective armed forces. Neither the Ethiopian nor the Nigerian case allows us to generalize facilely about organizational variables and their consequences for defining communal identifications and models for national integration.

Similar care must be taken when examining the impacts of civilian bureaucracies on formulas for integration and group identifications. Professional bureaucrats claim that power springs from the office, not the man, and that codified rules and regulations govern the operation of bureaucracies. Hiring and promotion are supposed to be done according to merit and examination systems. A large literature has grown up which explores the gaps between the values held by bureaucrats and the stated definitions of professional conduct. Social scientists and novelists have shown us the multiple pressures on bureaucrats from kin, patrons and clients. They have painted a picture of bureaucrats with fragmented value systems who operate in different ways in various contexts and they have explored the latent functions of bureaucracies.[20] Corruption, nepotism and cynicism have been detailed. Case studies on African countries have told us that where formal decision rules proscribe taking bribes in practice the watchword, as in Zaire, is "yibana mayele" -- steal cleverly.[21]

It has been argued that the central normative focus on interpersonal relationships is closely tied to the corporate family basis of African societies. Price maintains that the corporate nature of African cultures creates a situation in which the role pressures placed on officials by members of their social systems are overwhelmingly particularistic. Pressures come not only from a person's kinship group, but from other members of his society.[22] Society places social value on the status but not the role aspects of organizational positions.

These arguments should hold for militaries as well as civilian bureaucracies. And no doubt they do. We do not have much in-depth research comparing military and civil service attitudes toward organizational roles and toward communal ties. Thus we are left with comparisons of behaviors and with comparisons of statements or formulas meant to govern behavior toward communal groups and statements about ethnicity in societies. Certainly, we can find examples of particularistic behavior within militaries and between armed forces and society. However, constant interaction between militaries and citizens is less common than between civil servants and citizens. Moreover, civil servants are more spread out through society. The armed forces define themselves in stronger corporatist terms than do civil bureaucracies. That is, the smaller size, greater centralization and less functional specialization within armed forces, and especially within African armed forces as compared to civil bureaucracies, lead to a greater stress in public on the organization's corporate definition. Part of that definition is a claim to be above ethnicity. While formal codes exist which define civil service behavior, and training and recruitment stress bureaucratic roles, less emphasis is placed on socialization than in the armed forces. It is not that members of the armed forces in Africa behave more in terms of their formal role orientations than civil servants do. Rather, it is that they are somewhat more cut off from society and they have a more apolitical ethos, one that is frequently antipolitical. Compared to civil servants, there are greater tensions for the military between behavior and the formal values espoused with regard to groups and conflict.

Political Parties

Of all the institutions of rule in Africa, it has been political parties that have been most sensitive to cleavage and conflict and have tried to design formulas for handling communal and class cleavages. These formulas have had to do with internal party arrangements such as primaries, appointments to committees, and party membership criteria. And they have had to do with constitutional arrangements for contesting elections and establishing mechanisms for representation through assemblies at national and/or subnational levels, councils of state, and ministerial positions. There have also been direct party manifestoes and pronouncements on national integration in which national leaders have elaborated on their ideas concerning ethnic identification in society.

Here follow a few examples concerning electoral arrangements. In 1970, debates were held in Uganda over electoral proposals for parliament and presidency. President Obote proposed in what was called Document No. 5 that every candidate for parliament should stand in one basic and three national constituencies. The basic constituency would be in the candidate's home region and the three national constituencies would be in other regions of the country. The percentages won in each constituency would count as electoral votes, and the winning candidates would be those with the highest totals for the four constituencies.[23] The stated idea behind the proposals was that a candidate would have to appeal beyond his local constituency and to reach out to different ethnic communities. For presidential election, the same document proposed that parties would propose candidates. Each voter would cast a vote for or against the candidate. Electoral votes would be tabulated as constituency percentages for or against, and the candidate who had the largest total electoral votes would be elected. Again, the idea was to show widespread support across constituencies. At the same time, as the proposals were debated (and the Uganda People's Congress, the then ruling party, voted down the president's proposals) it became clear that some feared direct election would show that the UPC and Obote had little support among key ethnic groups, especially the Baganda.

In Nigeria during 1978-79, as the military moved to establish constitutional arrangements for a return to civilian rule, intense debates were held over proposals for electoral forms. The complicated provisions for winning election as president of the Nigerian Federal Republic included the insistence on a candidate's receiving a plurality, plus at least one quarter of the votes in two-thirds of the nineteen states.[24] If this did not occur, a run-off was required between the two top vote-getters with the national and state legislatures sitting as an electoral college and choosing between the candidates.[25] The clear aim was to force candidates to appeal beyond their own major ethnic blocs.

The Nigerian electoral provisions established staggered elections for governors, state legislatures, national legislatures, and the presidency. This meant that parties with ethnic bases would not be frozen out. The federal system provided them with the perquisites of executive and legislative office. Also, each state had to be represented in a president's cabinet. At the same time that ethnic representation was guaranteed, the military leaders insisted that only national parties would be allowed to contest elections. The constitutional drafting committee insisted that parties must make national appeals and have national outlooks and that members of a party's executive committee must be drawn from at least two-thirds of the states. New parties had to be created and a number of regulations were imposed to be policed by a federal electoral commission. This commission refused to certify all but five parties; it debarred parties it felt were not truly national in scope and support and it eliminated parties which were appealing to class and occupational categories.

The Nigerian formula for more states, competitive parties, and a strong center was one aimed at reconciling pluralism and ethnicity with the need for central authority and centrally determined allocations. In Nigeria, minority peoples, that is, non-Hausa, non-Ibo and non-Yoruba, saw the center as the protector of smaller ethnic-language groups. This perception depended, of course, on the center not being controlled by any one major ethnic group and on the perception that those in control would diffuse authority among different ethnic groups. In fact, as the elections made clear, each of the parties had strong ethnic support from one of the major ethnic groups, although Shehu Shagari's National Party of Nigeria had the broadest support.[26]

The reality in Nigeria was the continued success of leaders who could mobilize major ethnic constituencies and make alliances across ethnic groups. The appeals of ethnicity were so strong that Chief Awolowo, leader of Yoruba-speaking peoples, was able to win in Ibadan city, where his party had never been able to win elections, in the first civilian period when it ran against other Yoruba-backed parties. However, the formula was one where ethnicity was implicitly recognized and diffused through complicated electoral arrangements. Ethnic representation was guaranteed but diluted. It was a very different system from that of the national list of many African parliamentary elections which eliminated the geographic and constituency basis for representation. In national list systems, the ethnic component of representation was not necessarily abandoned as the list might be juggled by central authorities to take account of ethnic affiliations. But the direct tie between constituencies and representatives was broken and it was impossible to register dissent as an ethnic community except where an opposition party was allowed to contend. In the mid-1960s in the Ivory Coast and Mali, for example, only the candidates of the ruling party stood and they did so on a national list basis.

During this period, some observers of the African political scene made an argument for the utility of the political machine. The argument, in essence, was that Africa had a low level of political capacity at the center, and had ethnically heterogeneous populations. Thus, it was important to have political organizations congruent with the dispersion of power, the diversity of communal life, and the low level economies. Given the restricted economic resources available to African leaders, many payoffs would have to be symbolic and communal groups would have to have a feeling that they were not frozen out. Political organizations need to have a grass-roots existence. Effective political institutions would need to be decentralized and competitive whether within a one-party or multiparty framework.[27]

In the United States, the machine was a vehicle for upward social mobility for immigrant groups who saw politics as a vocation. Political machines were organizations whose norms were congruent with traditions of personalized action. Observers of African parties often painted a picture consistent with the description of those organizations as political machines. Normatively, some saw the machine as particularly suited to govern small communities in transition and as providing both flexibility and stability. Rewards distributed by machines tended to go to members but the membership basis was broader than it would be in more hierarchical systems.[28]

Many party machines were replaced by military regimes. Of course, some parties continued to exist into the 1970s and 1980s although they underwent evolution too. KANU in Kenya remained a rather decentralized organization with much internal factional competition and many ethnic alliances. It continued to hold primary elections to sort out distributions of power within the organization as did TANU in Tanzania. Although the latter's leadership tried during the late 1960s and 1970s to centralize authority and insisted on more commitment to stated ideological goals after 1967, TANU remained an open, mass party.

Lucy Behrman has argued that in Senegal the political forces in the government and in the Union Progressiste Sénègalais remained fundamentally the same in many respects despite the new constitutional formula first announced in 1974 by President Senghor whereby Senegal adopted, through constitutional amendment, a three-party system. The ruling UPS filled the slot for a left-social democratic party and the Parti Démocratique Senègalaise was designated as the right liberal party while the Marxist-Leninist position was filled by the Parti Africain d'Indépendance. "Senghor's machine was dependent then on the support of the more powerful marabouts, and the same basic relationship may be assumed to have existed during the past decade since there has been no realignment of power within this political structure."[29] Certainly, Senghor continued to deal with major Moslem Brotherhood leaders directly and through his own agents in an extremely clever fashion. The national leadership continued to reach local constituencies by bargaining with community leaders.

In Senegal, Ivory Coast and Kenya, we can find varying degrees of ethnic cleavage and tension. These are not societies which are homogeneous or cohesive by language, culture, or ethnic identification. But the ethnic lid has remained on. One can make arguments pro and con for the utility of the party machine in terms of growth strategies and equity considerations. But I think we have to give it relatively high marks in terms of ethnic bargaining and representation and to some extent in terms of elite circulation, at least at middle and lower levels. Moreover, the party machine seems, in Kenya and Senegal, to have handled the transition problem fairly well. This looks like it may be the case in Ivory Coast too.

One could make an argument that in Tanzania, too, despite the emphasis on ideology and building socialism, TANU also partook of many features of a political machine. It remained decentralized with lots of local autonomy in its regions and districts. Many of its cadres seemed to remain motivated by material interests, much to the chagrin of critics from the left who saw TANU as being too much in the hands of a self-serving petty bourgeoisie, with its top leadership constituting part of a bureaucratic bourgeoisie.[30]

Perhaps above all, Julius Nyerere continued to insist that TANU's claim that it speaks for the people must rest on its representative quality rather than on its ideology.[31] It is true that Nyerere never saw this representation in terms of specific communal groups. Quite the contrary, he stressed the importance of a national ethic and much of his written work and ideological formulations have been devoted to filling in the content of the ethic.[32] From the start of competitive elections within a one-party framework in 1965, TANU prohibited appeals to race, tribe or religion and there was general compliance with such strictures, at least in public meetings.[33] While candidates have been chosen in TANU elections when they were not local to an area or did not belong to an ethnic group indigenous to that area,[34] in the close analysis that was done for the 1965 general election in Tanzania, it was shown that most successful candidates had deep roots in their constituencies and usually had strong ethnic ties. In areas where local ethnic cleavages were severe, elections sometimes heightened cleavages, as they did in other African countries.[35]

Thus Tanzania, a country often considered to have fewer severe ethnic cleavages than most African countries, one which has a lingua franca and a widely shared Swahili political culture, is not without its ethnic differentiation and concern for representation of local and perhaps even ethnic communities. Indeed, Nyerere's continued emphasis on the need to distinguish between TANU and government lies in an understanding of the distance between state power and the people, in a recognition that bureaucrats work to stipulated procedures, and in the view that government is a top-down organization and one inherently coercive.[36] The party humanizes authority to the people and creates two-way links. It is proximate to and representative of explicitly social categories, workers and peasants, not to ethnic communities. Nonetheless, by being responsive to local concerns it cannot avoid ethnic issues where and when they arise.

Questions as to how open the ruling party in Tanzania is can be debated. This party is now the Chama Cha Mapinduzi, formed by the merger of TANU and the Afro-Shirazi Party of Zanzibar in February 1977. Whatever the judgment on these matters, CCM is not a vanguard party. It still retains, or at least its leader Nyerere retains, commitments to remain a relatively open party. One cannot use it to assess how a Marxist-Leninist party handles questions of ethnicity and applies public formulas to the identification of groups. One can say that the operative categories in theory and mostly in practice, for TANU and then the CCM, have been broad social formations, not communal ones. Nonetheless, TANU, like other African political parties, must take account of ethnic cleavage and must have mechanisms for coping with communal differentiation in a noncoercive way.

This same argument can be made also for parties that define themselves as Marxist-Leninist. Indeed, insofar as Marxist-Leninist parties are committed to more internal centralization than machine-type organizations, and more committed to effectively centralizing authority within the political system, the ethnic maps and formulas that the rulers of those parties hold should be even more crucial.

How have African Marxist-Leninist systems handled ethnicity? Before coming to grips with this question, we must recognize that Marxist-Leninist regimes in Africa are self-proclaimed. Structurally speaking, these regimes do not have much in common beyond their self-labeling. A military junta declared Congo-Brazzaville to be a Marxist-Leninist state; there was no communist party. This was true in Benin, Somalia, Madagascar and Ethiopia. With the collapse of Portuguese colonialism after the fall of the old order in Portugal, parties came to power in Angola and Mozambique that had participated in armed struggle. But only in 1977 did the MPLA and FRELIMO declare themselves as Marxist-Leninist. The historical experiences of these different states were not especially similar: the evolutions of their parties were not the same; and in some places real parties have not yet been created, or barely have been designated, as in Ethiopia.

In a tentative fashion, and in a few pages, I would like to discuss ethnic formulas under the Marxist-Leninist states, bearing in mind the relative paucity of current empirical work on a number of these countries.[37]

Much has been written on Marxism and the nationalities question, on Leninist and Stalinist solutions to ethnic heterogeneity, on the nationality question within the USSR, and on Marxist-Leninist approaches to nationalism and ethnicity in the industrial and nonindustrial countries. Certainly, nationality policy has changed over time in the USSR and the treatment of nationalities has differed among the USSR, China and Vietnam. Thus, given the wide divergences in size, history, levels of mobilization, organizational structure, and degrees of ethnic polarization, we should expect that ethnic formulas developed by African regimes would differ. But with this said, it is hard to find much of a Marxist-Leninist nationality component in theory or in practice for these regimes. Frequently, Marxist categories have not been applied to discuss ethnicity. And, since there has been no operative Marxist-Leninist party in Benin, Ethiopia, Madagascar, Somalia, or Congo-Brazzaville during most of their Marxist-Leninist phases, it has not been possible to centralize authority and penetrate across and into ethnic communities via the party vehicle. Nor has there been a developed proletariat which might, in theory at least, have been available to confront ethnic categories with a class base for power.

Indeed, if Marxist-Leninist regimes in Africa had chosen to base their power on a working class, it might well have created severe communal tensions because members of the urban working class have tended to be disproportionately recruited from certain ethnic groups. City and industrial growth have both taken place in the home areas of particular communities, as was the case with the Lari, whose abode is around Brazzaville. Although migration to large African cities has tended to make urban populations heterogeneous, there has been a great deal of industrial work force recruitment through ethnic networks and thus not only working class membership but specific occupations frequently are dominated by one or two ethnic communities. Decalo has noted that "labor has usually judged all regimes in Cotonou [Benin] according to their ethnic base."[38] In Ethiopia,

Eritreans have formed a disproportionately large share of the urban working class, which would have created a problem for any Ethiopian regime determined to prosecute a war against Eritrean autonomy if that Ethiopian regime had been serious about having a working class base for a communist party regime.

The same argument can be made about the creation of a Marxist-Leninist party in at least some states. Where African parties have remained relatively open and mass in nature, they have tended to recruit from many ethnic communities, as in Tanzania. These mass parties may have strong ethnic components within them, but the old CCP in Ghana, TANU in Tanzania, and even KANU in Kenya could not be said to have been dominated by one ethnic community. Insofar as Kikuyu influence was so strong during the Kenyatta regime, and even after Kenyatta's death, that influence has been exerted through the civil service and through commercial and entrepreneurial elites, with a growing influence in the armed forces. Where institutions have been so riddled with ethnic and factional strife, as they have been in Congo-Brazzaville, Benin and Madagascar, the creation of a narrow, "vanguard" party is going to lead to that party being seen as a vehicle for ethnic supremacy. The perception is likely to accord with reality. As Decalo notes for Congo-Brazzaville, ethnic animosities emerge in ideological guise and (for Benin) ethnic opposition is defined in terms of ideological deviations. Ideological dialectics become jargon to ward off personalistic, ethnic, and factional challenges.[39]

The military coups that brought regimes to power which declared for Marxism-Leninism in Benin and Congo-Brazzaville had personal, corporate and ethnic causes. There may well have been an ideological component to political events in Congo-Brazzaville, which is one of the most urbanized, industrialized and best-educated African countries. Ideological motivations seem to have been absent in Benin. But even for Congo-Brazzaville, as one politician in exile has said, "_Who_ the leader of the Congolese Revolution is matters as much as what his specific policies are. Socialism under a petit-nordist is simply not socialism to us."[40] Neither in Congo-Brazzaville nor in Benin were ethnic politics handled differently than they had been prior to the coming of Marxism-Leninism.[41]

If Benin and Congo-Brazzaville's ethnic politics were business mostly as usual, although garbed in ideological raiment, Somalia is perhaps not the best test for new ethnic formulations because it is one of the most linguistically, culturally and ethnically homogeneous countries in Africa, although there are regional distinctions between north and south, and clans and subclans have political importance. It was one year after the Somali coup which brought Siyaad Barre to power that he declared Somalia to be a socialist state and another six years later that a socialist party was created.[42] Barre has stressed that scientific socialism is consistent with Somali traditional values and with Islam as a state religion. He has eschewed the idea of class struggle in Somalia. Laitin argues that few analyses of economic problems there are undertaken from a socialist perspective.[43] More telling, for our purposes, is the regime's emphases on Somali language and on Islam, which appear contradictory, at least in the first instance, to the Marxist ideal of universalism and in the second to the Marxist ideal of materialism.[44] Moreover, the regime has emphasized a pan-Somali gathering of Somali-speaking peoples, and it has proved impossible to reconcile those two socialist and Marxist-Leninist regimes, Ethiopia and Somalia, on the basis of their socialist internationalism, proletarian internationalism, or membership in the community of Marxist-Leninist party-states!

I have already referred to ethnic formulas in Ethiopia in the discussion of military regimes. Because Ethiopia shares features of what have been understood in the West as multinational empires, it is perhaps the most interesting case of all because the regime must find a formula to reconcile diverse nationalities and ethnic communities with its concerns for revolutionary restructuring of Ethiopian society.[45] Indeed, it might find support for some of those goals from important Eritrean, and other, elements, were it not for communal strife. So far, the Ethiopian military -- and it has been a military, not a working class or communist party, which has defined ethnic formulas -- is still perceived by many within and without Ethiopia as a Marxist minority with a Shoan center of gravity, ruling by terror.[46] Nonetheless, while it has been prosecuting wars to re-establish central authority, albeit that this authority is seen as the traditional authority of a Shoan Amharic state, the regime's formulas are faithful to Marxist-Leninist theory in that religious and nationality discriminations are said to stem from capitalism, feudalism and imperialism.[47] Thus the argument that ethnic allegiances arise from lack of class consciousness, and the view that the regime will take into account the needs of nationalities as equal participants in economic and social processes and will recognize cultural diversity, are consistent with Marxist-Leninist nationalities policy. Since Marxist-Leninist nationalities policy has stressed in practice the maintenance of the centralized state, whatever its multiethnic composition, it is policy consistent with the unity promulgated by the old imperial regime.

Angola and Mozambique have regimes which came to power out of anticolonial struggle. In both countries armies fought, and parties must be understood in terms of civil-military relations of guerrilla fighting and, in the case of Angola, in terms of severe party and factional struggles within the anti-Portuguese forces. Angolan factions from the inception of the Angolan insurgency against the Portuguese, were, in the words of John Marcum, "led by competing elites with dissimilar social backgrounds. They reflected vertical and horizontal cleavages inherent in Angolan nationalism, that is, differences grounded in ethnic genesis, and cultural, class and racial stratification."[48] Marcum notes that while Angolan movements sought to transcend their origins, managed to attract some representation from the other ethnic-linguistic communities, and presented themselves as multiethnic, they were each seen as being exclusively ethnic and in fact received their support from primary ethnic segments.[49] Moreover, most of the Movemento Popular de Libertação de Angola (MPLA) was Catholic while substantial Protestant elements existed in the União das Populaçoẽs de Angola (UPA) (which later merged with other groups to form the Frente Nacional de Libertação de Angola, or FNLA).[50] More important has been the well-known association of MPLA leadership with mestiços and assimilados. Some of these people spoke no African language, knowing only Portuguese. UNITA was to stress its uni-racialism while the MPLA was to stress its multiracialism.

As civil war unfolded in Angola, the liberation movements were based in their ethnic areas. Although the MPLA won out, with Cuban support, in the military struggle, UNITA has been able to maintain itself as a fighting force -- with critical South African support -- in areas where it has ethnic support among the Ovimbundu, Ovambo, Chokwe and Lwena.

The late President Neto said: "Our fundamental task is that of transforming the tribalized man, full of racial and class complexes into a truly free man."[51] However, resentment within the MPLA, as well as without, exists against the mestico community. It is not clear to what extent the MPLA has been transformed from a movement into a vanguard party of militants. It does appear that internal dissidence has been so strong that at least one coup has been attempted from within the MPLA, and that this coup, put down with Cuban assistance, was motivated by an antimestiço animus.[52]

The MPLA's capacity to integrate Angola is conditioned by its capacity to cope with racially based dissension within its own ranks.[53] So far, the MPLA has not been able to transcend its own ethnic divisions and establish ethnic formulas which would persuade the Ovimbundu and the others that they will be accommodated in socialist Angola. Since the MPLA's own ethnic base is a minority, and its leadership is a minority within that minority, the prospects for its victory and for peace through force of arms are not bright.

It is Mozambique, of all the African countries that have declared themselves to be Marxist-Leninist, which has had the best-organized political party. FRELIMO was formed in 1962 by the merging of three exile organizations. Each of these organizations was, in the words of Alpers, "perhaps as much an ethnic as a nationalist union."[54] But unlike Angola's, the nationalist movement was not split into distinctly ethnic-based groups until after 1962. Two exiled leaders, Mondlane and Dos Santos, did not associate themselves with ethnic constituencies. Inside the country there were splits within FRELIMO over ideological issues and over what were designated as regionalism and tribalism.[55]

But by the early 1970s, FRELIMO had become more consensual and, through internal conflicts, more disciplined. How should we account for the development of FRELIMO as a dominant and relatively centralized party as compared to the factional struggles and continued ethnic splits elsewhere and especially in Angola? Was the structure of ethnic cleavage different in Mozambique? There were important ethnic communities, especially the Makonde, who had seen themselves in distinctive terms at various times. And there have been mesticos and assimilados as well as people who are not of African descent at all who have played leadership roles in FRELIMO and in the government. Yet FRELIMO has not been riven by racial issues since independence, as has been the MPLA.

It can be argued that leadership has been critical. It can also be argued that Tanzania played an important role in giving aid and comfort to certain factions and suppressing others. And it can be argued that through protracted struggle in Mozambique a party emerged which was able to raise class consciousness and to discipline itself in the process by linking with the masses.[56] But why struggles led to one outcome in Mozambique and to another in Angola is not so clear. Whether FRELIMO has had success in rural transformation has been much debated. Some observers are arguing that in Mozambique a new African society is being formed on the basis of the accumulated experiences FRELIMO acquired in its administration of liberated zones during the anticolonial struggle. Revolutionary theory and pragmatic policy are said to be linked so as to enable a Marxist-Leninist party to be truly transforming. The formula for coping with communal cleavage would be described as building class consciousness. The only viable and true formula, it has been argued, involves the surmounting of the false consciousness of ethnicity so that people see that the real relationships which define their existence, their cooperation and conflict, are those of class.

Parties, then, including Marxist-Leninist parties, also show a wide range of variation in the ways they try to deal with communal cleavages. By default perhaps, in most African societies, mass parties with the least tension between their self-definition and the structure of cleavage emerge as ruling institutions. Where parties are narrowly based in religious, ethnic, or ideological terms, the conclusion is different. However, even so-called Marxist-Leninist parties in Africa have in practice shown flexibility in creating formulas for dealing with ethnicity and national integration although they, too, have frequently become captured by particular ethnic communities.

Conclusion

Ethnicity and class must be understood situationally, that is, in specific social, political, economic and spatial contexts. Political and economic groupings which center on occupational and income differences may prove transient. They may get subordinated to identities of region and nationality. But communal identities may change and/or become politically subordinate to income and class differentiations. Or, class and ethnicity may fuse in particular situations. Communal identities may be eroded but they also may be displaced temporarily. They may also be hardened, or created anew. The formulas that elites put forward are important elements in these processes.

Where patronage and recruitment issues are seen as crucial, as they are in much of Africa, it is no surprise that communal orientations are at the forefront of politics. Under these circumstances, machine-type and patron-client relationships often govern. Patronage and recruitment issues get infused with communalism. Sometimes the political institutions of machine and personalistic relationships may cope with communalism; sometimes they get overwhelmed or exacerbate the tensions. This applies to militaries and civil services also.

All studies of politics in cities or rural areas must come to grips with the relationship between national and local politics. Local differentiations are played out in the context of the centralization of national resources. More generally speaking, political behaviors are affected by authoritative institutions through which community influence is channeled. That is, outputs of government affect patterns of competition and participation. Government's performance is an independent factor affecting support for or disaffection from the regime. Thus, in the future, as in the past, government's own policies and performance, which are at the same time a response to pressures from below and an expression of elite preferences, will themselves be consequential for the development of politics in class or communal directions.

The African revolutionary thinker who perhaps best understood the independent role of the state, who avoided broad formulas for conceptualizing class and ethnic relations, and who insisted on the defining importance of specific situations and contexts, is Amilcar Cabral.[57] While clearly operating with many Marxist categories, Cabral avoided calling himself a Marxist and resisted designating his party in Guinea-Bissau, the PAIGC, as Marxist-Leninist. He tried to give specific content to categories such as tribe, ethnic group, worker and peasant, and he was wary of general formulas.

For Cabral, revolutionary transformation of the wielders of state power comes about through class struggle. The level of productive forces is for him the essential determining element in the content and form of the class struggle. He also argued that colonialism violently usurped the freedom of development of productive forces in Africa, and that this phenomenon was the principal characteristic of imperialist domination. Thus for Cabral, cultural resistance and Africanization are critical undertakings practically and conceptually.[58] At the same time, he distinguished between colonialism -- direct foreign control expressed through state domination -- and neo-colonialism, where native elements have political power and develop either state capitalism or capitalism, and he gave the state a central place in his analysis. Cabral discussed ethnicity more than have Debray or Guevara or Mao, and it was partly his sensitivity to this question that made him dwell on the specifics of Guinea, directly colonized by a relatively backward European country and itself extremely underdeveloped.

Cabral described the proletariat in Guinea as embryonic and noted the creation by the colonial regime of a workers' aristocracy. The physical force of most importance in the liberation struggle was the peasantry. But for now neither of these two classes could distinguish true national independence from fictitious political independence.[59] Cabral wrote that we speak of peasants, but the term "peasants" is vague. The peasant who fought in Algeria or China is not our peasant.[60] In Guinea, unlike Angola, there were no plantations and agricultural companies, and no expropriation of land, which remained basically the cooperative property of the village. Cabral argued that exploitation of the peasantry was indirect, not direct. One could not mobilize the rural countryside with the slogan of "the land belongs to the people," nor simply on the basis of the struggle against colonialism. Mobilization had to take place by organizing around detailed and concrete demands.

One of the most interesting aspects of Cabral's writings is his treatment of class and ethnicity as interlocked phenomena:

> As we conceive it, the tribe exists and it does not exist. When the Portuguese came to our country the tribal economic system was already disintegrating. Portuguese colonialism contributed further to that disintegration, although they needed to maintain some part of the superstructure. As far as we were concerned it was not so much the economic base

> that led us to respect the tribal structure as a mobilizing element in our struggle, but its cultural aspects, the language, the songs, the dances, etc.[61]

Cabral argued that one could build on ethnic feelings to mobilize people. At the same time there were contradictions between tribes, albeit of secondary importance, that had to be taken into account. He disagreed with those African nationalists who maintained that there were no tribes, as well as with those who said that such distinctions were fundamental. In Guinea, indigenous ruling classes had ethnic expressions. Thus "semi-feudal" elements were described in tribal terms by Cabral, but he also looked at internal differentiation within ethnic groups -- for example, the relationship between the people and their chiefs.[62] He saw class contradictions as more basic than ethnic differences. And he treated tribalism, and even the existence of tribes, as something which was manifested generally only as a function of opportunistic attitudes on the part of detribalized individuals or groups.[63]

Cabral discussed "peoples" in terms of their contribution to agricultural production: "peoples" varied in many respects, including material possessions, but overwhelmingly shared the fact that most of them were farmers.[64] Their relationship to farming transcended ethnic and cultural diversities, since agricultural activities and politico-social situations were common. Thus the cultural and material realms were affected by common modes of production.

Cabral located his class analysis in a concern for what he called the "colonial situation" -- whether direct or indirect -- and argued that there were no revolutionary groups to start with in Guinea. He believed that there was, as yet, no fundamental conflict between town and countryside because of the extremely close connections that were still maintained through networks of relatives, migration back and forth, and the fact that some individuals held both rural and urban occupations. The petty bourgeoisie was not a property class in Africa, but a collection of people who served the state and benefited from controlling it. Cabral argued that occupation and class were not synonymous, although there was a tenuous link between certain types of work and class consciousness.[65]

Thus Cabral, more than others, has rooted his discussion of class and ethnicity in the specifics of time and place. He is sensitive to the historical evolution of African elites and to the critical role of the state apparatus in forming class and ethnic consciousness.

The elites who have come to power in Africa, whether through parties or armies, are usually not established elites. That is, they are not of high status by birth. Also, they are politically insecure. Because they do not have developed economic bases, their political positions rest on control of the state bureaucracies or party institutions they inhabit. Control of private property has not brought them to their political and institutional positions. The economic stake they create for themselves is insecure; it cannot be guaranteed if they lose political power or access to power. Some have understood the adoption of new political and ideological designations as means to provide elites with internal bonds and external boundaries.[66] However, ideology has not provided a firm cement in

Africa, and this is going to be true for Marxism-Leninism as it has been for varieties of African socialisms.

Moreover, the invocation of a particular ideology which both says that ethnic ties are manifestations of a lack of class consciousness and tries to invoke that consciousness cannot by itself deal with communal cleavages. And most central elites, including those in self-designated Marxist-Leninist states, are too weak to impose policies which try to centralize authority over ethnic units or to ignore ethnicity in the allocation of resources. These same elites, usually with a military base, fear creating the very parties that might have grass-roots organizations and might be able to channel centrally determined policies downward and outward. Viable Marxist-Leninist parties might be created as institutional vehicles for trying to cope with communal diversity but we have not seen such parties in practice in Africa for any length of time; if at all. Surely the rhetoric and symbols of Marxism-Leninism, whether wielded cynically as in Benin or Congo-Brazzaville, or apparently with commitment as in Angola, or with idiosyncratic interpretations as in Somalia, cannot restructure communal cleavages or reformulate issues posed in ethnic terms for large numbers of people.

Footnotes

1. Audry C. Smock, *Ibo-Politics: The Role of Ethnic Unions in Eastern Nigeria* (Cambridge, Mass.: Harvard University Press, 1971), p. 5.
2. Crawford Young reviews this literature in his well-documented and important study, *The Politics of Cultural Pluralism* (Madison: University of Wisconsin Press, 1976).
3. Cynthia H. Enloe, *Ethnic Soldiers: State Security in Divided Societies* (Athens: University of Georgia Press, 1980), p. 187.
4. Leonard Binder, "National Integration and Political Development," *American Political Science Review*, 58 (September 1964): 622-633.
5. Jean Gallais, "Signification de groupe Ethiopie et Mali," in *L'Homme* (Paris), 2 (May-August 1962): 106-129, as found and cited in Aristide Zolberg, "Patterns of National Integration," *Journal of Modern African Studies* (1967): 452.
6. Ronald Cohen illustrates this idea in his discussion, "Social Stratification in Bornu," in Arthur Tuden and Leonard Plotnicov, eds., *Social Stratification in Africa* (New York, 1970), especially pp. 253-263. "We must also view class differences as another status distinction in and of itself, since it is such a broadly inclusive feature and is, therefore, in a number of senses a multifaceted variable instead of a dichotomous category when it is actually utilized by people themselves to judge and generalize about someone else's or one's own high or low status in the society," pp. 255-256.

7. For a very insightful discussion of ethnicity see Nelson Kasfir, "Explaining Ethnic Political Participation," World Politics, 31 (April 1979): 365-388. Kasfir analyzes the meaning of "Nubian" in Uganda on pp. 378-385. Other discussions include Aidan Southall, "General Amin and the Coup: Great Man or Historical Inevitability?" Journal of Modern African Studies, 13 (March 1975): 85-105; and Dennis Pain, "The Nubians: Their Perceived Stratification System and Its Relation to the Asian Issue," in Michael Twaddle, ed., Expulsion of a Minority: Essay on Ugandan Asians (London: University of London, 1975).
8. An interesting recent case arose with the deportation of Alhaji Abdulraham Shugaba, who was the majority leader in Nigeria's Bornu State House of Assembly. Shugaba was born to a Nigerian mother in Maiduguri; his father was from Chad, where he was deported. The matter was tangled up with party politics and security questions, but a central issue -- who is Nigerian? -- was posed for people living in border states.
9. See, among others, Young, Politics of Cultural Pluralism, pp. 149-156; David Apter, The Political Kingdom in Uganda (Princeton: Princeton University Press, 1975); F.B. Wellbourn, Religion and Politics in Uganda, 1952-62 (Nairobi: East Africa Publishing House, 1965); Victor A. Olorunsola, ed., The Politics of Cultural Sub-Nationalism in Africa (Garden City: Doubleday Anchor Books, 1972); Nelson Kasfir, The Shrinking Political Arena: Participation and Ethnicity in African Politics with a Case Study of Uganda (Berkeley and Los Angeles: University of California Press, 1976); and Donald Rothchild and Michael Rogin, "Uganda," in Gwendolen Carter, ed., National Unity and Regionalism in Eight African States (Ithaca: Cornell University Press, 1966), pp. 337-445.
10. See Carl G. Rosberg, Jr. and John Nottingham, The Myth of 'Mau Mau': Nationalism in Kenya (New York: Praeger, 1966).
11. See Henry Bienen, Kenya: The Politics of Participation and Control (Princeton: Princeton University Press, 1974), and Cherry Gertzel, The Politics of Independent Kenya (Nairobi: East Africa Publishing House, 1970).
12. See Henry Bienen, Armies and Parties in Africa (New York: Africana Publishing Co., 1978).
13. See Robin Luckham, The Nigerian Military (Cambridge: Cambridge University Press, 1971).
14. The Provincial Military Administrative Council (PMAC) did announce its recognition of the right of any nationality in Ethiopia to self-determination while it was declaring the unity of Ethiopian nationalities based on common struggle against feudalism, bureaucratic capitalism, imperialism, and reactionary forces. The Dergue had in mind for Eritrea internal autonomy but not as a special region. The formula was similar to that articulated by the revolutionary Bolshevik government after 1917 in the USSR. See the Ethiopian Herald, April 21, 1976, as cited in Marina and David Ottaway, Ethiopia: Empire in Revolution (New York: Africana Publishing Co., 1978), especially pp. 151-162.

15. *Ibid*., p. 158.
16. David Laitin and Drew Harker, "Military Rule and National Secession: Nigeria and Ethiopia," draft paper, May 1979.
17. Ottaway and Ottaway, *Ethiopia*, p. 131.
18. Laitin and Harker, "Military Rule and National Secession."
19. *Ibid*., p. 50.
20. See, among others, Arnold J. Heidenheimer, ed., *Political Corruption: Readings in Comparative Analysis* (New York: Holt, Rinehart & Winston, 1970); Bert F. Hoselitz, "Levels of Economic Performance and Bureaucratic Structures," in Joseph LaPalombara, ed.,, *Bureaucracy and Political Development* (Princeton: Princeton University Press, 1963), pp. 168-198; and Fred Riggs, *Administration in Developing Countries* (Boston: Houghton Mifflin, 1964). The Nigerian novelist Chinua Achebe has dealt with these themes in *Things Fall Apart* (London: Heineman, 1958; repr. 1964).
21. David J. Gould, *Bureaucratic Corruption and Underdevelopent in the Third World: The Case of Zaire* (New York: Pergamon Press, 1980).
22. Robert Price, *Society and Bureaucracy in Contemporary Ghana* (Berkeley and Los Angeles: University of California Press, 1975), p. 29.
23. James Mittelman, *Ideology and Politics in Uganda: From Obote to Amin* (Ithaca: Cornell University Press, 1975), pp. 125-126.
24. Just what getting one quarter in two-thirds of the states meant was to be argued and decided in the electoral commission and special tribunal, and the supreme court after the election in 1979.
25. The military government amended the constitution after the election to allow for a run-off to be carried out, if necessary, through a second and, if necessary, a third direct election.
26. The Nigerian presidential election results of 1979 have been reproduced in *West Africa*, August 27, 1979, p. 1573.
27. Henry Bienen, "Political Parties and Political Machines in Africa," in Michael Lofchie, ed., *The State of the Nation* (Berkeley and Los Angeles: University of California Press, 1971), pp. 195-216; Aristide Zolberg, *Creating Political Order: The Party-States of West Africa* (Chicago: Rand McNally, 1966).
28. Zolberg, *Creating Political Order*, pp. 160-161.
29. Lucy Behrman, "Muslim Politics and Development in Senegal," *Journal of Modern African Studies*, 15 (1977): 271.
30. See Issa G. Shivji, *Class Struggles in Tanzania* (London: Heinemann Educational Books, 1976); John Saul, "The State in Post-Colonial Societies -- Tanzania," in Ralph Miliband and John Saville, eds., *The Socialist Register 1974* (London: The Merlin Press, 1974), pp. 349-372.
31. Bismark U. Mwansasu, "The Changing Role of TANU," in Bismark U. Mwansasu and Cranford Pratt, eds., *Towards Socialism in Tanzania* (Toronto: University of Toronto Press, 1979), p. 172.
32. See Julius Nyerere, "Importance of a National Ethic," in Julius Nyerere, *Freedom and Unity* (Dar es Salaam: Oxford University Press, 1966), pp. 174-175 and other works in this collection, and in his *Freedom and Socialism* (Dar es Salaam: Oxford University Press, 1968).

33. Lionel Cliffe, "The Campaigns," in Lionel Cliffe, ed., One Party Democracy: The 1965 Tanzania General Election (Nairobi: East Africa Publishing House, 1967), p. 240.
34. A person might be a local or have local roots but not be from a dominant ethnic group in the area. See Cliffe, "The Campaigns," pp. 308-320.
35. See Crawford Young, Politics in the Congo (Princeton: Princeton University Press, 1967), and Young, Politics of Cultural Pluralism, pp. 163-215.
36. Mwansasu, "Changing Role of TANU," pp. 177-178, citing Julius Nyerere, Majadliano ya Mkutano Mkuu, Taarifa Rasmi, June 1969, pp. 708-709; and Julius Nyerere, "Maana TANU Kushika Hatamu" ("The Meaning of the Supremacy of TANU"), Uhuru, July 7, 1975.
37 I have benefited from reading work in progress by Crawford Young on Afro-Marxist regimes.
38. Samuel Decalo, "Ideological Rhetoric and Scientific Socialism in Benin and Congo/Brazzaville," in Carl G. Rosberg and Thomas M. Callaghy, eds., Socialism in Sub-Saharan Africa (Berkeley: Institute of International Studies, 1979), p. 243.
39. Ibid., pp. 249 and 246. Of course, in the USSR and in China, ideology has been used as a vehicle for expressing and warding off factional challenges. This is not a phenomenon peculiar to African Marxist-Leninist regimes.
40. As quoted in ibid., p. 251.
41. Historically, Madagascar was a society ethnically polarized between highlands Merina and various lowlands groups. It has caste-like features, religious diversity, and a language given by the Merina cultural core. But I cannot find a secondary literature on the development of ethnic patterns after Ratsiraka headed the military regime and the Revolutionary Socialist Malagasy Charter was proclaimed in 1975.
42. I rely heavily on the work of David Laitin on Somalia. See Laitin and Harker, "Military Rule," and Laitin's "Somalia's Military Government and Scientific Socialism," in Rosberg and Callaghy, Socialism in Sub-Saharan Africa, pp. 175-206.
43. Ibid., p. 197.
44. Ibid., p. 199. Laitin himself does not argue this point but rather sees the use of Somali as an official language of the state as enabling and even inducing people to debate what socialism means. This may be true, and it also may be true that Siyaad has fought reactionary Islamic leaders, and, like many others, has argued that Islam and socialism are compatible. But this said, the emphasis on language and religion cannot be an ethnic formula derived from Marxism-Leninism.
45. For a discussion of the ethnic composition of Ethiopia, see Donald Levine, Greater Ethiopia: The Evolution of a Multiethnic Society (Chicago: University of Chicago Press, 1974).
46. P.T.W. Baxter, "Ethiopia's Unacknowledged Problem: The Oromo," African Affairs, 77 (July 1979): 295.
47. John Harbeson, "Socialist Politics in Revolutionary Ethiopia," in Rosberg and Callaghy, Socialism in Sub-Saharan Africa, p. 365.

48. John A. Marcum, _The Angolan Revolution: Exile Politics and Guerrilla Warfare 1962-1976_ (Cambridge, Mass.: MIT Press, 1979), p. 46.
49. _Ibid_., p. 48. For Holden Roberto's FNLA, that support was Bakongo; for what came to be Jonas Savimbi's UNITA it was Ovimbundu; and for MPLA it was Mbundu.
50. There was, however, significant Methodist presence in the MPLA leadership. _Ibid_.
51. From an MPLA pamphlet of 1974 or 1975, quoted in Kevin Brown, "Angolan Socialism," in Rosberg and Callaghy, _Socialism in Sub-Saharan Africa_, p. 289.
52. Brown, "Angolan Socialism," pp. 314-315.
53. _Ibid_.
54. Edward A. Alpers, "The Struggle for Socialism in Mozambique, 1960-1972," in Rosberg and Callaghy, _Socialism in Sub-Saharan Africa_, p. 267.
55. See _Mozambique Revolution_ (official organ of FRELIMO) 28 (May 1967) and 36 (October-December 1968), as cited by Alpers, "Struggle for Socialism," p. 408, fn. 31.
56. Alpers, "Struggle for Socialism," argues this, and so does John Saul, "FRELIMO and the Mozambique Revolution," in Giovanni Arrighi and John Saul, eds., _Essays on the Political Economy of Africa_ (New York: Monthly Review Press, 1973), pp. 378-404. The latter, writing before FRELIMO's victory and not _ex post facto_, is more guarded.
57. I am borrowing below from my "State and Revolution: The Works of Amilcar Cabral," _The Journal of Modern African Studies_, 15 (December, 1977): 555-568.
58. Brown, "Angolan Socialism," p. 302.
59. Amilcar Cabral, _Revolution in Guinea: Selected Texts_, trans. by Richard Handyside (New York: Monthly Review Press, 1969), p. 105.
60. _Ibid_., p. 158.
61. _Ibid_., p. 160.
62. _Ibid_., p. 64.
63. _Ibid_., p. 104. All this is from Bienen, "State and Revolution," pp. 560-561.
64. See Amilcar Cabral, "On the Contribution of the 'Peoples' of Guine to Agricultural Production in Guine," in Ronald H. Chilcote, ed., _Emerging Nationalism in Portuguese Africa: Documents_ (Stanford: Hoover Institution Press, 1972), pp. 352-355.
65. _Ibid_., p. 562.
66. Kenneth Jowitt, "Scientific Socialist Regimes in Africa: Political Differentiation, Avoidance and Unawareness," in Rosberg and Callaghy, _Socialism in Sub-Saharan Africa_, p. 141.

7
The Manipulation of Ethnicity: South Africa in Comparative Perspective

Heribert Adam

Two contradictory perspectives and conclusions prevail in the vast social science literature on South Africa. (1) History is seen as the unfolding of predetermined forces. For Marxists, irreconcilable antagonisms inevitably clash in a globally spreading class war. Its outcome is certain. For orthodox black and Afrikaner nationalists alike, a race war is in the offing. Preparation and determination will decide the winner. In this Social Darwinist version of eternal fitness competition, efforts to avoid an escalating conflict are doomed to fail, at best delaying, at worst obfuscating the inevitable struggle. "Suffer the Future"[1] is the almost unanimous resignative warning for South Africa from liberal academics while the majority of African nationalists agree that the current white system can deliver only one thing: the timetable for conflict. Perhaps there is one irrefutable truth in such prophecies of doom: the insight that there no longer can be any winners but only losers in a South African escalation is hardly sufficient to stop at least some antagonists from trying.

(2) On the other side, guarded optimism prevails that views history as basically open-ended. A dialectic of action-response-counteraction is stressed, because there are always unintended and unanticipated effects of conscious intervention. The structural forces in motion are seen as susceptible to conscious redirection albeit with unforeseen consequences.

However, this policy-oriented social engineering frequently overestimates the likelihood of its success, an attitude found among constitutional experts. Constitution-making easily becomes a self-deceptive exercise when, in good faith, blueprints for the future -- be they partition, confederalism or consociationalism -- are offered as magic formula, if only the combatants would convert to their propagated merits. It is precisely the neglected obstacles to the well-intentioned grand designs that deserve the most but get the least attention. The de facto collapse of almost all highly praised Westminster constitutions after a few years of independence in postcolonial Africa counsels caution. Constitutions reflect rather than alter power relationships. The institutionalized rules of the political game are observed, because all competing groups consider compliance to be in their interest. However, the most well-designed constitutions are ignored or set aside where they do not tally with a political reality, where they are imposed against vital interest groups, or where constitutions are transplanted from one political culture without much regard for the historical uniqueness of a different society.

This analysis will try to avoid the naiveté of social engineering, and the equally dangerous paralysis of mere onlookers to historical causes.[2] It will attempt an interdisciplinary, comparative, and policy-oriented critical analysis of current strategies for conflict solution in South Africa.

Comparing South Africa

In order to better understand the nature of the South African state as well as the chances and direction of its possible transformation, in comparative terms, a rough categorization of the more than 150 contemporary sovereign states in the world may be helpful. While most of the literature distinguishes between military and civilian governments, democracies and dictatorships, one-party and multiparty states, four types of current regimes could be singled out more meaningfully in terms of how they maintain power: (1) traditional autocracies; (2) totalitarian/revolutionary regimes; (3) authoritarian/corporate states; and (4) democratic welfare states.

(1) In *traditional autocracies* a single person or family holds absolute power, with no provisions for accountability or legal limitations. The ruler amasses personal fortunes in the midst of extremes of poverty and wealth (Mobutu, Bokassa). A clique of sycophant associates, based on kinship ties or personal loyalties, participates in expropriating general revenue. In exchange for arbitrary administrative favors, foreign companies add to the general corruption and graft of an official "kleptocracy."[3] Social scientists have used the Weberian concept of *patrimonialism* to describe many of the modern African autocracies.[4] In a patrimonial regime, the power of the ruler depends on his ability to secure the cooperation of the support staff in his administration. The ruler extends patronage and the rewards of office in exchange for loyalty and tribute. Since the state coffers constitute the only major source of wealth, the struggle for access among competing cliques is fierce. Attempted coups d'état occur frequently. An adroit ruler will play factions off against each other and politics takes the form of elite rivalry. More inept despots revert to tyranny and establish an arbitrary government, based on fear (Idi Amin), including genocide of suspected opponents, usually people of different ethnic origin. Gross violation of civil rights characterizes all autocratic regimes, since the concept of legal protection exists in theory only and dissent is considered treason. Crucial for the stability of the regime are the attitudes of strategic officer cliques and whether the ruler succeeds in using the "military against itself" (Sadat).[5] Most autocrats also possess private armies who have pledged their allegiance to the ruler personally rather than to the state. When the autocrat loses office or dies, many institutions are therefore in danger of disintegrating since their rationale is tied to the deposed patriarch. Classical examples of autocracy outside Africa are the royal household of Saudi Arabia, the Samoza family dictatorship in Nicaragua, or "Papa Doc"

Duvalier's tyranny in Haiti. Needless to say, South Africa does not belong in this category nor in the second type in which she is frequently lumped under the label fascism.

(2) Totalitarian/revolutionary regimes are characterized by mobilization of the masses in the name of an ideology. In traditional autocracies the rulers hardly take account of politically uninvolved masses as long as they can exact taxes and services from them. Unlike the static autocracies, the revolutionary regimes politicize the people for a transformation or restoration of traditional institutions. Such regimes emerge at a higher stage of economic development and after a severe crisis of the previous rule. In terms of ideological orientation three types of totalitarian mobilization can be distinguished: (1) fascism (Nazi Germany); (2) Stalinism and its offshoots (North Korea of Kim Il Sung); and (3) theocracies or messianic states (Iran of Khomeini, Libya of Qaddafi).

All three types use indiscriminate, arbitrary terror against ideologically defined outgroups and perceived foreign agents in and outside the polity. This scapegoating serves the function of uniting a heterogeneous populist movement and exhorting it to make sacrifices for economic and military goals. A regimented populace provides an audience for constant purges in the leadership, which unmasks subversives and demands vigilance. The infallible supreme leader tolerates a personality cult or poses as indispensable guardian and interpreter of the sacred doctrine. His power is based on a symbiotic relationship of army and party. Both cooperate in parallel political structures in controlling the state with either one the senior protector of revolutionary heritage. Where socialist-revolutionary regimes have been established in Africa (Ethiopia, Angola, Mozambique, Congo-Brazzaville, Guinea), their totalitarian tendencies were greatly ameliorated by lack of ideological cohesion and the immediate problems of underdevelement. The leadership in totalitarian/revolutionary regimes is less prone to ostentatious consumption or the personal corruption of autocrats. Revolutionary leaders also rely more on collegial executives ("democratic centralism") although there are no general institutional controls of the accession to and the succession of the supreme posts. The leadership operates in a strictly hierarchical system, camouflaged with spurious elections or fabricated mass support. No public dissension is tolerated and political life is reduced to acclamation. The ruling elite faces the problem of how to keep the ideological zeal alive and prevent spiritual withdrawal and ideological apathy.

The Afrikaner ideology of apartheid never aimed at mobilizing the majority black population and, on the contrary, arranged for its depolitization through Bantu education. Unlike other ideologies, apartheid as an ideology never did achieve a hegemony because of the constant challenge in the country by English liberalism, embraced by many blacks and business alike. Afrikaner "civil religion"[6] gave racial segregation a strong ideological underpinning, but was primarily directed at remobilizing a defeated Afrikaner nationalism. It did not contain a convincing eschatology but merely a rationale for the exclusive use of the state to further sectional interests. Apartheid as a blueprint of expedience, in contrast to the sacred doctrine of revolutionaries, has now been largely discarded by the Afrikaner intelligentsia itself. Ideological confusion rather than religious intransigence reigns. Unlike Northern Ireland, the South African conflict is not perceived in absolutes of good and evil, but in terms of more or less privilege and security, to be open to compromise and bargaining, if need be. The pragmatic commitment to power maintenance by a threatened ethnic oligarchy[7] puts South Africa in the category of authoritarian states.

(3) The *authoritarian/corporate state* displays a limited pluralism, legitimized by the state. It lacks a clear ideology and justifies exclusion of the many from political participation on traditional or technocratic grounds. "At the same time, the authoritarian regime is not one of arbitrary power...neither tyranny nor despotism, and [it] does not grant absolute power."[8] Corporatist authoritarianism became a standard concept in the analysis of Iberian and Latin American societies which did not fit the fascist one-party model.[9] Corporatism corresponds to delayed capitalist development and dependency. In combination with international capital, paternalistic elites use the state machinery to modernize. State agencies and private interests interpenetrate. Social security systems are established for the first time to forestall a growing class consciousness. The Latin American struggle for independence, for instance, eventually gave rise to populist mass movements in nonindustrialized urban centers -- corporatism amounts to their forced demobilization. In the interest of middle classes, the military directed political participation into licensed channels which were not subversive for the system. Brazil, Argentina and Peru are the classical examples where lower-class protest is neutralized by a mixture of right-wing repression and left-wing reform. In the tightly linked coalition of military technocratic and bureaucratic elites, the actual head of the government is chairman among equals. Officers and civilian technocrats and managers have become interchangeable. The "role expansion" of the military as economic modernizers makes the army dependent on recruitment from outside its ranks and prevents the military from ruling alone. Unlike the military in autocracies, the army in corporate states is not divorced from the community and represents a highly professional section of the executive-administrative center. In Africa, the Nigerian army resembles most closely the Latin American examples with the additional task of preserving the unity of the country. In South Africa, too, the military has taken over the role of political warfare without having a formal institutionalized role in the civilian political center. It has not initiated society-wide reforms, except for increasingly integrating its own ranks so that blacks now amount to 20 percent. However, political policing by the military can hardly guarantee industrial peace, so crucial in a modern economy where voluntary identification with the system of production is the basis for its growth.

(4) *Democratic-welfare states* in advanced industrial societies have so far succeeded better in securing mass loyalty and legitimacy. Minimal material security, universal citizenship and equal access to political participation guarantee, at least in theory, a model for the happiness of the greatest numbers. In reality, to be sure, democratic grassroots participation has given way to corporate and bureaucratic power in a spreading executive state. While the classical liberal model of constitutionalism undoubtedly needs revision in light of the various new problems, particularly in divided societies

such as South Africa, important elements remain valid and exert great appeal.

The Case for South Africa's Uniqueness

South Africa can be viewed as an African country only in a geographic and demographic sense. In most other respects the republic is so different from any other African society that it must be conceptualized as a qualitatively different case. This distinction rests above all on a different historical development.[10]

Unlike the rest of Africa, whose weak native middle classes have failed to carry out an autonomous capitalist development, South Africa has long experienced strong indigenous economic growth. For a variety of historical and geographical reasons the country underwent an early capital accumulation. In South Africa, unlike other colonial societies, national interests ensured the reinvestment of accumulated capital in the local economy instead of its being drained to the metropole. Because of their permanent rather than temporary ties with the colonized land, the independent settlers, strengthened by a constant stream of skilled European immigrants, developed an elaborate infrastructure for South Africa's expansive extractive industry and commercial farming. After the forced isolation during World War II, this was supplemented by a rapidly growing manufacturing sector, aided by protectionist policies and an abundant labor supply.

Moreover, this development to an industrial empire took place in conjunction with an early anticolonial movement, that is, the bitter struggle between English capital and the indigenous Afrikaner farming group over state control. All these factors combined to create socioeconomic conditions different from the traditional colonization by a merchant class acting under the umbrella and in the interest of metropolitan capital. South Africa's leading class now compromises indigenous entrepreneurs in their own right. It may be called "an outpost of monopoly capitalism" in the global economic system, as Marxist analysts suggest,[11] but it certainly constitutes a rather self-sufficient and increasingly invulnerable bridgehead.[12] As the stronger, not the weaker partner in terms of subregional control, South African state interests can dictate the conditions for the operation of foreign capital. South Africa's intransigence on Namibia in the face of outside pressure further underscores this view.

In this respect, South Africa's position differs fundamentally from Israel's, with which it otherwise can be compared in many ways. Serving as a strategic-political outpost of its U.S. sponsor, Israel is economically and politically dependent, while the Western influence on South Africa is increasingly limited. Both countries, however, have become, for different reasons, integral parts of the Western political system. Therefore, they can be expected to be defended at high costs if they seem about to fall under Eastern bloc control.

In addition, the continued economic dependency on colonial trade links and on a preindependence infrastructure severely limits the leverage of socialist governments in southern Africa. The entire subregion has developed historically as one system, feeding the center on the Rand with labor and providing outlets for exports, regardless of national boundaries. The cooperative economic policies of socialist Mozambique and Zimbabwe toward South Africa have demonstrated how much the outer periphery is locked into an interdependent economy, in which white South Africa holds most of the cards. It is on this basis that Pretoria can pursue an internal geopolitical fragmentation policy contrary to the rest of the continent. While in the rest of Africa the state elite struggles to maintain the postcolonial, fragile unity, official South African policy aims at dismantling the unitary state into confederations under white economic hegemony. It is probably safe to say that the majority of people in independent Africa would welcome greater self-determination at the periphery, including secession from existing states, while the black majority of South Africans clearly favors preserving the unitary state.

However, a just and genuine geopolitical partition of the country which would be negotiated and mutually acceptable is highly inconceivable in South Africa. The major reason lies in the country's economic development and interdependence, which would be thwarted by population resettlements and artificial racial boundaries. Where partition has been implemented (Palestine, Ireland, Cyprus, Pakistan), it has occurred in agricultural societies and has hardly proved a lasting solution to communal strife.[13] Despite the unequal distribution of wealth and uneven development internally, South Africa constitutes an economically integrated society. This has shaped some common outlook and common heritage for South Africa's population groups, regardless of their enforced place in the racial hierarchy. Partition is only conceivable in a stalemate, coinciding with frozen battle lines after both antagonists have become exhausted. There would be little left to divide between the survivors.

Government Strategies of Change and Persistence

Any analysis of so controversial an issue as current South African politics rests on implicit assumptions that color the conclusions. For a meaningful discussion it seems imperative to make these assumptions explicit. The evaluation of government strategies in this assessment is based on the following seven propositions:

(1) Meaningful change is likely to emerge in South Africa only with the collaboration of decisive sections of the white ruling group. In anticipation of imminent challenges, the apartheid state has strengthened itself economically and militarily, buttressed by control over strategic resources and aided by foreign allies, to such an extent that no revolutionary movement on its own is able to defeat this powerful state machinery, except in its own propaganda. Short of an unlikely regional war with massive superpower interference, South Africa's racial minority monopoly is certain to continue for the time being, its universal repugnance notwithstanding. An unjust regime is not necessarily an unstable one.

(2) To be sure, the costs of racial privilege maintenance are also increasing through heightened internal unrest, escalating guerrilla incursions and international ostracism. However, unlike the Zimbabwean situation, since blacks cannot force whites to negotiate

their capitulation, the conflict remains an uneasy stalemate. In this fragile situation, Pretoria now tries to cut its costs by policy adaptations to new exigencies. The skilled manpower shortage in a booming economy has led to abolitions of certain training and mobility restrictions for blacks. A growth-oriented policy demands modifications of traditional apartheid regulations. The Afrikaner ruling elite has offered a new alliance to English business to achieve an optimally mobilized war economy in their mutual interest. In this endeavor an emerging black petty bourgeoisie in the Bantustans and urban townships is actively courted to form a conservative bulwark against an increasingly politicized youth.

(3) It would seem false to dismiss these policy changes as merely cosmetic. To be sure, they do not tally with black expectations. But neither do they indicate inflexibility. They represent adaptations of a sophisticated control system that rearranged the switches and rewired the bureaucratic machinery to make it more efficient. If the system so far did not go further in reforming itself, it is because there was no pressing need, not because it was incapable of doing so.

(4) These reformist policies and the accompanying rhetoric strained the cohesion of the Afrikaner ruling group. Deep cleavages and heated controversies about strategies and tactics of privilege maintenance reveal conflicting interests and reflect different class positions of whites. English and Afrikaner business alike, particularly export-oriented big capital, show themselves much more willing to accede to black accommodation than do white unions and a vast civil service, dependent on state protection. Heated ideological debates about the essence of white survival mirror conflicting perceptions of identity in a group that is often falsely portrayed as a monolithic bloc. Conventional wisdom views groups under threat as closing ranks. In the South African case the opposite is happening: a pressured ruling group splits because crucial sections perceive their material and ideal interests quite differently under new circumstances. It is only at this junction of a serious split of the Afrikaner nationalist movement (beyond the breakaway of ultra-right fringe groups) that a meaningful reformist policy would have a chance. A disintegration of the Nationalist Party would open the possibility of new political alliances within the white camp and across the color line. It is this leverage for blacks which can be considered the most promising hope for change in the short run, despite the spreading dreams of revolutionary change among young blacks.

(5) It would seem false to characterize the dominant white outlook as one based on fear. An entire genre of literature dwells on the high anxiety level of South Africans. Neurotic maladjustment to a threatening reality is said to have resulted in a "national psychopathology."[14] Fear or anxiety, however, presupposes the perception of being threatened. It is doubtful whether the majority of whites, particularly Afrikaners, do in fact feel threatened by blacks. The deep-seated conviction of their own superiority blocks a perception of threat. The success of the government in quelling racial disorder, the tough, no-nonsense image of the authorities, regardless of the damage to South Africa's standing abroad, reinforce an experience that everything is under control. As a foreign

correspondent perceptively observed:

> So far, 20 years after the main resistance movements went underground, black insurgency remains something that most whites encounter only in their newspapers, and then usually when the guerrillas are caught.[15]

(6) If fear of blacks is not the dominant white outlook, neither can it be said to be hate. Emotional rejection presupposes a measure of perceived equality. As empirical studies have revealed, South African whites are not more prejudiced toward outgroups than comparative groups in Western Europe.[16] Conformity pressure among a cohesive ethnic group (Afrikaner), not different personality predispositions, accounts for the distinct South African authoritarianism.[17]

(7) At the risk of overgeneralization, one can say the majority of whites treat blacks as a statistical category only. From this point of view, they constitute a problem among others, to be managed properly like a flood or similar natural disaster. A vast bureaucracy takes care of what otherwise could develop into a costly burden, because of their numbers and economic usefulness. Besides their indispensable labor, blacks mostly constitute a nuisance for whites, neither feared nor hated, human beings outside the volk, not necessarily inferior, but also not fit for equal admission to the community. This outlook allows for perfectly normal elite contact, albeit confined to the formal or ceremonial level. Beyond consultation about problems that are better solved with the cooperation of black elites, the ruling group has imposed upon its subordinates what Paulo Freire called the "culture of silence."

However the need for internal pacification in light of a growing black militancy accounts for a new emphasis on previously neglected policies. Recent strategies of co-optation aim at lubricating the system of white control beyond coercion and neglect by the majority population.

The most important implication of South Africa's extraordinary economic development, and accumulated wealth, lies in the vast leverage it affords the regime to manage internal conflict. The dominant focus on coercion overlooks the rich incentives that a government of a wealthy state has at its disposal to diffuse, cajole or simply buy off dissent. With an 8 percent real GNP growth in 1980, the South African budget can easily absorb the elimination of racial wage differentials in the public sector. The state can afford to bribe a hesitant Bantustan leadership into accepting Pretoria's designs for independence. In the latest case of Ciskei, Minister P.G.J. Koornhof is reported to have said: "Independence: we'll make it attractive!" This includes payrolling an entire client bourgeoisie. In the view of a local political scientist: "The rapidly expanding middle class of politicians, chiefs, civil servants, teachers and merchants are all dependent on the state for salaries or loans; and the Ciskeian state is dependent on South Africa. These people form a privileged class tied to South Africa's chariot wheels by bonds of common economic interests."[18] The official white reaction to the Soweto upheavals of 1976, too, was a promise of material improvement for urban blacks, but not political inclusion. Such strategies aim at containment rather than reformist transformation.

Co-optation of urban blacks as a method of conflict regulation, however, encounters two contradictions. (1) The collaborators want to be part of the ruling bourgeoisie, not a separate one. As long as the distinguishing criterion is an imposed racial one, it will be invidious. (2) Collaboration discredits participant elites among their constituencies, particularly when the benefits remain symbolic only. Even for an imaginative "collaborative opposition" (the so-called Buthelezi strategy) which must be distinguished from opportunistic collaboration,[19] it is imperative that there is a clear dissociation from government policy.

Why has noncollaboration such widespread appeal, particularly among those urban blacks who are relatively better off? Insisting on noncollaboration is the most effective means to restore a violated dignity, to assert identity, and to establish what could be called the <u>symbolic sovereignty</u> of the powerless. Like the suicide of political prisoners, it remains the ultimate act of self-determination in the face of its denial.

In light of the high costs of old-style coercion and the limits of co-optation of traditional elites, the Nationalist Party now practices a third strategy which may be called the creation of counter-elites. The nomination of Coloureds and Indians to the President's Council, without any pretense of their being community representatives, illustrates this approach. An election would mean abandoning Nationalist Party control over the personnel of subordinate elites. The election experiment backfired when the elected majority abandoned the Coloured Council in protest, while Indian South Africans never had the opportunity of any community-wide election. The appointed representatives are now lavishly courted and presented as community spokesmen in the hope that the exposure will act as a self-fulfilling prophecy. Appointed political leaders, however, share in many ways the role of "yellow" company unions in the economic realm. Even the Federated Chamber of Industries has recommended to its members that they should deal with an unregistered trade union if it represents a worker majority. In the same way, nominated subordinate spokesmen in politics will be questioned as long as they represent only themselves. Without the legitimacy of demonstrated representativeness, counter-elites deceive their sponsors.

In short, Afrikaner Nationalist hegemony is indeed willing to negotiate, but on its terms. It is in the process of extending state patronage to collaborating ethnic elites, as long as Afrikaner Nationalism has the decisive say in selecting the new allies. It aims at cross-ethnic alliances and is even prepared to share power and abandon apartheid for its black counterparts, as long as Nationalist Afrikanerdom retains the ultimate control.[20] In this process race has, indeed, become irrelevant for the power holders. What has not been jettisoned is the commitment to retain Afrikaner sovereignty, if necessary on a much smaller scale. It is for this purpose alone that all kinds of modifications of traditional apartheid are contemplated, including possible concessions to regional autonomy (Natal) and other political experimentation at the periphery.

How far did the white election results of April 1981 indicate support for these more sophisticated attempts of regime management of conflict? Since both the right-wing and liberal factions gained considerable support, the answer is qualified. From the results two conclusions about the likelihood and methods of change can be drawn. First, major reforms cannot be achieved within the present political system, in which the ultimate power rests in the white electorate. In order to overcome this obstacle, a reform-minded leadership has two options. It could choose to abolish the electoral system in favor of a technocratic-military regime in the name of emergency conditions. This is likely only after a severe crisis (effective strike, credible military threats and domestic chaos) but could also be precipitated by the clear loss of electoral support, either through further defections to the right or through a Nationalist Party split into Transvaal and Cape wings. The sketched events, while possible, are not likely at present. The second option remains an enlargement of the electorate. The President's Council could, as its much heralded new constitutional proposal, recommend that Coloured and Indian voters be placed on a common voters roll for a multiracial parliament. Whether this option of an expanded electorate was adopted would depend on whether opinion surveys indicated Indian and Coloured majority support as a kind of reciprocation for late admission to a reformed National fold. However, since the majority of the hypothetical new voters would be likely to support the liberal Progressive Reform Party (if it were interested at all any longer in a parliamentary system which still excluded the majority), this alternative is easily ruled out, since it would entail the loss of control of the ruling group. Both options, but particularly the second one, would also encounter great opposition from the Afrikaner right wing. Therefore, they could be carried out only in a much closer alliance between reformist nationalists and liberal progressives which would mean sharing of power with and more policy concessions to the official opposition. The concessions to English business interests would pose no problems since all South African reformists of whatever background share a high degree of consensus about the essence of political devolution (group rights, mutual veto, proportional representation, some form of federalism). However, the present Nationalist leadership would still be very reluctant to relinquish exclusive Afrikaner control and share political power with their former ethnic adversary. The demobilization of ethnic consciousness, even in the face of perceived worse alternatives, remains a slow process indeed.

In the absence of a dramatic crisis, and with the present comfortable parliamentary majority of the Nationalist Party, more likely in the short run is a course that may be called the _administration of immobility_. In the words of Lawrence Schlemmer: "The NP may well be immobilized, forced to pursue status-quo politics, yielding only to crises, stumbling from precedent to precedent and attempting to appease reality with promises."[21] Most Nationalist Party majorities were cut in half in the Afrikaner heartland in the 1981 election. With up to 30 percent support for the ultra-right, the parliamentary incumbents must tread carefully indeed, and not lay themselves open to charges of "selling out." Thirty percent may well be the ceiling of ultra-right support in terms of the size of its dominant constituencies, that is, blue-collar Afrikaners and the lower echelons of the civil service. But in the emotionally charged atmosphere of ethnic politics in South Africa, a minor move of dramatic symbolism can tip the scales.[22] In order to be re-elected, Nationalist caucus members with strong right-wing constituencies can hardly afford to pose as champions of radical reform. Nor by inclination would they want to. It has to be remembered that the policy modifications of the Nationalists were reluctantly adopted because of the failure of previous policies, particularly in the economic realm. This remains a crucial difference between Nationalist and liberal reformers, who believe in change for additional reasons beyond self-interest. All these factors make the pursuit of status quo politics the most likely alternative at present, despite the rhetoric and genuine intentions of some Afrikaner intellectuals that "something meaningful must be done soon" in order to avoid the predicted crisis. Even sophisticated attempts at regime management of ethnic conflict, however, will prove unstable if they lack the crucial legitimacy among the ruled. Legitimacy is the key to successful conflict regulation.

Legitimacy and Minority Rule

Legitimacy refers to the degree of acceptance of their rulers by the governed. Social scientists usually distinguish three types of legitimacy, based on (1) divine origin, inheritance, or similar dynastic vocation of rulers, (2) the personal charisma and performance of rulers or (3) civil legitimacy, based on democratic-representative election of authority holders according to generally accepted bureaucratic procedures. Obviously the last type of legitimacy is of the main concern in the South African context. There is a direct correlation between the degree of legitimacy and the degree of coercion and stability: a regime based on a high degree of popular consent needs to rely less on force and is, therefore, more stable because of the common trust of its citizens. A system based on racial domination by a minority is now considered the most illegitimate foundation of rule. Such a regime may have power but no legitimate authority.

Some authors go so far as to assert the corresponding principle of ethnic proportionality in the name of "racial sovereignty." Unlike the liberal concept of "majority rule" -- for which ethnicity does not matter as long as the rulers represent the will of the majority -- racial sovereignty means that "the rulers of each society should as far as possible be racially or ethnically representative."[23] Black Africa, Mazrui and Gordon argue, is not united in a commitment either to human rights or to free elections for majority rule, but only to racial representativeness of the leaders. In the absence of this ethnic correspondence between rulers and majority, political power is viewed as alien, colonial domination. "Foreign rule is not merely rule by a nation-state from abroad, but rule by a foreign racial or ethnic minority. White rule in Southern Africa is illegitimate partly because it violates this principle of racial sovereignty."[24]

The concept of "racial sovereignty" runs counter to all hitherto acceptable liberal ideals of colorblind, individual universalism. It enshrines re-ethnicization as a necessary prerequisite for legitimate authority.[25]

On the other hand, there has always been a fairly wide consensus among black South Africans of all political persuasions that whites are not colonialists who could or should be driven out or repressed to seek revenge. Inkatha, the ANC, as well as Black Consciousness spokespersons, have always emphasized that they did not wish to replace white racism with a black one, although such sentiments may inevitably change in the process of ethnic countermobilization. In the words of the chairman of a popular Soweto committee:

> In South Africa, Whites and other non-Africans [*sic*] make up nearly a third of the population, as opposed to figures like one percent and four percent elsewhere, and nobody suggests that they are just colonists or settlers with no role in the future.[26]

The growing self-perception of former "Europeans" as "white Africans" together with the mutually binding high economic stakes in South Africa make analogies with processes of decolonization elsewhere in Africa dubious. South Africa is not a problem of colonialism, but of coexistence under conditions of mutually acceptable nonracial group politics. To suggest white emigration as "the most desirable and humane solution to the Southern African problem"[27] is off the mark from the perspective of all protagonists, both practically and morally.

Legitimacy presupposes a minimum of politicization. Where there is neither rejection of nor assent to rule, the question of legitimacy does not arise. Such situations may frequently exist among illiterate peasant populations who merely tolerate different types of rulers without attaching a notion of "just" or "fair" to a distant control. Taxes are traditionally paid but few other impacts of a remote government are perceived by an isolated periphery. What kind of ruler occupies the center is of the least concern to peasants in a subsistence economy. While such acquiescence is more widespread than frequently recognized, the apolitical ignorance is decreasing with the rapid incorporation of rural areas into the central economic system, the expansion of the market economy, and the penetration of communication technology into the remotest hut. Migrant laborers, even though illiterate, have, nevertheless, a keen awareness of what constitutes a fair wage and proper treatment. The proletarianization of the peasantry creates the need for civil legitimacy of rule over an atomized and increasingly politicized populace. Invidious racial distinctions have politicized South African society for the past decades in a way no other African country has experienced. This makes consociational rule particularly difficult since an unquestioned, apolitical loyalty to communal leaders can no longer be taken for granted. "Deferential attitudes to the segmental leaders," which Lijphart stresses as a favorable condition for consociationalism, are waning among all groups, and so is the "political inertness of the non-elite public."[28] Judging from the 1981 election, this change now applies even to such a cohesive model ethnic group as Afrikanerdom, whose traditional tribal solidarity is eroding over acrimonious controversies as to how to preserve it. In other words, mobilized ethnicity, both of the ruling white and the excluded black kind, develops its own dynamic which easily escapes the control of its originators.

Rulers often deceive themselves by construing mere acquiescence as tacit support. The supposed loyalty of the subjects indicates that the daily struggle for survival leaves no other option than to accept any tolerable domination, whether legitimate or not. In situations of absolute control and total despair, victims may indeed identify with their masters, as the reports about behavior in Nazi concentration camps have revealed. This "identification with the oppressor" then remains the only way of keeping psychologically alive. However, wherever a minimum of security exists and alternatives are available, people establish their identity in opposition to oppressors as the first step to real resistance. They resist by delegitimizing an intolerable burden. They define themselves by creating what could be called their _ideological sovereignty_. The Black Consciousness Movement, at least in its initial phase, has thus far been the most successful attempt at establishing a black counter-sovereignty, albeit confined to the symbolic realm.

It is also useful to distinguish between _internal_ and _external legitimacy_. The Bantustans lack international recognition, but not necessarily domestic acceptance, among certain sections of their population. The South African government itself, ostracized as a pariah in the assembly of nations, finds more support among its white constituency at home the more it is delegitimized abroad. On the other hand, the morale of the black population and its motivation to resist may well be strengthened by the external rejection of Pretoria.[29] Thus, external legitimacy influences internal attitudes and, in turn, depends on the domestic conditions. However, external legitimacy is grossly overrated with regard to the stability of a regime. As long as a minimal degree of domestic legitimacy exists -- or at least the perception that a government is strong and effective in maintaining its power and providing essential services, regardless of its discriminatory nature -- the absence of external legitimacy hardly affects the day-to-day functioning of an authoritarian regime, as the South African case clearly demonstrates.

Internal legitimacy, on the other hand, is crucial for the costs of racial domination. How people react in the absence of popular consent is hard to foretell for all circumstances. One author writing without South Africa in mind states: "[People] may transfer legitimacy sentiments to revolutionary movements (internal or international), to ethnic or similar internal groups, or even to international agencies. Or they may relapse into cynicism and alienation, punctuated by outbreaks of rioting, crime waves, terrorism or similar expressions of dissatisfaction."[30] The diverse expressions of black resistance in South Africa would seem to confirm this view. Even more foreboding for rational conflict regulation is a growing search for irrational solutions among the victims. While black political leaders, both in and outside prison, still command a strong following,[31] there is a growing longing for a messianic salvation. Market research reveals that the articulation of grievances and reform politics by black politicians strikes a diminishing chord. In the words of a researcher: "It was discovered that no single leader enjoyed the support of a wide cross-section of blacks. What those questioned yearned for was not a politician to express their views, but a messianic figure who would come forth to unite them and lead them to freedom."[32]

Messiahs do not usually display the moderation and pragmatism expected from the grand coalition of an elite cartel in a consociational democracy. Such prospects would seem to confirm a prediction made by virtually all analysts of South Africa: "The longer that South Africa's whites keep blacks in peonage the more superhuman will be the statesmanship required to stop blacks from someday turning the tables."[33]

Majority Rule or Consociationalism

A realistic discussion of constitutional "solutions" in South Africa rests on several crucial assumptions: (1) Any fundamental change can only be achieved with the cooperation, and not against, the dominant section in the ruling group. This follows from a second assumption, namely that (2) neither side can defeat the other, i.e., impose its solution on its opponent; and (3) even if a unilateral victory of one antagonist were possible, the price would be too high to make it worthwhile. Only if one accepts the evidence for these normative assumptions does it make sense to talk about constitutional conflict regulation.

It follows from the above that neither the white unilateral designs of neo-apartheid, nor the black preference for one person one vote in a unitary state, have a chance of relatively peaceful realization. Short of partition, this makes the consociational model of group politics the only serious contender. It is a second-best option from the view of committed partisans but clearly the only realistic alternative to escalating strife. It amounts to an institutionalized truce instead of open warfare.

As is known from the pioneering work of Gerhard Lehmbruch[34] and Arend Lijphart,[35] the consociational model consists of (1) a grand coalition of the political leaders of all population groups, (2) mutual veto power, (3) proportionality, particularly in the allocation of resources and civil service appointments, based on (4) a high degree of internal autonomy for each segment. Van den Berghe asserts: "A consociational democracy exists when the class interests of the ruling elite in preserving a unitary multiethnic state prevail over countervailing interests to break the state down into its ethnic components. The C.D. is thus a special case of 'bourgeois democracy.'"[36] It must be added that the class interests of the various subordinate communal elite groups must also coincide to reject partition. This is the case for leading sections of the white and black bourgeoisie in South Africa. However, quite unlike all other plural societies, the present South Africa distinguishes itself by the absence of three decisive preconditions to consociationalism: (1) voluntary association instead of imposed group membership; (2) lack of restrictions of segmental leadership; and (3) equal distribution of resources.

The consociational model is based on group politics, not on direct, individual participation in the political process. What is decisive is (1) who defines group membership and, intimately related, (2) what is the nature of the group's cohesion. In other words, is group membership imposed without the consent of the organized subjects or does their membership reflect voluntary identification with the group's existence? Is the consociational arrangement based on racial or communal divisions with ascribed membership of people who remain in their inherited segment without choice for life? Does the group-based conflict regulation reflect nonracial geographic cleavages, class or regional interests, ideological or cultural preferences with the option for individuals, at least theoretically, to choose alignments, to opt out and cross group boundaries?

If the group membership is imposed, consociationalism will not work, because it will be considered institutionalized racialism. If the group affiliation is freely chosen, consociationalism could work, even if it mainly coincides with phenotypical distinctions. That those distinctions are not offensive and do not stigmatize people is crucial.

Assuming such voluntary group identification were to be allowed in South Africa, three main ethnic or what could be called cultural heritage groups could emerge in place of the present four racial groups: (1) the African heritage stream, joined by some so-called Coloured people and even some whites, but most likely split at least along Zulu and Xhosa language lines; (2) the Afrikaner heritage group, joined by part of the Coloured population; and (3) the English heritage line, joined by most Indians, some Coloureds and some liberal-minded black Africans. The invidious category of Coloureds would have disappeared completely, since it has no distinct cultural base. Indian affiliations would most likely assert themselves at some local levels but not at the national level. There would be several ideologically based and class-based parties, vying with more ethnically oriented associations for membership. It is most likely that a mixture of both would emerge, for example a Zulu workers association. All groups would require a certain percentage of the popular vote (say a minimum of 10 percent) in order to gain representation.[37] Again, such a system would distinguish itself from the Westminster-type winner-take-all provisions in that (1) it would not allow one group to dominate the rest; (2) it would ensure proportional representation at all levels; and (3) it would include veto rights, particularly on central education/cultural issues. Veto rights introduce the danger of immobility and status quo politics. However, if carefully defined and judiciously exercised, vetoes would only be used as an emergency brake. The constitution, with a strong enforceable bill of rights, could also be underwritten by outside third parties (Western Five) with the threat of specific sanctions against violators.

A second precondition for a functioning elite cartel remains the free political activity of all leaders, provided they adhere to the constitutional rules. Leaders whose representativeness is doubted because their competitors are restricted destabilize consociationalism. In practical terms, this would mean a general amnesty for political prisoners and exiles, provided they abstain from extra-constitutional activity. Only through the participation in the political competition of all persons who claim a following can their representativeness be tested.

All analysts of consociationalism stress the importance of a relatively equal distribution of resources among the competing groups. This equality does not exist in South Africa. However, the real economic interdependence partially compensates for existing power differentials. If those differentials were addressed at the political level -- through proportional revenue-sharing and equalization payments for less-developed regions or institutional sectors, including affirmative action programs to compensate for past deficiencies -- unequal development would not necessarily present an insurmountable obstacle in a fast growing economy. In the same way as class conflicts are reduced in progressive West European countries through institutionalized bargaining and compulsory arbitration with ever fewer strikes and lockouts, conflicts about material privileges in South Africa could be settled without escalating violence. The sophisticated bargaining procedures in modern industrial relations may be well worth probing for possible adaptation to the political level.

An important facilitator of successful consociationalism is the _interpenetration of elites_ in nongovernmental institutions. If elite accommodation is confined to the political realm only, without the social basis of mutual contact within crucial support groups (universities, professional and business organizations, unions, churches, army, voluntary associations, service clubs, etc.), a much more politicized distrust is the likely consequence. Although South Africa has made considerable advances in reintegrating segregated institutions, particularly with regard to universities and sports clubs, the black presence often amounts to a token and does not nearly approach "elite proportionality." Moderation on the part of all leaders in a consociational model cannot be expected to prevail without the institutional support structures which create mutual trust in place of fear. The arduous process of reaching an accommodation often proves as important as the actual result.

Most analysts paint the alternative crisis scenario in rather apocalyptic terms of ultimate white defeat. In the words of a South African commentator:

> The conflict would not end unless the morale of the whites was so eroded by call-ups, by the impoverishment of war, by casualties, by disruption of social and family life, by emigration, and by bloodshed that resistance collapsed.[38]

Other academics engage in unrealistic thinking based on false analogies:

> The [apartheid] system is so unequal and iniquitous that it is not amenable to concessions and compromise. Past experience with decolonization elsewhere in Africa, especially in Zimbabwe (which was in almost every respect a miniature version of South Africa), seems to indicate that the end of white domination is in sight. The only question is whether it will take the form of a prolonged civil war, a negotiated partition or a frantic white exodus.[39]

However, Zimbabwe as an agricultural country is far from a "miniature version" of highly industrialized and urbanized South Africa, apart from all the other crucial differences between the two cases. People locked into manufacturing jobs for sound reasons tend to be far more reluctant to risk an uncertain all-out confrontation than peasants in remote parts of tribal lands. Even at the height of the war, the few urban centers were oases of relative peace in Zimbabwe. In South Africa, the frustration and anger among the black urban proletariat has certainly reached unprecedented heights. At the same time, the vulnerability of an industrial economy to disruption, sabotage and paralysis remains far greater than in rural settings. David Halberstam has illustrated this vulnerability most convincingly:

> Transport? Every lorry on the highway is driven by a black. Access? A black can go to any place in Johannesburg as long as his pass is in order, he is dressed poorly and he is servile in manner. Perhaps the black armored cars of this war will be the garbage trucks driven by black men through the streets of the rich. Technology and weaponry? One black in an old VW, armed with a small missile, can drive by a major airport, and one man can knock out a plane or a train. Vulnerability? Every home is vulnerable, every skyscraper, every gold mine. A society which lives off black labor cannot exist in the foreseeable future without it. Where black labor goes, terrorists can go too.[40]

On the other hand, unfulfilled expectations in a too slowly liberalizing society may constitute a necessary but not a sufficient precondition for successful revolution. The vast objective scope of the rulers for both repression and concessions can be used effectively to undermine the "unity of the oppressed" and shape different expectations. The deep cleavages among the black apartheid opposition do not reflect so much personality idiosyncrasies as this relative deprivation: the differential placement in the hierarchy of exclusion and exploitation. Hence, different black priorities and strategies emerge which often culminate more in fighting each other than in fighting the common source of the malaise. In this vein, middle-class blacks are mainly concerned with asserting their violated dignity through acts of symbolic liberation (Black Consciousness); migrants and peasants focus, above all, on material improvements by working within the system (Inkatha, unions); for exiles confrontation and disassociation remain crucial (ANC); finally, the liberal white apartheid opposition (PFP) aims primarily at negotiated security and economic stability. As long as a highly adaptive, manipulative capacity of the rulers to manage these concerns prevails, comparisons with the self-destructive nature of Nazi fascism or the fumbling of a weak and illegal regime in Rhodesia are, at best, misleading. South Africa may be drifting into a civil war situation à la Northern Ireland, and later perhaps even Lebanon, but this does not inevitably signal a reversal of political power, as it did not in Belfast or Beirut. Presumptuous and casual predictions that "the end of white domination is in sight" do not only constitute ill-informed, wishful thinking, but are counterrevolutionary because they lead to dangerous underestimation of the opposing forces.

What seems more plausible is an erosion of racial domination far less dramatic and on another front than frequently foreseen. The inevitable penetration of the economy by skilled subordinates, combined with a growing trade union militancy, could undermine and ultimately reform South African society far more effectively than call-ups. It is because the stakes are so high for those in control that they could be expected to compromise far more readily than the prophets of doom envisage.

Footnotes

1. Robert I. Rotberg, Suffer the Future, Policy Choices in Southern Africa (Cambridge, Mass.: Harvard University Press, 1980).
2. It may be worth speculating on how much the popular slogan that "history is on our side" has paralyzed rather than inspired political involvement. After all, why should one take risks now when "victory is certain" later?
3. Stanislav Andreski, The African Predicament (London: M. Joseph, 1968), pp. 110-133.
4. See Guenther Roth, "Personal Rulership, Patrimonialism, and Empire-Building in the New States," World Politics, 20, 2 (January 1968): 194-206; Victor T. LeVine, "African Patrimonial Regimes in Comparative Perspective," Journal of Modern African Studies, 18, 4 (December 1980): 657-673.
5. So Amos Perlmutter writes in an informative, comparative survey, "The Comparative Analysis of Military Regimes," World Politics, 4 (1980): 96-120.
6. For a comprehensive analysis of Afrikaner ideology in the pre-1948 phase see T. Dunbar Moodie, The Rise of Afrikanerdom (Berkeley and Los Angeles: University of California Press, 1975). For the later period, see the authoritative account by Hermann Giliomee in Heribert Adam and Hermann Giliomee, eds., Ethnic Power Mobilized: Can South Africa Change? (New Haven: Yale University Press, 1979), particularly chapter 4. The most sensitive author writing from a Marxist perspective on Afrikaner nationalism is Dan O'Meara, "Analysing Afrikaner Nationalism," African Affairs, 77, 306 (January 1978): 45-72, and "The Afrikaner Broederbond 1927-1948," Journal of Southern African Studies, 3, 2 (1977): 156-186.
7. For a full elaboration of the theme, see Heribert Adam, Modernizing Racial Domination (Berkeley and Los Angeles: University of California Press, 1971).
8. Ralf Dahrendorf, Society and Democracy in Germany (New York: Doubleday, 1967), p. 59.
9. See particularly Juan J. Linz, "Totalitarian and Authoritarian Regimes," in F.I. Greenstein and Nelson W. Polsby, eds., Handbook of Political Science, vol. 3, Macro-Political Theory (Reading, Mass.: Addison-Wesley Publishing Co., 1975), pp. 175-411; James M. Malloy, ed., Authoritarianism and Corporatism in Latin America (Pittsburg: University of Pittsburg Press, 1977); Guillermo A. O'Donnell, Modernization and Bureaucratic

Authoritarianism: Studies in South American Politics (Berkeley: Institute of International Studies, 1973); Fredrick B. Pike and Thomas Stritch, eds., The New Corporatism: Social-Political Structures in the Iberian World (Notre-Dame: University of Notre-Dame Press, 1974); Gino Germani, Authoritarianism, Fascism and National Populism (New Brunswick, N.J.: Transaction Books, 1978). For an informative overview of the debate, see Robert A. Manson, "Perspectives on Corporatist Approaches to Political Change in Latin America," Plural Societies, 10, 3/4 (Autumn/Winter 1979): 21-39.

10. As will be seen, the vast literature on dependency is hardly applicable to the South African case in the world economy. It is, however, meaningful internally, where average incomes in the most developed 10 percent of South Africa are twelve times greater than those in the least developed 10 percent. (Jil Naltrass, Financial Mail, March 6, 1981.)
11. Ann Seidman and Neva Makgetla, Outposts of Monopoly Capitalism: Southern Africa in the Changing Global Economy (London: Zed Press, 1980).
12. When an area with only 6 percent of Africa's population generates 50 percent of the continent's electricity, manufactures 74 percent of its railway trucks, 42 percent of its motor vehicles, 94 percent of Africa's books and newspapers, employs more medical doctors and engineers than in the rest of the continent -- then such a highly developed economic power can hardly be compared with the prevailing underdevelopment everywhere else. It can be shown that despite the professed boycott of the apartheid state, the developed South increasingly expands its economic influence over the region far beyond neighboring hostage countries and satellite states. Business Week, September 22, 1980, reports that South African export to black African areas outside the Rand bloc "is exploding at an exponential rate," from manufactured goods (78 percent in 1979) to vital food. According to this report, Nigeria, which leads the call for sanctions against Pretoria, nonetheless receives South African meat and other products through such devices as double invoicing and false certificates of origin, while much of the oil imported by South Africa originates in Nigeria's offshore wells. The African states benefit from quicker delivery and lower freight costs as well as some specific African designs. The reliable reports of contacts between Anglo-America and the Soviet Union, which share common interests in diamonds and gold marketing, add an ironical aspect to the public rhetoric.
13. The best comparative analysis of partition applied to South Africa is N.M. Stultz, "On Partition," Social Dynamics, 5 (June 1979): 1-13.
14. For example, H.I.J. van der Spuy with D.A.F. Shamley, eds., The Psychology of Apartheid: A Psychosocial Perspective on South Africa (Washington D.C.: University Press of America, 1978).
15. John F. Burns, New York Times, September 21, 1980.
16. Theodor Hanf, Heribert Weiland, Gerda Vierdag, Südafrika: Friedlicher Wandel? (Munich: Kaiser, 1978).

17. Thomas F. Pettigrew, "Personalty and Sociocultural Factors in Intergroup Attitudes: A Cross-National Comparison," Conflict Resolution, 2, 1 (March 1958).
18. Nancy Charton, "Ciskei Independent?" Reality, January 1981, pp. 10-11. For similar ties on a larger scale and its contradiction, see Newell M. Stultz, Transkei's Half Loaf (New Haven: Yale University Press, 1979).
19. On this important distinction see the perceptive essay by André du Toit, "Emerging Strategies for Political Control: Nationalist Afrikanerdom," in R.M. Price and Carl G. Rosberg, eds., The Apartheid Regime, Political Power and Racial Domination (Berkeley: Institute of International Studies, 1980), p. 7.
20. Presumably this is the major reason why the Nationalist Party refused to participate in the Buthelezi Commission, which originated as a black initiative outside Pretoria's control.
21. Lawrence Schlemmer, "How the Afrikaner Split Can Free the Nats," Sunday Tribune, May 3, 1981.
22. Even a thoughtless blunder, quite outside the racial context, may swing white votes now more than policies on race relations. In the 1981 elections this was apparently the effect of a statement made by a cabinet minister (Munnik) that elderly people can live healthily on R20 a month.
23. Ali A. Mazrui and David F. Gordon, "Independent African States and the Struggle for Southern Africa," in John Seiler, ed., Southern Africa Since the Portuguese Coup (Boulder, Colo.: Westview Press, 1980), pp. 183-194.
24. Ibid., p. 185.
25. Assuming the presently inconceivable were to occur and a black American or native Indian were to be elected as a U.S. president, his or her rule would have to be considered as foreign and illegitimate because the incumbent would not be racially representative of the majority. By the same token many African presidents who come from minority ethnic groups in the country, such as Nyerere or Kaunda, should have doubts about their legitimacy! Moreover, "racial sovereignty" corresponds perfectly with Afrikaner notions of ethnic pluralism and partition. If only rulers who are racial equivalents of the majority can be legitimate by definition, then the political fragmentation of South Africa in such a way that somewhere Afrikaners or whites constitute a numerical majority is indeed their only way for sovereignty as non-foreigners. "Separate development" has long grasped this message. "Racial sovereignty" merely represents the other side of the apartheid coin.
26. N. Motlana, Star Weekly, August 30, 1980, p. 8.
27. Pierre L. van den Berghe, ed., The Liberal Dilemma In South Africa (London: Groom Helm, 1979), p. 62. Van den Berghe bases his "solution" on the assumption that "there is no room left for compromise," p. 56, which is, I believe, false.
28. Arend Lijphart, Democracy in Plural Societies (New Haven: Yale University Press, 1977), p. 169.
29. Afrikaner Nationalists are fond of pointing out that the start of major unrest in the country usually coincides with some dramatic movement in South Africa's external relations, such as the Vorster-Kissinger meeting in Austria shortly after the

outbreak of the Soweto upheaval in June 1976. The implied conspiracy theory, where foreign agitators instigate riots for their own end, of course, detracts from the real causes of unrest. Ironically, the myth is aided by exile movements, falsely claiming undue credit for underground organizations in order to enhance their standing with sponsors. On the causes of the 1976 Soweto events see the official, remarkably frank report by Judge P.M. Cillie, Report of the Commission of Inquiry into the Riots at Soweto and Elsewhere from the 16th of June to the 28th of Feburary 1977 (Pretoria: Government Printer, 1978). The literature on Soweto is swelling. The most informative accounts are by John Kane-Berman and, from a neo-Marxist perspective, Baruch Hirson, Year of Fire, Year of Ash -- the Soweto Revolt: Roots of a Revolution? (London: Zed Press, 1979). Hirson mainly blames the "petty bourgeois limitations" of the students for the failure of the revolt. The Black Consciousness Movement, he argues, was not tuned to the problems of the black working class, which, he hopes, the ANC will attend to more effectively. Richard Rathborne, "Review Article: The People and Soweto," Journal of Southern African Studies, 6, 1 (October 1979), aptly criticizes the neglect of the social texture (religion, crime, drugs, etc.) in such mechanical class analysis.

30. John H. Herz, "Legitimacy: Can We Retrieve It?" Comparative Politics, 10, 3 (April 1978): 321.
31. See the "Freiburg survey," Theodor Hanf et al., Südafrika: Friedlicher Wandel?
32. B. Kubeka of Bates Wells Rostron Ltd. as quoted in Daily News,, October 8, 1980.
33. Adam Hochschild, "Enlightened Despotism," Harpers, January 1981.
34. Gerhard Lehmbruch, Proporzdemokratie (Tubingen: Mohr, 1967).
35. Lijphart, Democracy in Plural Societies. There is a growing literature on consociationalism. For South Africa in particular, see: R.I. Rotberg and J. Barratt, eds., Conflict and Compromise in South Africa (Lexington, Mass.: Lexington Books, 1980); F. van Zyl Slabbert and David Welsh, South Africa's Options, (Cape Town: David Philip, 1979); John A. Benyon, ed., Constitutional Change in South Africa (Pietermaritzburg: University of Natal Press, 1978); Nic J. Rhoodie, "Federalism/Confederalism as a Means of White-Black Conflict Resolution: Conceptual Dissonance in White Nationalist Ranks," Politikon, 7, 2 (December 1980): 101-110.
36. Pierre L. van den Berghe, The Ethnic Phenomenon (New York: Elsevier, 1981), p. 188.
37. The mechanisms of official recognition could be devised along the lines of party registrations in the present white system. A relatively high number of signatures would be required before a group is officially admitted. People could declare their group affiliation during voter's registration and change it, if they so desired, before the next election.
38. Ken Owen, Sunday Times, April 26, 1981.
39. van den Berghe, The Ethnic Phenomenon, p. 174.
40. David Halberstam, "The Fire to Come in South Africa," Atlantic, May 1980, pp. 81-95.

8
A Response to Heribert Adam, and a Rebuttal

Pierre L. van den Berghe
Heribert Adam

I disagree with practically everything Heribert Adam has said except that South Africa is not a fascist state. We agree about what South Africa is not, but we fundamentally disagree about what it is. Heribert has a tremendous advantage over me in that he goes to South Africa every year, and I haven't been in South Africa for twenty years.

Basically, I think the fallacy in Heribert's arguments is that he takes at face value the seeming reasonableness, flexibility, and pragmatism of the technocratic/academic/bureaucratic ruling class to whom he talks when he goes to South Africa, and loses sight of the constituency which this class represents. I have characterized South Africa as a "herrenvolk democracy," namely a nationalist movement which is internally democratic and externally repressive and dictatorial. South Africa is both a Western-style democracy of sorts and a colonial regime, combined into a single state. Unlike Heribert, I do not reject the colonial analogy. The colonial analogy is truly apposite and appropriate to understanding South Africa. South Africa has a classical colonial system, certainly a classical system of internal colonialism combined with a parliamentary democracy of sorts within the herrenvolk.

In effect, the present South African government since 1948 represents a constituency which is almost exactly 10 percent of the total population (about two-thirds of the white population), a constituency which is defined in the first instance in terms of being a nationalist movement of white Afrikaners. It is a genuine nationalist movement, which, in addition, is a racist movement because of the colonial context in which it finds itself. Secondly, because of the position of this ruling group within the South African economy, the ruling party represents in the first instance the white artisan working class, those whom Marxists call the aristocracy of labor. The Nationalist Party represents and entrenches the interests of a privileged sector of the working group, namely the white working class. This was very clearly the case in the incipient phase of Afrikaner nationalism in the early part of the twentieth century. It is somewhat less the case in the recent decades since Afrikaners have made considerable progress in the control of the economy that was hitherto English-dominated. Still, to a very considerable extent, the Nationalist Party speaks to a constituency which is only 10 percent of the total population and which includes a very substantial working class section. It entrenches not only the political privileges, but also the economic privileges of that small minority, and in such a way that, unless it maintains a monopoly of power, the economic privileges are threatened too.

So, in effect, the ruling regime of South Africa is a captive of a system in which a monopoly of political power is a sine qua non of maintenance of those economic privileges of the white working class and petty bourgeoisie; and that is why, however pragmatic, flexible, intelligent, and rational the ruling circles of the Nationalist Party might want to be, they simply cannot get away with making substantial concessions. Assuming that they should want to, they would be facing an immediate danger of dissidence within their own movement.

Now, let me examine briefly Heribert's thesis of co-optation. He believes in the capacity of the regime to co-opt such substantial numbers of blacks as to transform South Africa into a consociational democracy along European lines. The first problem is that the theory seems to see consociation as an extension of co-optation. One begins with co-optation and ends up with consociation. The fallacy there, I think, is that co-optation is the very opposite of consociation: co-optation means giving the shadow of power without the substance; consociation means sharing the substance of power. To see co-optation as evolving into consociation is a contradiction in terms. All the basic preconditions of consociation which have been spelled out by Lehmbruch, Lijphart and others in the European context are totally missing in South Africa. The odds against successful consociation, even in the European context, are very long indeed. Consociation requires demanding and seldom-met conditions to work successfully, even in societies in which groups start up with fewer economic, demographic, and political disparities than in South Africa. To expect that South Africa might move sufficiently in that direction to make anything resembling consociation possible is an extraordinarily long shot.

While consociation is not, in my view, within the scope of possible outcomes for South Africa, co-optation is indeed, to a limited extent, a possibility of the regime. I think it is true that the regime does have some potential for co-optation -- not that it has made much of it, but co-optation is at least theoretically feasible in South Africa. Even there, though, the prospect is not very good. Whereas I would put the odds against consociation at a million to one, I would put the odds against successful co-optation in South Africa at perhaps a thousand to one. Let us look at a country where conditions of co-optation are much more feasible, namely the United States. The policies of "affirmative action" in the United States are clearly co-optation techniques, and they have had very limited success. I mean, limited success in the sense that they did defuse politically the upper third of the American black population, who could be considered middle class. Affirmative action certainly did not take any steps toward alleviating the misery or defusing the situation of the black lumpenproletariat of the American cities which is, by almost every index, considerably worse off today than it was ten or fifteen years ago when these policies were put into operation.

The United States has many times the resources of South Africa, a more advanced state of industrialization, and a much more favorable demographic ratio: nonwhites represent less than 20 percent of the U.S. population compared to 83 percent of the South African population. Given the extremely modest success of co-optation-type policies in the United States, where nearly all conditions were extremely favorable and where the economic resource basis was considerably greater, I think it is extremely far-fetched to expect co-optation to work to any appreciable extent in South Africa. It is even more far-fetched to expect it to be in any sense satisfactory or acceptable to the oppressed population to whom the policies would be directed. So I do think co-optation is not quite as far-fetched as consociation, but still very unlikely in the South African context.

Let's put Heribert's thesis of the flexibility of the South African regime to several tests. Heribert asserts that South Africa is, in fact, deracializing its system of domination. There are some signals of token, cosmetic desegregation which may look impressive in the South African context, but I am suggesting that South Africa is slowly moving to a situation where, perhaps, in another ten years, it might be at the same point in deracialization as Portugal was in her African colonies some forty years ago. To suggest that this pace of change is in any sense going to defuse the revolutionary situation in South Africa is, I believe, totally unrealistic. South Africa today is still much more explicitly racist than Angola or Mozambique were in the 1940s or 1950s. To my mind it is doubtful that one can even speak of a policy of deracialization of domination in South Africa. For example, the government does not even bring itself to abolish such totally gratuitous vexations as the Prohibition of Mixed Marriages Act or the Immorality Act, which make no contribution to the regime's stability or control. Another test of the regime's ability to backtrack and to reflect some pragmatism at no cost to itself would be the assimilation of the Coloureds, or Browns. Its failure to do so is an example of pure folly on the part of the South African government. About 50 percent of the Afrikaans-speaking people in South Africa are "Coloureds." About 90 percent of the Coloureds are Afrikaans-speaking. In fact, Coloureds are as much brown Afrikaners in South Africa as Afro-Americans are "Afro-Saxons" in the United States. By all linguistic, religious, and other cultural criteria, Coloureds would be easily assimilable; they were begging for assimilation into the white group until about ten years ago. By now, there is some indication that the Coloureds have been kicked so often that they seem to have become radicalized: for 300 years those people have been begging to be allowed to assimilate into the white group, yet the government has not taken any meaningful steps in that direction, even though by doing so it could have doubled its demographic basis of support. There are about as many Coloured Afrikaners as there are white Afrikaners. Assimilation of the Coloureds could instantaneously double the size of the _volk_. There was everything to be gained by doing so, and really nothing substantial to be lost except racial hang-ups. Yet, the opportunity was never taken and, even now, the government shows no meaningful intention of doing so, thereby demonstrating its continued unwillingness to deracialize its policies. All the elements of "flexibility" are missing in the case of the Coloureds. I can take another example of the lack of disposition of the South African government to deracialize and to compromise. A government that assassinates such an extremely mild "bourgeois liberal" as Steve Biko is hardly a government ready to co-opt and compromise.

The last point I would like to make is on the possibility of partition. Unlike Heribert, I do see partition as an alternative. If I can put the odds on consociation at one in a million, and those on meaningful co-optation one in a thousand, then the odds on partition are perhaps one in ten. I think partition in the form of the establishment of a residual "Whitestan" in the Western Cape is, indeed, an interesting possibility. This would create a white-plus-Coloured state, with a 15 percent black minority, in the western part of the Cape Province. That I see as a possibility after a drawn-out civil war. Partition would probably not come about through peaceful accommodation and negotiation. There are indeed few, if any, precedents for amicable partition, although Canada might be a candidate.

Pierre L. van den Berghe

Van den Berghe's main evidence for the intrinsic irrationality of white rule rests on the exclusion of the Coloured middle-group. However, from the perspective of maintaining Afrikaner power, this policy can be considered "rational," however immoral: the 1948 parliamentary victory of the Nationalist Party was narrow. Only the disenfranchisement of the Coloured voters consolidated this unexpected grasp of political power which would have been jeopardized by the then Coloured vote for the liberal opposition. Nowadays, when the fragmented Nationalists expect Coloured support in the conflict with the extreme right, the Coloureds are re-enfranchised again, albeit, one would suspect, with limited success. This manipulative co-optation exercise proves again that the present technocratic regime is hardly guided by irrational racism, but by a ruthless, shrewd survival policy.

Incidentally, South Africa's unfortunate treatment of her brown Afrikaners constitutes a telling case against sociobiological explanations of ethnic relations. The exclusion of so close a genetic and cultural kin group in favor of nonrelated English and other (white) European immigrants disproves sociobiologists, like van den Berghe, who emphasize preferential kin selection as the root of ethnic nepotism. Paramount for the inclusion or exclusion of ethnic groups -- decisive even for their very existence, their perpetuation, mobilization or disappearance -- are situational criteria such as political considerations, and not primordial givens of genetic make-ups or psychological dispositions such as racial antagonisms.

Heribert Adam

9
Modernization, Ethnic Competition, and the Rationality of Politics in Contemporary Africa

Robert H. Bates

Theories of modernization imply the demise of ethnic competition. This is true of sociological theories, in which specific, differentiated, "rational" interests are held to displace generalized, diffuse, "primordial" ties. It is also true of Marxist theories, in which horizontal class cleavages are held to displace vertical segmentary ties as markets broaden and as social relations become organized about the capitalist means of production. A consensus exists, then, that ethnic competition belongs to the premodern era; insofar as it persists, it is an irrational form of behavior or a form of false consciousness.

Despite the predictions of these theories of social change, ethnic competition strongly endures. It is a feature of politics even in the most modern of nation-states. More relevant to the subject of this essay, in contemporary Africa, the levels of ethnic competition and modernization co-vary. Awareness of these facts must provoke a reappraisal of modernization theories, and this essay joins with my own earlier work and the recent work of others in stressing the weakness of classical expectations concerning ethnic behavior.[1]

Where this essay differs from the work of others is in emphasizing the rational basis for ethnic competition. For what I argue is that ethnic groups represent, in essence, coalitions which have been formed as part of rational efforts to secure benefits created by the forces of modernization -- benefits which are desired but scarce.

Definitions

The major terms of the argument are "modernity," "ethnic competition," and "ethnic group."

In keeping with conventional usage, I define modernity operationally and call those societies more modern which attain higher levels of the following variables: education, per capita income, urbanization, political participation, industrial employment, and media participation. In practice, I will restrict my attention to the first three of these variables. I feel justified in using a single term -- "modernity" -- to refer to these distinct variables, for it has been repeatedly demonstrated that they are highly interrelated and that their interrelation derives from their tapping a single underlying dimension.

By ethnic competition, I mean the striving by ethnic groups for valued goods which are scarce in comparison to the demand for them.

The definition of an ethnic group is, perforce, complex. Like all groups, ethnic groups are organized about a set of common activities, be they social, economic, or political; they contain people who share a conviction that they have common interests and a common fate; and they propound a cultural symbolism expressing their cohesiveness. The primary factor that distinguishes ethnic groups from other kinds of groups is the symbolism which they employ. The symbolism is characterized by one or more of the following: collective myths of origin; the assertion of ties of kinship or blood, be they real or putative; a mythology expressive of the cultural uniqueness or superiority of the group; and a conscious elaboration of language and heritage. In addition, ethnic groups differ from other groups in their composition; they include persons from every stage of life and every socioeconomic level.[2]

It is important to note that ethnic groups need not be tribes. The term "tribe" denotes a group, generally rural, which is bound by traditional political structures to which people are linked by the mechanisms of traditional political obligation. Ethnic groups need not be based on traditional political institutions; rather, many are based upon newly created organizations, forged in the competitive environment of modern nation-states. And the ties that bind the members of ethnic groups are often material interests, and not traditional obligations. Ethnic groups may expand into the rural sector and gain the backing of tribes; this in fact can be a politically dangerous stage in their evolution, and this paper will examine some of the circumstances that can promote such urban-rural linkages. But, nonetheless, ethnic groups should be distinguished from tribal groups, and the origins and dynamics of the former should be considered independently of what is known and asserted about traditional political behavior in Africa.

The Argument and Evidence for Its Plausibility

Modernity is a cluster of desired goods. This is not to state that it is uncritically accepted, nor that African people do not decry the costs of modernization. The development of such philosophies as Negritude, Humanism, and the multitudinous versions of African Socialism by African intellectuals, and the spread of urban prophet churches and antiwitchcraft movements among some of the African masses, suggest the sensitivity of many to the costs of modernization.[3] Nevertheless, it is obvious that the components of modernity are strongly desired.

Not only is modernity desired, but the goods it represents are scarce in proportion to the demand for them. The inevitable result is that people compete. This competition is best illustrated in the struggles over income and for several of the resources which create it: land, markets, and jobs.

Land. In the agricultural societies of Africa, particularly where the population is dense, the penetration of a money economy gives rise to an intense competition for land. As Colson states: "By themselves such changes had an impact on local systems of land rights as men began to evaluate the land they used in new ways. They also led to an increasing number of legal battles over land; for men were encouraged to establish long-term rights in particular holdings either for immediate use or for subsequent gain."[4] While much of the competition for land is intraethnic, much of it is interethnic as well. Hill has published accounts of several major interethnic land disputes in Ghana.[5] The dispute between the Kikuyu and Masai over control of the former White Highlands has created a major cleavage in the political life of Kenya. And a major source of urban conflict is the tension between those indigenous ethnic groups who have alienated their lands and those immigrant groups who have benefited from the occupation of urban real estate.[6]

Markets. The competition for control of markets is equally intense. One of the best analysts of this phenomenon is Cohen, who documents the intensive rivalry between the Yoruba and Hausa for control over trading routes to the interior along the southern coast of Nigeria.[7] Lloyd documents the rivalry between Itsekiri and Urhobo for marketing facilities in Warri. Such conflicts characterize Eastern Africa as well.[8]

Jobs. Equally as pervasive is the competition for jobs. Parkin discusses the rivalry between Luo and Baganda for employment in the industrial and service sectors of Kampala; Grillo, in his analysis of the East African Railways, notes a similar rivalry between Luo and Abaluhya.[9] Competition for employment has been noted between the Bamileke and Douala in the Cameroun, between Nyanja and Bemba speakers on the copper belt of Northern Rhodesia, between indigenous Africans and strangers in Abidjan, and between Kasai Baluba and Bena Lulua in Kasai province in the Congo.[10] Exacerbating these tensions has been the expansion of the production of educated employables at a rate in excess of the expansion of job opportunities, a phenomenon that has been studied in Ghana, Nigeria, and elsewhere.

Modernity and Stratification. Not only does modernization thus create competition; but because the elements of modernity are valued and scarce, they form the basis of a new stratification system in Africa. This is not to state that traditional criteria of social ranking are totally relinquished; indeed, all evidence is to the contrary, and we shall later argue that the interplay between the two stratification systems is in fact crucially important. Nonetheless, those who possess the attributes of modernity can more successfully claim higher social rank in contemporary African society than can those who do not.

The creation of new stratification systems is documented in studies of occupational prestige in Africa. These studies find that modern roles, such as those of the teacher or clerk, are given high prestige and that in fact they are generally ranked higher than the roles of traditional societies, such as those of craftsman or hunter. The results of these studies go beyond suggesting that modern occupations are prestigious, however, to emphasizing that modern conceptions of stratification are being utilized by African peoples. Thus, studies report a close correspondence between the ranking of occupations by Africans in the Congo and those produced in the more developed countries.[11] Hicks, in a study of occupational prestige in Zambia, finds the criteria for the rankings to be similar to those reported in industrialized societies.[12] The prestige of an occupation can largely be accounted for, he indicates, by the degree of responsibility, service value, income, education and the nature of the working conditions associated with it.

The numerous studies of elite formation in Africa also suggest that those who possess the attributes of modernity can successfully lay claim to high status in many indigenous societies. Thus, Lloyd in a series of articles notes the rise of wealthy traders and educated clerical workers as a new elite in Yoruba society; Austin discusses the same phenomenon among the Ashanti and, like Lloyd, documents the conflicts between the new elite and the traditional ruling classes.[13] The same pattern has been reported in East Africa. Studies of the Gisu, Chagga, Luo, and Abaluhya describe how the spread of education and cash cropping generated new groups of literates and wealthy traders and how these new segments of the population successfully lay claim to elite status.[14]

Stratification and Competition. Crucial to the emergence of ethnic competition is that societies as well as individuals tend to be evaluated along the dimension of modernity. Those groups which are wealthier, better-educated, and more urbanized tend to be envied, resented, and sometimes feared by others; the basis for these sentiments is the recognition of their superior position in the new system of stratification.

In Calabar, for example, the indigenous Efik took readily to education, while the immigrant Ibo lacked both education and the wealth which would follow; the result was tension and hostility, sentiments which were exacerbated by the Ibos' attempts to close the gap.[15] This case finds its parallel in the famed rivalry between the Ibo and the Yoruba. So too with the northern peoples of Nigeria: their fear of the superiority of the southern peoples in the modern social and state system led them to an explicit policy of "northernization" whereby they gave privileged access to educational and employment opportunities to residents of the north. Their unequal status led to another point of conflict with the southerners, this time over the date of self-government in Nigeria. As related for the Birom of northern Nigeria, "Birom leaders expressed unease at the thought of rapid achievement of selfgovernment -- in fact at the idea of selfgovernment before the Birom have produced enough professional men and traders and artisans to be able to claim all the occupations of control which exist in their Division."[16] As a result, the Birom aligned with the Northern Peoples Congress instead of with one of the several Southern parties which were competing for their allegiance, for the Congress favored a later date for self-government.

This pattern is also found in East Africa. In Kenya, for example, the less modernized pastoralists feared the perpetuation of the disparity between themselves and the more educated, urbanized, and wealthy agriculturalists which would result were Kenya to become independent under the political control of the agriculturalist peoples. As a result, the pastoralist groups less fervently pressed for self-government in Kenya than did the agriculturalists; and when they saw that their efforts were to fail, they sought to fragment power through a federal constitution. This conflict provided a major basis for the political competition between KANU and KADU in Kenya.[17] As Zolberg succinctly states, in the case of the Ivory Coast: "Many [of the changes introduced under colonial rule] have reinforced old differentiations between tribes by adding to them new ones based on modern attributes, such as wealth and education."[18] And as one Nigerian commentator states, ethnic competition breaks out when "groups must compete for places in the class, status and power systems of the new nation. In a manner of speaking...it is a form of social indecision regarding the strategy of equitable distribution of...advantages available to people in the new African nations."[19]

Basis for the Formation of Ethnic Groups

The basic question that arises from this discussion is: why should the competition for the components of modernity and for status positions as defined by modernity involve ethnic groups at all? At least three answers can be given to this question. The first is that the distribution both of modernity and of ethnic groups tends to be governed by the factor of space. Where modernization takes place often largely determines who gets modernized. The second is that administrative and ethnic areas often coincide. And the third is that it is often useful for those engaged in the competition for modernity to generate and mobilize the support of ethnic groupings.

The Factor of Space. It is the geographers who most forcefully portray the spatial patterns of modernization. Originating in "nodes" or "central places," modernity then spreads or "diffuses" into the more remote regions of the territory, they report. They also demonstrate that the level of modernity slopes downward with distance, with the central places being the most modernized, the proximate areas being the next most developed, and the hinterlands lagging behind.[20] While there is considerable debate over whether territoriality is a required component of the definition of an ethnic group, there is no denying that the members of an ethnic group tend to cluster in space; nor can it be questioned but that colonial policy made every attempt to assign ethnic groups to stable and rigidly defined areas. One result of this correspondence in spatial orderings is that members of ethnic groups will tend to have preferences with respect to allocational decisions which are homogeneous and well defined. This is particularly the case with respect to siting decisions, i.e., choices as to where to locate specific facilities or projects. The benefits from such decisions accrue to those who are most proximate and they diminish monotonically with distance from the project. Insofar as a group occupies a specific area, then, its members would have a uniform preference with respect to sites; they would prefer that resources be devoted to constructing projects that are in or near "their" area and they would rank alternative locations in the order of their proximity.

Many ethnic conflicts in fact take the form of locational disputes. A major reason for the split between the Bemba-speaking and other factions in Zambia, for example, was that the Bemba-speaking areas appeared to derive disproportionate benefits from projects built in response to Rhodesia's UDI; a railway and pipeline were constructed in the northeast and the main road in the area upgraded to provide Zambia with a life line to the sea. Ethnic competition takes the form of regional conflict in other nations as well; and this, of course, is what one would expect, given the locational feature of allocational decisions.

Another result of the correspondence of spatial orderings is that groups are differently advantaged in terms of their attainments. Soja, in his analysis of "modernization" in Kenya, finds that the Kikuyu, being proximate to Nairobi and the highlands, are the most urbanized and educated, and among the wealthier, of the ethnic groups in Kenya.[21] Similarly, Coleman and Abernethy argue that the initial advantage of the Yoruba in Nigeria derived from their proximity to Lagos and from the early establishment of missions in Lagos and Abeokuta.[22] The Ibo, being more remote from these areas, were initially less exposed to the centers of modernization; and being less proximate to the locus of mission activity, they lagged badly in the attainment of education and well-paying jobs. The pre-eminence of the Baganda, and the tensions which have resulted, have also been explained in terms of their proximity to the administrative capital and largest town in Uganda. As one analyst reports:

> In Uganda (with the possible exception of the southeastern area) geographical distance from the capital city is sufficient to provide a rough indicator of the degree of modernity. Most Ugandans are fully aware that the Baganda profited more than others from their close proximity to the administrative center of the country.[23]

So pervasive a phenomenon does this appear that spatial proximity is sometimes offered as an alternative to classic notions of "cultural receptivity" in explaining differing rates of change. Thus, Kasfir, discussing Apter's structural-cultural theory of modernization in Buganda, notes that given the proximity of the Baganda to Kampala, the "argument...cannot be proved or disproved."[24] And Gugler, in discussing the general resistance of pastoralists to the forces of change, comments that "the underlying more general factor [is] probably that many of these are difficult of access to schools and administration alike."[25]

<u>Space, Administration, and the Incentive to Organize</u>. Local administration serves as one of the primary agents of modernization in Africa. And the colonial powers, by delineating administrative boundaries along "tribal" lines, made it in the interests of their subjects to organize ethnic groupings so as to gain control over the administrative mechanisms which themselves controlled the modernization process.

This assertion is best demonstrated in the studies of one of the primary sources of income in Africa -- land. Colson notes that colonial policy produced two contradictory developments in land law. On the one hand, the growth of the cash economy furnished an incentive for individual ownership; on the other, the dominant mythology of the colonial administration, that land was "communally owned," restricted permanent rights to land to the members of the local ethnic group.[26] A clear implication of Colson's analysis is that as the benefits of land ownership increased, as they did with the spread of cash cropping, so did the importance of retaining and affirming membership in ethnic groups. The political consequences of this rapidly became evident in the conduct of the local councils in which jurisdiction over land rights had in part been vested. The local councils began to function as ethnic organizations, legislating so as to restrict ownership to members of local ethnic groups and to divest "strangers" of rights of permanent tenure. In this way, the material benefits to be derived from land ownership were purposefully restricted to local residents.[27]

The power of the local administration over economic resources extended beyond the control of land tenure to such other matters as access to markets and market stalls, the regulation of crop production and animal husbandry, the construction of roads for the export of produce, and the like. At the behest of those who had the greatest stake in the modern economy, often organized in "improvement unions," many councils acted so as to bias the distribution of these programs for the benefit of the local population and away from immigrant strangers. La Fontaine reports that local leaders in Mbale sought to distribute roads and payments to coffee growers so as to benefit Gisu cash croppers exclusively. Lonsdale documents the attempts by the Kavirondo local councils to restrict access to markets to Abaluhya and Luo cash croppers. And Lloyd, in his discussion of Urhobo and Itsekiri rivalry in municipal elections in Warri, notes that "it was said during the 1955 election campaigns that whichever tribe won the election would restrict the lease of stalls to its own members and thus give them a monopoly of the trade in the town."[28]

Given that power over the distribution of many of the benefits of modernity is vested in the local administration, and given the correspondence between administrative and ethnic boundaries, it is natural that persons would create politically cohesive groups and utilize these to restrict the degree to which the administration could compel the sharing of benefits with others. The demand of ethnic groups for their own districts and councils represents a logical continuation of this process, for by securing this demand they could more perfectly exclude others from such benefits and thereby reserve to themselves a larger portion.

The Behavior of the Moderns. A third major class of reasons for the formation of ethnic groups is that in the competition for the benefits of modernity, it has been in the interests of the most modern elements to sponsor the growth of "traditional" consciousness in Africa.

Before explaining why this is so, it is instructive to indicate the extent of the evidence for the major role that moderns have played in organizing "traditional" groupings. The "educateds" often were the founders of ethnic unions. Thus Lonsdale speaks of "the Christian establishment" of mission-trained literates who helped to form the Kavirondo Taxpayers Welfare Association.[29] Twaddle writes of the "'new men' created by missionary education" who helped to form the Young Bagwere Association.[30] And the role of the former mission students in organizing the independent schools among the Kikuyu and promoting ethnic consciousness in that tribe has been discussed by many authors.[31] In terms of income, it is often those who are better-off by dint of their occupations in the modern sector -- the clerks, cash croppers, and traders -- who form ethnic unions. Ottenberg, for example, notes that it was those who "work as clerks for the local British Administration, who teach in local schools, who work for traders, or who are traders themselves" who founded ethnic unions among the Afikpo Ibo.[32] Sklar notes that both the Egbe Omo Oduduwa and the Ibo State Union "were created by representatives of the new and rising class -- lawyers, doctors, businessmen, civil servants."[33] And Bennett notes that the Bahaya Union was led largely by relatively prosperous cash croppers and members of cooperative societies whose economic interests were threatened by the government's policies toward coffee cultivation and land management.[34]

The role of urban dwellers in the formation of ethnic unions stands out most clearly in the literature from West Africa. There, ethnic unions were most often formed in urban centers and only later exported to the rural areas whose names they often bore. As stated by Offodile: "It is significant to observe that almost all the tribal unions now existing in Nigeria were found, not in the very towns of the tribes represented, but outside their own villages, and sometimes outside their own tribal territories."[35] Thus it was in Lagos where the Ibo came into competition with the Yoruba and where Ibo tribal consciousness was formed. It was in Leopoldville that Abako was founded; only later was it exported to the majority of the Bakongo in the surrounding rural territories.[36] And even in the case of such minor tribal unions as the Afikpo Town Welfare Association, the origin of the organization lay in the city: "An educated Afikpo man working at a trade-union post in Aba, who had traveled widely in the course of his work...realized the need for a protective union to aid Afikpo people.... At his own expense he had 400 membership cards printed to organize Afikpo people."[37]

In explaining the behavior of the moderns, we can note numerous motivations for the formation of these unions, but one stands out above all others: the perception by the moderns that they must organize collective support to advance their position in the competition for the benefits of modernity. To this we now turn.

<u>Social and Economic Competition and the Formation of Groups</u>. We have noted that there are good reasons for one ethnic group to be more advanced than another; in part, the spatial diffusion of modernization makes this inevitable. The members of an advantaged ethnic group are motivated to defend their leading position; they devise methods for retaining their privileged positions, such as biasing local council legislation in the ways we have described. Moreover, because modernity is desired, the less favored members of a privileged ethnic group place immense pressure on their more advantaged brothers to share the benefits derived from their advanced positions. Thus, family loyalties are activated to secure jobs; the income of the more prosperous is claimed by kin, often to meet school fees that will in turn secure future prosperity; and urban dwellers find their households being used by country folk in search of urban

employment.[38] Under the pressure of the less advantaged, a sense of obligation resembling that usually extended to immediate kin is thus broadened to include fellow village dwellers and even persons from other villages and districts; and the language of relationships, such as the use of putative kin terms, is broadened to suggest this expansion. The result of these pressures is that the more advantaged members of the group are forced to draw into their sphere others of their kind. And the social-climbing less advantaged generate a mythology of consanguinity in search of modern benefits. The initially advantaged group thus consolidates and comes to view itself as an ethnic grouping in the process.

A major result is to create among the members of other groups a sense of threat and disadvantage. In the competition for jobs, it is the more modern elements of these groups who most directly experience this threat and perceive it in ethnic terms. They come to understand that they are placed at a disadvantage by their inability to activate the sense of ethnic obligation so as to gain access to the modern sector. Moreover, they perceive that their individual progress is closely determined by the collective standing of their group; they therefore initiate programs of collective advancement in response.

The creation of ethnic support by the competitors for jobs has been noted by Grillo in his analysis of the Railway African Union in Uganda. High office in the union often led to promotion to more advanced jobs in the railway company, and so was much desired by railway employees. In order to enhance their mobility prospects by gaining union office, contenders for advancement would sometimes make appeals to tribal loyalty. As stated by Grillo, "Although those seeking office...may have little sense of tribal interest, tribalism may be one of the weapons used in the struggle."[39] A similar pattern has been noted in Kenya, where accusations are made regarding the "Kikuyuization" of the government services and ethnic pressures are mounted by the leaders of the less advanced groups in protest over job discrimination.[40] However, the most striking illustration of the creation of ethnic action is to be found in West Africa. Perceiving that their individual fates in the struggle for modernity were tied to the collective standing of people from their own areas, the most modern members of the less modernized groups organized large-scale programs of advancement among their people. Abernethy furnished the best discussion of the phenomenon:

> The struggle for employment was bound to produce frustration, and those not chosen for the best jobs found it easy to blame their plight on the advantages possessed by members of other groups. Of course, different groups clearly did have differential access to education, which in turn was the key to job mobility.... What was the best course of action open to the urban migrant who was acutely concerned lest his ethnic group fall behind others in the struggle...? Certainly the rural masses had to be informed of the problem. If the masses were not aware of their ethnicity, then they would have to learn who they really were through the efforts of "ethnic missionaries" returning to the homeland. These "missionaries" would also have to outline a strategy by which the ethnic group, once fully

> conscious of its unity and its potential, could compete with its rivals. Clearly the competition required enrolling more children within school, particularly at the secondary level for the graduates of a good local secondary school would be assured of rapid...mobility within modern society.[41]

In this manner, Abernethy accounts for the formation of ethnic unions among the Ibibio, Ibo, and Urhobo in Nigeria.

Political Competition. A similar pattern obtains in politics. In the political arena, it is not just power that is at stake, but also the benefit which power can bring: control over the distribution of modernity itself.

There can be no doubt but that electoral competition arouses ethnic conflict. The tensions arising from the 1964 elections in Nigeria were one of the precipitates of civil war in that country. The primary instruments of the virulent ethnic conflict in the Congo were the numerous political parties formed by politicians to contest the 1960 elections. And it was elections, or the anticipation of them, that precipitated ethnic conflict in Zanzibar, Rwanda, and Ghana.[42]

Perhaps the main reason for these conflicts is that in the competition for power, ethnic appeals are useful to politicians. Given that most constituencies tend to be dominated by the members of one ethnic group -- a result of the politics of apportionment and delimitation -- an ethnic appeal is an attractive and efficacious weapon in the competition for office. Moreover, because ethnic groups contain persons of all occupations, socioeconomic backgrounds, lifestyles, and positions in the life cycle, the appeal of common ethnicity can generate unified support where other issues would be divisive. As a result, in the competition for power, and for the benefits of modernity and the prestige which it confers, politicians stimulate the formation of competitively aligned ethnic groups. As stated by Sklar: "Tribal movements may be created and instigated to action by the new men of power in furtherance of their own special interests."[43]

Naturally, this is not to state that the politicians are alone to blame for the rise of ethnic conflict. Indeed, while they do instigate ethnic conflicts, they often behave like captives of the forces which they helped to create. This leads us to the last question with which we plan to deal in this paper: having accounted for the formation of ethnic groups, how can we explain their persistence?

The Persistence of Ethnic Groups

Ethnic groups persist largely because of their capacity to extract goods and services from the modern sector and thereby satisfy the demands of their members. Insofar as they provide these benefits to their members, they are able to gain their support and achieve their loyalty.

The capacity of ethnic groups to extract benefits from the modern sector is best demonstrated in their relationship with those who have achieved positions of prominence in that sector. Ethnic groups exert powerful social pressures upon the modern elite in order to satisfy the demands of their members. Perhaps the most persuasive evidence for these assertions is the reaction of the modern elite itself. Its members experience their positions not only as privileged but also as onerous; they feel that they are at the center of tremendous social pressures. As Uchendu states: "My town demanded leadership from me. But this leadership is a trying as well as a thankless experience. My town has a passionate desire to get up."[44] Chona, once vice president of the Republic of Zambia, argues the same point: modern leaders are subject to concerted pressures and forced to act as spokesmen for ethnic interests. Rather than blaming members of the national elite for instigating "tribalism," he states, the citizens of Zambia should blame the "local leaders in the villages and towns" who "travel...to Lusaka to meet leaders" in order to urge them to serve parochial interests. Chona concludes, "Unless the local leaders in the villages and towns stop being competitive against other groups and begin to regard top leaders as national leaders we shall not find a lasting solution to [the problem of ethnic conflict]."[45]

The demands upon the modern elite are predictable. Characteristically, they include demands for material resources: financial contributions from the moderns for the construction of new facilities and for the creation of educational funds. Some groups even levy taxes upon their more prosperous urban members.[46] The demands of ethnic groups are also for service: the use of the skills of the moderns, be they technical, educational, or political, on behalf of "their people."

The capacity of ethnic groups to extract goods and services from the modern elite derives from several sources. They control the allocation of strong inducements. For example, because elite skills are a desired commodity, ethnic groups are able to win the use of these skills by making their acquisition contingent upon ethnic service. Thus Lloyd reports for the Itsekiri that "several lawyer-politicians who were sent to England with community funds are expected on their return to repay the debt either in cash...or by winning tangible benefits for their people."[47] Other inducements include prestige: symbols of status are conferred or withdrawn by ethnic groups in recognition of services performed for the group. Plotnicov documents the conferring of status by ethnic groups in Jos; members of the modern elite, he writes,

> have the skills for dealing with government and the wider community. Their knowledge in legal and economic matters, secretarial and bookkeeping procedures, and their general sophistication in modern and urban affairs is indispensible. Their value to the ethnic group...[is] recognized through the granting of high...offices...and sometimes...titles as well, which further reinforce the modern elite's powers.[48]

Ethnic groups also possess strong sanctions, most notably the capacity to withdraw elite status. This is vividly revealed in the political sector, where, for example, the Luo, Sukuma, and Lozi turned out of office several of their most renowned political leaders, some of cabinet rank. The reported reason for the imposition of these sanctions was the elite's failure to serve local interests.[49]

There is another reason for the ability of ethnic groups to extract goods and services from the incumbents of the modern sector. For many moderns, what is prestigious is still defined in terms of traditional criteria. For many, modernity becomes a resource which they utilize to attain prestige within the traditional sector.

Evidence for the continued presence of traditional notions of stratification is contained in the very studies which emphasize the pre-eminence of modern stratification systems. Hicks, for example, finds that the standard deviation of the rankings of modern occupations by Africans is higher than that of the rankings by Europeans; he attributes this in part to the use by Africans of two sets of stratification criteria, one traditional and the other modern.[50] Similarly, Foster found that Ghanaians give higher ranks to traditional political offices -- chiefs and councilors -- than they give to many modern occupations "despite the fact that often holders of these [offices] had little education and received little...pay. Clearly the respondents in Foster's study were using at least two dimensions to rate the full list of occupations: one being the western dimension and the other a traditional dimension."[51]

The acceptance of traditional stratification patterns leads many moderns to convert their success in the modern sector into prestige in the traditional order by utilizing their wealth to obtain prestigious positions in their ethnic groupings. Many studies cite the purchasing of traditional titles by the successful entrants into the modern sector. Others note the use of wealth to purchase traditional offices. Still others record the use of income derived in the modern sector to practice clientage and to finance ceremonies so as to enhance social standing in the traditional order. As Balandier states:

> [A wealthy person] can...make "sociological investments"; in this case, he uses new economic conditions to achieve or to reinforce a traditional type of prominence. The size of his "clientele" and the extension of his generosity will reveal his degree of success; his profit will be expressed in prestige and authority...the economic "game" is still only a method to achieve goals determined by the old social and cultural system.[52]

Ethnic groups are thus able to extract investments from persons seeking access to elite positions in the "modern" order. Moreover, the "moderns" need, and seek to elicit, the support of ethnic groupings. Forces thus promote the supply and demand of services between the ethnic groupings and the most "progressive" elements of modernizing societies.

Conclusion

Theories of social change predict the demise of ethnic grouping. A major contribution of the study of African politics is to document the falseness of this prediction. Modernization and ethnic conflict do intersect, both empirically and intellectually. In this essay I have examined this point of intersection and tried to explain how the process of modernization can promote ethnic group formation.

It is precisely the rationality of ethnic behavior that has eluded the modernization theorists of both the sociological and political-economic persuasion. Because they failed to perceive the usefulness of ethnic organization, they failed to predict the efflorescence of ethnic competition. Ethnic organization is a means of organizing so as to attain the benefits of modernization; it is a form of coalition-building in the rational pursuit of specific objectives. In this conclusion I wish to amplify this argument with special reference to politics.

Political systems allocate resources. In particular, they make decisions regarding the location of wells, clinics, schools, roads, markets and other facilities. The location of such facilities determines who gets the benefits. If a clinic, for example, is located in district A, it is _not_ located in district B. In the parlance of game theory, the political process determining the allocation of such facilities is simple: the outcome is either 1 (if you win and get the project) or 0 (if you do not).[53]

In the competition for benefits, it is obviously useful for groups to be "bigger"; in most forms of political competition groups do better if they are larger in size. But it is also true that it is useful to limit the size of groups and to restrict the number of beneficiaries. Exclusion is desirable if the benefits are fixed in size; a budgetary allocation or a capital fund will produce greater benefits per capita if spread over fewer people. Exclusion is desirable even when the benefits are nonmonetary. If there are, say, a fixed number of classrooms or scholarships or jobs at a project site, then, once again, individuals do better if there are fewer claimants for these benefits. While there are forces which make it desirable to expand the number of people in a coalition competing for favorable allocational decisions, there are thus also strong forces promoting the restriction of these benefits. For the benefits are subject to dilution or crowding, and rational actors will therefore seek to confine them to as few people as possible.[54]

We have noted that because the incidence of the allocation of benefits and ethnic group membership are both a function of spatial location, it is natural that those seeking benefits should have preferences which are organized along "ethnic" lines. Further considerations promote such behavior. Institutions -- such as local governments, "traditional" political systems, kinship ties, markets and trading networks -- are likely already to exist in specific locations; their pre-existence reduces the cost of organizing. Moreover, the uniformity of language within groups and the difficulty of communicating across linguistic lines means that, for a limited set of resources, organizers will prefer intragroup organizing to organizing across groups. Not only the incidence of the benefits but also the incidence of the costs of organization therefore correspond to spatial boundaries.

As a consequence of these considerations, I argue, actors, in the rational pursuit of benefits, will organize competitive groupings and these will take ethnic form. Ethnic groups are, in short, a form of minimum winning coalition, large enough to secure benefits in the competition for spoils but also small enough to maximize the per capita value of these benefits.[55]

It should be noted that this analysis provides a rational grounding for one of the major interpretations of ethnic group behavior: that of situational analysis. Situational analysis stressed that ethnic groups often lack an "objective" basis. It also stressed that they are dynamic. Ethnic groups, it noted, were in fact sometimes "invented," forged out of cultural materials which had lain latent until mobilized in efforts to organize. Situational analysis also stressed that the boundaries of ethnic groups were subject to repeated redefinition. When opposed to another major group, then group membership would be defined inclusively; in other situations, the group would be internally divided and ethnic membership would then be more restrictively defined.[56] The analysis which I advance furnishes an explanation for the patterns highlighted by situational analysis. For depending upon the issue and the competitive arena, the size of the coalition which would insure maximum individual benefits to its members would vary; and different criteria of inclusion and exclusion would therefore be invoked as groups formed in efforts to reap maximal benefits in the competition for scarce resources. The dynamics underscored by situational analysis thus became predictable corollaries of an approach based upon the assumption of rational behavior.

What are the implications of this analysis? One, of course, is that ethnic competition will not disappear as modernization proceeds; the very purpose of this paper is to show the weakness of such reasoning. Nor will it decline as people are subject to higher levels of education; clearly, the behavior is not atavistic or "prerational" and so it will not diminish as people acquire more sophisticated training. Rather, it would appear, what is required is a form of correction which exploits the very properties which generate the phenomenon in the first place: the desire for benefits and the capacity to act rationally in pursuit of them.

The appropriate response, then, is one of institutional design. Efforts should be devoted to creating institutional environments which alter incentives so that persons organize coalitions of a different nature when in pursuit of their interests. Attempts should focus on exploiting the very nature of ethnic competition so as to channel and diffuse it.

Examples of such attempts come from Nigeria, where constitutional designers created new electoral rules in an effort to decouple the desire for office from the making of political appeals to ethnic consciousness. New states were created in an effort to reduce the expected value of attempts to organize regional blocs; as each state included a smaller proportion of the electorate and controlled a smaller proportion of governmental delegates, each had less reason to aspire to be winning in the competition for public benefits.

Moreover, the founding fathers of the new Nigerian republic required political parties to draw a fixed minimum proportion of votes in a designated proportion of separated and widely scattered jurisdictions. The purpose was to reduce the attractiveness of narrow ethnic appeals; such appeals, while mobilizing some, would alienate others and so reduce the capacity of the parties to fulfill their distributional quotas.

Current political events in Nigeria leave open the question of the effectiveness of these measures. Nonetheless, they underscore that those most affected by ethnic conflict recognize the rational wellsprings of ethnic behavior and seek to control it by designing institutional environments wherein persons, in the rational pursuit of their own best interests, will have reason to behave in ways consistent with the maintenance of political order.

In conclusion, in this essay I should note that I have taken what can be termed an instrumentalist view of ethnic behavior. Others, most notably Geertz, have established a consummatory interpretation.[57] Consummatory behavior would lead to ethnic conflict over issues other than material advantage. Moreover, it would lead to ethnic conflict even when it "didn't make sense," i.e., when the costs exceed the benefits to the parties concerned.

It is important to realize that both principles are at work. And it may well be the case that while instrumentalist behavior may lead to the formation of ethnic groups, consummatory behavior is important in explaining their persistence; as I have stressed, people who organize ethnic groups are often captured by the forces they set in motion. Nonetheless, for two major reasons, I feel that instrumentalist considerations are paramount. The first is that the consummatory model makes ethnic behavior constant, whereas as a matter of empirical fact ethnic behavior has been found to be variable. The second and related point is that ethnic behavior is controllable. Just as ethnicity can be organized, it can be _dis_organized. The rise and fall of ethnic consciousness marks the history of almost every nation-state; and efforts to create political institutions to contain, to ameliorate, and to defuse ethnic self-assertion mark the constitutional histories of many political communities. Theories of ethnic behavior based upon the power of primordial sentiments suggest the futility of such efforts. What should be suggested, rather, is their difficulty. For the rational component of ethnic behavior clearly exists and it has been and should be exploited so as to direct political choice-making into less explosive channels.

Footnotes

This paper is a further development of ideas initially developed in papers presented to the Program of Eastern African Studies of Syracuse University and published in _Comparative Political Studies_ (January 1974): 457-484.

1. See, for example, Stanley Greenburg, _Race and State in Capitalist Development_ (New Haven and London: Yale University Press, 1980); and Crawford Young, _The Politics of Cultural Pluralism_ (Madison: University of Wisconsin Press, 1976).
2. For discussion of the definitional problem of the term, see Robert Melson and Howard Wolpe, "Modernization and the Politics of Communalism: A Theoretical Perspective," _American Political_

Science Review, 64 (December 1970): 1122-1130; Pierre L. van den Berghe, "Introduction," in Pierre L. van den Berghe, ed., Africa: Social Problems of Change and Conflict (San Francisco: Chandler, 1965), pp. 1-11; Paul Mercier, "On the Meaning of 'Tribalism' in Black Africa," in van den Berghe, ed., Africa: Social Problems of Change and Conflict pp. 483-501; and Abner Cohen, Custom and Politics in Urban Africa (Berkeley and Los Angeles: University of California Press, 1969). To be noted is that to affirm the reality of ethnic groups is not to ignore their internal divisiveness. Conflicts between major segments and areal groups, between commoners and persons of royal blood, and between clans and villages -- all these cleavages do exist. Nonetheless, internal conflicts do not necessarily weaken the capacity of the larger ethnic groups to mobilize their members for collective purposes; the conflict-laden, internally divided Ibo are a case in point. The fact of internal division and conflict should therefore not be taken as evidence of the absence of effective ethnic collectivities nor discredit the validity of our enterprise.

3. One of the major social costs of modernization is the development of new stratification patterns, and thus inequality; another is the generation of new individual opportunities, and thus the promotion of individual self-interest at the expense of traditional social obligations. At the elite level, African Socialism, and its other variants, represent an attempt to speak to these problems. At the mass level, witchcraft and antiwitchcraft movements represent parallel attempts to deal with these problems. For analyses of the relationship between social change and attempts to control witchcraft, see the contributions in John Middleton and E.H. Winter, eds., Witchcraft and Sorcery in East Africa (London: Routledge and Kegan Paul, 1963).
4. Elizabeth Colson, "The Impact of the Colonial Period on the Definition of Land Rights," in Victor Turner, ed., Profiles in Change: African Society and Colonial Rule (Cambridge: Cambridge University Press, 1971), p. 194.
5. See Appendix 3 in Polly Hill, The Migrant Cocoa-Farmers of Southern Ghana (Cambridge: Cambridge University Press, 1963). See also the cases reported in L.T. Chubb, Ibo Land Tenure (Ibadan, Nigeria: Ibadan University Press, 1961); and Michael Twaddle, "Tribalism in Eastern Nigeria," in P.H. Gulliver, ed., Tradition and Transition in East Africa (Berkeley and Los Angeles: University of California Press, 1969), pp. 193-208.
6. See, for example, the discussion of the politics of Mbale contained in J.S. LaFontaine, "Tribalism among the Gisu," in Gulliver, ed., Tradition and Transition in East Africa, pp. 177-192; and the analyses of the politics of urban Nigeria contained in Richard L. Sklar, Nigerian Political Parties (Princeton: Princeton University Press, 1963).
7. Cohen, Custom and Politics in Urban Africa
8. P.C. Lloyd, "Tribalism in Warri," in West African Institute of Social and Economic Research, Fifth Annual Conference Proceedings (Ibadan, Nigeria: University College, 1956), pp. 78-87. See also J.M. Lonsdale, "Political Associations in Western Kenya," in Robert I. Rotberg and Ali A. Mazrui, eds., Protest

and Power in Black Africa (New York: Oxford University Press, 1970), pp. 589-638; and Dharam P. Ghai, "The Bugandan Trade Boycott: A Study in Tribal Political and Economic Nationalism," in Rotberg and Mazrui, eds., *Protest and Power in Black Africa*, pp. 755-770.

9. David J. Parkin, "Tribe as Fact and Fiction in an East African City," in Gulliver, *Tradition and Transition in East Africa*, pp. 273-296 and R.D. Grillo, "The Tribal Factor in an East African Trade Union," in Gulliver, *Tradition and Transition in East Africa*, pp. 297-321.
10. See Willard R. Johnson, *The Cameroon Federation* (Princeton: Princeton University Press, 1970); Chapter 10 of Gwendolen M. Carter, *Independence for Africa* (New York: Frederick A. Praeger, 1960); and C.W. Anderson et al., *Issues of Political Development* (Englewood Cliffs, N.J.: Prentice Hall, 1967).
11. N. Xydias, "Prestige of Occupations," in Daryll Forde, ed., *Social Implications of Industrialization and Urbanization in Africa South of the Sahara* (Paris: UNESCO, 1956), pp. 458-469; J. Clyde Mitchell and A.L. Epstein, "Occupational Prestige and Social Status among Urban Africans in Northern Rhodesia," *Africa*, 29, 1 (1959): 22-39; and J.C. Mitchell and S.H. Irvine, "Social Positions and the Grading of Occupation," *Rhodes-Livingstone Journal*, 38 (1965): 42-54.
12. R.E. Hicks, "Occupational Prestige and its Factors: A Study of Zambian Railway Workers," *African Social Research*, 1 (1966): 41-58.
13. Lloyd's articles are referenced and summarized in P.C. Lloyd, *The New Elites of Tropical Africa* (London: Oxford University Press, 1966); Dennis Austin, *Politics in Ghana 1946-1960* (London: Oxford University Press, 1964).
14. J.S. LaFontaine, *City Politics: A Study of Leopoldville, 1962-63* (Cambridge: Cambridge University Press, 1970); Lonsdale, "Political Associations in Western Kenya"; Twaddle, "Tribalism in Eastern Nigeria."
15. W.T. Morrill, "Immigrants and Associations: The Ibo in Twentieth Century Calabar," *Comparative Studies in Society and History* 5 (July 1963): 424-448.
16. T.M. Baker, "Political Control Amongst the Birom," in West African Institute of Social and Economic Research, *Fifth Annual Conference Proceedings*, pp. 88-94.
17. See, for example, the discussion in Fred G. Burke, "Political Evolution in Kenya," in Stanley Diamond and Fred G. Burke, eds., *The Transformation of East Africa* (New York: Basic Books, 1966); and Donald Rothchild, "Ethnic Inequalities in Kenya," *Journal of Modern African Studies*, 7 (December 1969): 689-711.
18. Aristide R. Zolberg, *One Party Government in the Ivory Coast* (Princeton: Princeton University Press, 1964; rev. 1969), p. 5.
19. A.A. Akiwowo, "The Sociology of Nigerian Tribalism," *Phylon*, 23 (Summer 1964): 162.
20. See, for example, the analyses by Peter R. Gould, "Problems of Structuring and Measuring Spatial Changes in the Modernization Process: Tanzania 1920-1963." Paper presented at the annual meeting of the American Political Science Association, Washington D.C., September 2-7; and by Edward W. Soja, *The Geography of*

Modernization in Kenya (Syracuse, N.Y.: Syracuse University Press, 1968).

21. Ibid.; see also Rothchild, "Ethnic Inequalities in Kenya," pp. 689-711.
22. James S. Coleman, Nigeria: Background to Nationalism (Berkeley and Los Angeles: University of California Press, 1958); and David B. Abernethy, The Political Dilemma of Popular Education: An African Case (Stanford: Stanford University Press, 1969).
23. Nelson Kasfir, "Cultural Sub-Nationalism in Uganda," in Victor A. Olorunsola, ed., The Politics of Cultural Sub-Nationalism in Africa (Garden City, N.Y.: Anchor, 1972), pp. 47-148.
24. Ibid., p. 76.
25. Josef Gugler, "The Impact of Labour Migration on Society and Economy in Sub-Saharan Africa: Empirical Findings and Theoretical Considerations," African Social Research, 6 (December 1968): 465.
26. Colson, "Impact of the Colonial Period on the Definition of Land Rights," p. 194.
27. See also the discussion in Coleman, Nigeria: Background to Nationalism, p. 59, and Chubb, Ibo Land Tenure, p. 25.
28. Lloyd, "Tribalism in Warri," p. 86; also, LaFontaine, City Politics: A Study of Leopoldville, 1962-63; and Lonsdale, "Political Associations in Western Kenya."
29. Lonsdale, "Political Associations in Western Kenya," p. 628.
30. Twaddle, "Tribalism in Eastern Nigeria," p. 197.
31. See, for example, Carl G. Rosberg and John Nottingham, The Myth of "Mau-Mau": Nationalism in Kenya (New York: Frederick A. Praeger, 1966), pp. 105-135.
32. Simon Ottenberg, "Improvement Associations Among the Afikpo Ibo," Africa, 15 (January 1955): 4.
33. Sklar, Nigerian Political Parties, p. 72.
34. George Bennett, "Tribalism in Politics," in Gulliver, ed., Tradition and Transition in East Africa, p. 80.
35. E.P.O. Offodile, "Growth and Influence of Tribal Unions," West African Review, 18 (August 1947): 937.
36. An excellent discussion is contained in LaFontaine, City Politics: A Study of Leopoldville.
37. Ottenberg, "Improvement Associations Among the Afikpo Ibo," pp. 15-16.
38. Relevant materials include Grillo, "The Tribal Factor in an East African Trade Union," pp. 297-321; Jean L. Comhaire, "Economic Change and the Extended Family," in van den Berghe, ed., Africa: Social Problems of Change and Conflict, pp. 117-127; and John C. Caldwell, African Rural-Urban Migration: The Movement to Ghana's Towns (New York: Columbia University Press, 1969). Particularly germane are the discussions of "the failure of class formation" in Africa which underscore the extent of the use of kinship networks to extract the benefits accruing to the more modern elements of society. Good examples are provided in P. Mercier, "Problems of Social Stratification in West Africa," in Immanuel Wallerstein, ed., Social Change: The Colonial Situation (New York: John Wiley, 1968), pp. 340-358; and in Arthur Tuden and Leonard Plotnicov, eds., Social Stratification in Africa (New York: Free Press, 1970).

39. Grillo, "The Tribal Factor in an East African Trade Union," p. 318.
40. Rothchild, "Ethnic Inequalities in Kenya," pp. 689-711. See also Gavin Kitching, *Class and Economic Change in Kenya* (New Haven and London: Yale University Press, 1980).
41. Abernethy, *Political Dilemma of Popular Education*, pp. 107-108.
42. See the discussion in Austin, *Politics in Ghana 1946-1960*; Crawford Young, *Politics in the Congo* (Princeton: Princeton University Press, 1965); Michael F. Lofchie, "The Zanzibari Revolution," in Rotberg and Mazrui, *Protest and Power in Black Africa*, pp. 924-967; and René Lemarchand, "The Coup in Rwanda," in Rotberg and Mazrui, *Protest and Power in Black Africa*, pp. 924-967.
43. Richard L. Sklar, "Political Science and National Integration -- A Radical Approach," *Journal of Modern African Studies*, 5 (May 1967): 6.
44. Victor G. Uchendu, *The Igbo of Southeast Nigeria* (New York: Holt, Rinehart and Winston, 1965), p. 9.
45. Mainza Chona, "Who is Responsible for Tribalism?" *Zambia News*, May 5, 1968.
46. For example, see Offodile, "Growth and Influence of Tribal Unions," p. 937, for a description of the levying of educational funds by the Ibo State Union; for an analysis of the contribution of urban dwellers to the construction of civic facilities in rural villages, see A.F. Hershfield, "Ibo Sons Abroad: A Window on the World." Paper presented at the annual meeting of the African Studies Association, Montreal, October 15-18, 1969. See also the discusson of the impact on rural development of financial contributions from urban dwellers in Caldwell, *African Rural-Urban Migration*, pp. 161 ff.
47. Lloyd, "Tribalism in Warri," p. 86.
48. Leonard Plotnicov, "The Modern African Elite of Jos, Nigeria," in Tuden and Plotnicov, eds., *Social Stratification in Africa*, p. 289.
49. *Africa Report*, 15, 4 (1970): 8-9; *Africa Diary*, 10, 3 (1970): 4788-4789; Ruth S. Morgenthau, "African Elections: Tanzania's Contribution," *Africa Report*, 10, 11 (1965): 12-16; and Ian Scott and Robert Molteno, "The Zambian General Elections," *Africa Report* 14, 1 (1969): 42-47.
50. R.E. Hicks, "Occupational Prestige and Its Factors: A Study of Zambia Railway Workers," *African Social Research*, 1 (1966): 213.
51. *Ibid*., p. 219.
52. Georges Balandier, "Traditional Social Structure and Economic Changes," in van den Berghe, ed., *Africa: Social Problems of Change and Conflict*, pp. 392-395.
53. More formally, a game is simple if V(B) = 0 or 1 for all B P(N), where B is a coalition or subset of N, the set of all players; where P(N) is the power set of N; and where V is the characteristic function.
54. The process, in short, is subject to decreasing returns to scale. More formally, if for all $T \subseteq S \subseteq N$, then a game is

subject to decreasing returns to scale if $\frac{V(S)}{|S|} < \frac{V(T)}{|T|}$.

55. More formally, where the competition for spoils can be represented as an N person game in characteristic function form, where we restrict ourselves to the space of coalitionally rational payoff configurations, and where payoffs are subject to decreasing returns to scale, then, if a coalition is winning, it is minimally winning. See Richard McKelvey and Richard Smith, "A Comment on the Debate Over Riker's Size Principle," unpublished, and William Riker, *The Theory of Political Coalitions* (New Haven and London: Yale University Press, 1962).
56. See Anderson et al., *Issues of Political Development*; also, Young, *Politics of Cultural Pluralism*.
57. Clifford Geertz, "The Integrative Revolution: Primordial Sentiments and Civil Politics in the New States," in Clifford Geertz, ed., *Old Societies and New States* (New York: Free Press, 1963).

10 Collective Demands for Improved Distributions

Donald Rothchild

Group disparities and unequal exchange are, in and of themselves, insufficient to explain the course of interethnic conflict. Collective inequalities may indeed produce inconsistencies and conflict situations, but they tell us little about the manner in which public choice is conditioned. To learn more about the political dynamics of conflict and conflict management, then, it is necessary to gain an understanding of the felt dissatisfactions of ethnic and subregional interests as well as the claims these groups make upon public authorities.

The way in which public authorities respond to these group demands is critical to the well-being -- and even the survival -- of the political system. Where geoethnic or racial interests are frustrated in their ability to channel demands effectively or gain minimal satisfaction for their legitimate claims, they may turn inward to despair, migrate to a new environment, or go outside the political system to articulate their pent-up anger through violent behavior. Yet, as shown by the 1979 race riots in Liberia, such shock treatment may come to be considered as an alternative means of articulating demands which the authorities will have to deal with over the long term.

Thus collective demands for political, economic and social equity must be viewed as a natural and indispensable part of the political process, not as evidence of a malfunction or a negative attitude. In this, it is possible to make a critical distinction between negotiable and nonnegotiable demands, i.e., between those demands, substantial or not, which can be accommodated by the political system, and those threatening to the system, which exceed what its decision elite is prepared to consider seriously. The former are elastic and realistic in expectations, accepting of the legitimacy of the political order and working within it to effect change. The latter are inelastic and possibly unrealistic in expectations, questioning -- and possibly rejecting -- the validity of the state's authority and seeking extrication from some aspects, at least, of state regulation. Because nonnegotiable demands pose a fundamental challenge to the integrity of the state, such claims, particularly those to autonomous authority, appear as unreasonable and highly menacing, possibly bringing on a refusal by state authorities to negotiate. The response of federal Nigerian official spokesmen to the secessionist demands of Biafran leaders is indicative. "The disintegration of Nigeria," they asserted, "is not negotiable."[1] Some secessionist demands may, of course, be no more than rhetorical statements; they may well represent a tough stance which, in reality, is little more than a prelude to resolute bargaining. Yet, to the extent that secessionist leaders (in, for example, Katanga, Eritrea, Biafra) have pressed ahead in a determined manner with nonnegotiable demands, inconsistent objectives, ideological polarization, and a destructive process of conflict escalation have ensued.[2]

Though the potential for destructive conflict is ever-present in interethnic relations, it is vitally important at the same time to recognize the full potential for constructive social interaction. This point is often left in obscurity. In particular, those focusing on ethnic self-determination tend to view ethnic group claims in inelastic terms. Demands are described as fundamental in nature, requiring separate statehood in most instances for full satisfaction.[3] Such an externalist view of the self-determination process seems to me to be much too restricted with respect to the process of formulating and communicating ethnic demands,[4] for the ethnic claim has been reduced primarily to a nonnegotiable one. Noting that ethnic groups are frequently content to urge the granting of reasonable demands is no mere quibble. To view ethnic conflict in terms of nonnegotiable exactions is to make the issues at stake, and hence the solutions, needlessly extreme and fundamental.

Clearly, the possibilities for the management of ethnic conflict are significantly enlarged by a knowledge of the wide range of demands pressed upon public authorities. In order to illustrate the breadth of these demands, the stated claims of elites will be supplemented by survey data which compares the felt dissatisfactions and demands of different subregions within the same state. A focus on the statements of elites places great stress on cultural autonomy, political power sharing protection, and territorial separatism, as well as distributive demands. On the other hand, a focus upon the attitudes of the general public shows claims largely emphasizing the distributive side, especially in relatively open political systems. Moreover, it is apparent, from an examination of public opinion surveys I conducted in Kenya, Zambia and Ghana, that subregional demand and expectation patterns vary significantly between the relatively advantaged and relatively disadvantaged subregions.

After outlining the six categories of major demands evident in middle Africa, this chapter will concentrate upon general public dissatisfactions over distributive issues and upon the critical roles that political elites play in shaping these dissatisfactions into effective demands for communication to the center. As Jonathan S. Barker observes, "Rural, urban, and ethnic discontent do not spontaneously unite";[5] the task of organizing these dissatisfactions for presentation to central decision elites must necessarily fall on their spokesmen. This examination of reasonable claims within the system should be read in conjunction with the chapter in this volume by Crawford Young on "Comparative Claims to Political Sovereignty," which takes into account the nonnegotiable dimension.

The Range of Demands

Although negotiable and nonnegotiable demands represent two basic tendencies in the policy process as it affects interethnic relations, it is not always easy, in real-life situations, to characterize specific collective demands under such rubrics. The extent to which claims upon central authorities can be accommodated must be determined separately for each situation and time period. For example, a demand for favorable public allocations, say on the siting of new industries, may be negotiable in one time-place context; in another, where conditions of scarcity or values on proportionality and justice change the context, the problem alters and the dimensions of conflict deepen. As subregional stakes increase and the interest struggle intensifies, the spirit of accommodation over this seemingly negotiable issue on distributing resources becomes stretched, and it passes, almost unnoticed, from the category of negotiable to nonnegotiable demand. Similarly, autonomy demands, which could lead to the unraveling of the political community if not handled in a statesmanlike manner, have become the subject of exchanges in the Sudan and Nigeria, without entailing a challenge in either case to the system's identity or survival. In these instances, agreement to establish separate subregions within the state had the effects of reducing the likelihood of secession, of strengthening central leverage and capacity, and of redirecting conflict along more manageable lines. Moreover, population transfers may be a negotiable matter in one context (Soviet-West German exchanges affecting the ethnic Germans); in another, they may prove nonnegotiable (the expulsion of Asians from Uganda). In brief, then, the dichotomy between negotiable and nonnegotiable demands is present in all types of interethnic contacts and management strategies, requiring us to supplement them with some crosscutting categories.

A wide range of demands is indeed convergent with these two basic tendencies of negotiation and nonnegotiation: these include demands for enhanced distribution, representation and participation; cultural autonomy; political and economic protection; territorial separatism and independence; and counterdemands, that is, opposing claims which grow out of and are responsive to all the various demands for change on the part of ethnically defined interest groups. When these demands are examined, it is apparent that distributive demands, built largely upon the drive for collective benefits, are generally distinguishable from the others, grounded as they are upon a call for various forms of collective self-determination and human rights. As Richard Rose and Guy Peters assert, the language of collective rights is less the language of negotiation than that of obligation[6] and hence applies more meaningfully to demands for change which are fundamental in nature.

Here a further distinction -- between claims to rights within the state (the internal dimension) and rights outside of the state (the external dimension) -- is pertinent. The distinction has important implications for the intensity of conflict and, consequently, for the relevant policies on managing conflict. The negotiable possibilities of those internalist demands for representation and participation, cultural autonomy, and political and economic protection make reconciliatory outcomes a realizable objective. Representation and participation demands can be met, to a degree, by the adoption of various power sharing mechanisms; calls for cultural autonomy can be partially satisfied by placing restraints on assimilationist policies and by enhancing the permissible area for group participation (as in the Sudan); and claims to political and economic protection can be conceded, temporarily at least, to ensure political stability and economic growth (Sierra Leone's constitution of the late 1960s reserved twelve of seventy-eight seats in parliament for chiefs, and Zimbabwe's independence constitution allotted ten of 100 parliamentary seats for election by voters on the white voters' roll).

In the case of such externalist demands as those for territorial separatism and independence, however, appeals for change are not easily negotiable. Where the geoethnic group clusters around a given territory within the state, the possibility that its demands for change may prove nonnegotiable and lead to the splitting of the original state remain very real. Quite possibly, these demands for full separation and autonomy will be resisted by the counterdemands of smaller minorities to be included in the newly created state. This became quite evident in the two non-Ibo areas of the former Eastern Region of Nigeria. Yet the logic upon which the majority collective of the seceding geoethnic section operates is the uselessness of any reconciliatory efforts to protect its interests within the present state set-up. Precisely because the demand for secession is a "radical" claim involving the destruction of the established state, it "must be based upon definite and substantial grievances," and can only be invoked in earnest after all other compromises have proved futile.[7] Conflict theorists may argue cogently for greater tolerance toward "political divorce,"[8] but in practice Third World decision-makers are not likely to be more prepared to preside over the dissolution of their countries than their Western (United States or Canada) or Eastern (USSR) counterparts have been.

The catch-all category of counterdemands applies across the board to all types of demands made upon the political authorities. These can be distributive and involve a resistance to pressures for redistributive public policies. They can also take the form of opposition to more basic changes within the state structure. The determination to maintain a system of unequal recruitment, extractions, or allocations rests upon a will to counteract contending forces insisting upon transformed intergroup relations. More fundamentally, these counterclaims can entail a defense of political, economic, or cultural subjection. Such counterclaims, most striking in the case of South Africa's apartheid, may include both "a negative defensive aspect" that urges retention of the current structure of domination, and a "positive revolutionary aspect" that seeks to restructure society in such a fashion as to perpetuate racial or ethnic subjection, for example, by policies of displacement (expelling, repatriating, or regrouping populations).[9] Finally, counterdemands cover those claims, public as well as private, which resist the dismemberment of the state. It was in this vein that the federal government of Nigeria and the central regimes of the Sudan, Ethiopia, and Chad jealously guarded against efforts by third parties to negotiate their disputes with geoethnic secessionist movements at the periphery. They considered a willingness to enter into an exchange relationship tantamount to a move toward the recognition of the seceding party's legitimacy. The consequences of such a stance did not bode well for the peaceful management of conflict. As three social scientists conclude pessimistically, "Disputes caused by efforts at territorial secession [in Africa]...have never been managed successfully just because mere OAU intervention would suggest endorsement of some of the rebels' claims."[10]

In sum, then, it is not accurate, as some social scientists are inclined to do, to describe ethnic demands solely in terms of intense, nonnegotiable relationships. When these collective demands are disaggregated, what emerges, in fact, is an array of claims striking in its diversity. And not only do these demands vary widely in their intensity and scope, but the groups advancing these demands exert different kinds of pressures as circumstances alter over time. Nigeria's Ibo-led Eastern regional government changed its attitude between 1958 and 1964 on the guidelines to be employed in allocating revenue, largely as a reflection of economic activities, especially in the oil industry. Whereas Eastern regional spokesmen urged the criterion of subregional need on the Raisman Commission in the earlier period, they placed a higher priority on derivation as petroleum revenues rose markedly in the 1960s. By 1964, as production trends in the oil industry became increasingly obvious, the government of eastern Nigeria sought to enhance its controls over revenues generated in its area; thus it submitted a memorandum to the 1964 fiscal review commissioner stating that the system of revenue allocation then in operation had proved "extremely unreasonable, unfair and inequitable" to the subregion and that all revenue from royalties and rents on oil should be returned to the subregion of origin.[11] In another instance, changing attitudes in Kenya's dominant European minority became apparent in the 1920s on the issue of East African federation; the feelings of insecurity which gave rise to European championship of territorial enlargement subsequently led to an abandonment of the idea by the dominant Europeans when the British government rejected the interlinked demand for an unofficial European majority on Kenya's legislative council.[12]

Like the fluid collectivity of persons interacting to make up an ethnic people, then, the claims advanced by such interest-defined groupings seem likely to shift as the groups' perceptions of interest turn about. Where such switches in demand patterns lead to a polarization between groups and to conflict escalation, then the process of managing conflict will likely be complicated; contrariwise, and this is the point so often missed, these switches can also lead to an acceptance of the prevailing rules of the political system and to a perception on all sides that the greatest benefits are likely to follow from a struggle to maximize group interest from within that system. In the latter case, destructive conflict gives way in time to constructive conflict, and collective demands are adapted to the workings of a "bargaining culture."[13]

The General Public's Felt Dissatisfactions

In most middle African countries, scarcity gives rise to a struggle over how public rewards and resources are to be distributed. In this competition, the geoethnic and racial collective, as a distinct category of society united by a sense of common fate, plays an important part in, and is itself validated by, its usefulness as an instrument of such action. "Ethnic groups," as Robert H. Bates observes, "persist largely because of their capacity to extract goods and services from the modern sector and thereby satisfy the demands of their members. Insofar as they provide these benefits to their members, they are able to gain their support and achieve their loyalty."[14] Irrespective of ideology or political structure, the ethnic connection is likely to survive adverse criticism and exhortation -- so long, at least, as such a linking process produces the kinds of economic and political as well as psychological benefits sought by group members.

It is as one examines collective demands more closely that noticeable distinctions emerge between the "class" interests of group members and elite. If Richard L. Sklar is on sound ground in describing class formation as "determined by relations of power" -- political as well as economic[15] -- then it is possible to make an important distinction between group members expressing their felt dissatisfactions and elite brokerage elements who actually communicate the general public's discontents into the political process. As will become clear from an examination of the political practices and statements of elites and the author's survey data on the public's feelings of deprivation, the two constituencies differ in terms of associations, access, and interests. Members of the dominant class act as constituency ambassadors or "delegates" to the central government "patron,"[16] their status being dependent upon continued local support. Such local backing certainly requires a basic degree of sensitivity on their part to the felt dissatisfactions of local constituents -- particularly those in the relatively advantaged areas close to the primary political centers. Spokesmen in such areas tend to be the more active and persistent communicators and to be relatively effective in gaining preferred access to governmental decision-makers. Irrespective of an area's level of development, however, the extent of the gap between the public's felt dissatisfactions and the dominant elite's selection of which dissatisfactions to press upon government is critical to the system's general responsiveness, and, in a related way, to its long-term stability.

How, then, is one to acquire a comprehensive picture of public needs and wants? If the existence of hegemonial decision systems in many parts of middle Africa tends to complicate the communications process and to give observers a somewhat incomplete record of publicly expressed demands, then it becomes necessary to turn to other techniques to gain an insight into the general population's priorities and expectations. Hence, the social scientist makes use of survey data. In this instance, I am fortunate to have secured access to data from Kenya (1966), Zambia (1969-70), and Ghana (1973).[17] To be sure, these surveys suffer at least as much as any administered in Third World settings from the drawbacks of insufficiency, lack of precision, and inadequacies of interpretation and validation; yet, for all these limitations, the data they yield -- imperfect and tentative as they may be -- represent a distinct advance over untested notions too facilely advanced regarding the contemporary African scene.

Although the specific public expressions of neglect and dissatisfaction were largely situational, reflecting the particular problems and constraints encountered by each of the countries surveyed, a broadly similar emphasis on immediate conditions seems evident. Not surprisingly, in light of colonial disregard and contemporary poverty, respondents concentrated on the need for concrete measures for improving their areas. Yet two general findings do emerge which seem of particular relevance to the examination of intergroup conflict. First, there was a tendency, especially pronounced in the Ghanaian and Zambian findings, for relatively disadvantaged areas to insist _less_ intensely upon the allocation of expensive public goods and services than was the case for their relatively advantaged counterparts. And second, there was evidence of a latent sympathy for a redistribution of scarce resources on the part of the relatively advantaged as well as the relatively disadvantaged subregions, most notably in the Kenya survey. The implications of these findings for the management of intergroup conflict will be discussed after we first set out the data on patterns of collective discontent.

The Kenya survey, based on 653 interviews conducted in all parts of the country, was administered during the period that the Kikuyu, the largest, most economically advantaged, and most politically powerful people, were consolidating their hold on the life of the country. If critics assailed "Kikuyu-Luo" domination around the time of independence in 1963, the focus of attack by 1966 was more and more centered on the Kikuyu alone. In fact, wariness of Kikuyu hegemony became so pronounced that a new term, "Kikuyuization," came into use during this period.[18] A look at comparative data on the distribution of social services by province in 1970, moreover, shows (predominantly Luo) Nyanza province considerably behind (mainly Kikuyu) Central province. Whereas Nyanza province, with 19.4 percent of the total population, had 13.1 percent of total secondary school enrollments, Central province, with 15.3 percent of the total population, had 22.9 percent of total secondary school enrollments in the country. Nearby Nairobi, with 4.4 percent of the population, had 18.7 percent of secondary school enrollments (many of those Kikuyu). Nyanza province received 1.2 percent of all National Housing Corporation expenditures in 1970; Central province secured 15.1 percent of these loans (and neighboring Nairobi was the recipient of 65.2 percent of these expenditures). Finally, there were 1,269 people per hospital bed, and 2,219 people per medical practitioner in Nyanza province, while in Central province and adjacent Nairobi, there were 766 and 152 people respectively per hospital bed, and 1,287 and 84 people respectively per medical practitioner. In addition, except for access by the people of Coastal province to hospital and medical facilities (no doubt, partly attributable to the presence of many elite Europeans and Kikuyus), all the other provinces were like Nyanza in the way that they trailed Central province's access to resources.[19] As a consequence, it is not unreasonable to classify the Kikuyu (36 percent of the survey sample) as the relatively advantaged and to include the following groups as among the relatively disadvantaged: the Luo (20 percent of the total sample), the Kamba (9 percent), the Luhya (17 percent), the Embu/Meru (8 percent), the Coastal people (3 percent), and the Kisii (5 percent).

Survey findings were quite revealing as to the specific nature of Kenyan dissatisfactions, but less so as to differing perceptions on the part of the relatively advantaged and disadvantaged. When respondents were asked to describe the ways in which smaller ethnic peoples had been neglected in the past, they were clear in placing primary emphasis on the role of poor educational facilities (59 percent of the total sample of interviewees). Less significant areas of neglect included the following: poor medical services and facilities, 27 percent; lack of communications or transport, 24 percent; failure of the government to give or to develop the land, 11 percent; inadequate provision of agricultural or land management training, 10 percent; and insufficient representation in government jobs and political representation, 9 percent. In light of public reports circulating in Kenya at the time, these felt dissatisfactions on the part of the public regarding political recruitment appeared generally on the low side, possibly reflecting a perception of priorities on the part of the general public which differed from that of the dominant class.

Somewhat surprisingly, however, responses with respect to the neglect of the smaller ethnic peoples of Kenya did not show any statistically significant differences at that time as among ethnic peoples. In fact, in a number of instances (i.e., the availability of medical care, education, land management training, and land development), the relatively advantaged Kikuyu respondents showed more sensitivity to neglect in the periphery than did the less advantaged Luo, Kamba, Luhya, Embu/Meru, Kisii, and Coastal peoples. To be sure, a higher proportion of Kikuyu than non-Kikuyu respondents rejected the entire proposition that the smaller peoples had been neglected in any way; yet even in this somewhat unique instance, the differences between the relatively advantaged and disadvantaged were by no means as great as might have been anticipated. As the Luo, Luhya, and other ethnic groups advanced and became more assertive in the 1970s, however, observers perceived a shift in attitudes upon the part of the dominant Kikuyus toward defensive counterdemands; even so, the general sensitivity of their responses to the 1966 survey is a possible indication of a political culture conducive to a process of tacit exchange, and, quite possibly, governmental policies of subregional distribution.

To contend that the relatively advantaged respondents of the 1966 Kenya survey do not express deep opposition to reallocative policies is not the same thing as maintaining that the less advantaged will press for added public distributions for themselves or that government will provide a much-needed lead in this direction. What, then, does the Kenya survey reveal about public attitudes on the desired course government should pursue in facilitating corrective equity among subregions? Certainly there was considerable sentiment throughout Kenya in support of government action to equalize subregional disparities: 71 percent of the total sample agreed to reduce, if necessary, the allocations to the more advantaged in order that government might spend more money on building secondary schools for the less advantaged peoples. Support for increased government spending on industries, water, and electricity in the less advantaged subregions was less intense, but even in this instance, 60 percent of respondents concurred with the need for such reallocative measures.

Again, no statistically significant variation emerged when these responses were cross-tabulated by ethnic group.

The responses to another question in the Kenya survey give an indication both of the public's felt dissatisfactions and its priorities as to government action. Replying to an open-ended question as to what government might do to help the less advantaged geoethnic peoples to improve their conditions, respondents from all areas put the following in the foreground: improving health facilities; building adult literacy centers and sending teachers to the less advantaged localities; initiating self-help schemes (i.e., cooperatives and trade centers); assigning experts to the less advantaged areas and residents of these parts to the more advantaged areas for learning purposes; starting irrigation and land improvement schemes; improving roads, communications, and transport; and training people in land development and agriculture generally. Although cross-tabulations on the responses by ethnic group again demonstrated broadly similar attitudes among the relatively advantaged and disadvantaged, it was nonetheless apparent that the relatively disadvantaged were more insistent than were the relatively advantaged that the government should do more to help the smaller peoples gain greater access to education and to economic and political positions. Whereas 33 percent of the relatively advantaged Luo/Kamba/Luhya/Embu-Meru/Coast/Kisii sample of 403 respondents pointed to the need for government to build adult literacy centers and send teachers to their areas, only 20 percent of the relatively advantaged Kikuyu sample of 213 did likewise; similarly, 6 percent of the respondents from the relatively disadvantaged areas spoke of a need to give their people more political and economic representation, while a mere 3 percent of the relatively advantaged mentioned this matter. The appeals of education here were clearly a recognition of its central role in the presentation and articulation of collective interests in the political process. "In the modern world," as the _KANU Manifesto 1979_ affirms in Kenya, "education must remain the key to human fulfillment."[20]

In the case of Zambia, the survey showed the respondents of a relatively disadvantaged subregion also to be broadly concerned over their immediate conditions. The particular disadvantages labored under in the rural areas is indicated by a survey of rural nutritional needs and priorities conducted by the National Food and Nutrition Commission of Zambia in the Northern Province in 1969-70, and coded and tabulated by this author. Fourteen village interviews were conducted in each of the following areas: Lake Tanganyika, Bangweulu Swamp, Chambeshi Valley, Luangwa, Isoka, and Plateau (Mporokoso). After a short pretesting, interviewing teams were sent to the villages, chosen on the basis of the population and random sample. Contact was made with the local ward councillors and district secretary before the survey was conducted in each village. Each team was composed of four Zambians, one Zambian team leader, and one survey supervisor. The village report was written up by the survey supervisor in conjuction with the team leader and village headman.

A picture of isolation and neglect, carried over as part of the colonial inheritance, emerges. Bad roads made many villages inaccessible to governmental services and facilities as well as to essential marketing outlets. In addition, a quarter of the villages were found to be between three and nine miles from such a road. Only 5 percent had an operating market within two miles, making trade difficult and costs high. Children were forced to walk long distances to school (36 percent of respondent villages were five or more miles from the nearest school), and even the most hardy and determined pupils found their opportunities limited by the fact that schools often went no further than Grade Four. Those (particularly among the girls) fortunate enough to have parents willing to finance boarding elsewhere often had to travel long distances to find adequate facilities. And finally, the most rudimentary clinical facilities were a long walk for the average villager; roughly half the villages were ten or more miles from any kind of medical care. The figures become more poignant when it is realized that they are calculated on the basis of the most direct routes (not necessarily roads), that, in a number of cases, they involve travel over bad roads and by canoe, and that the nearest clinic can frequently administer only the most rudimentary kinds of treatment, necessitating further travel to a better-equipped and better-staffed dispensary or hospital. Such lack of opportunity in the rural areas has increasingly gained recognition among villagers, and, not surprisingly, they have voted with their feet by migrating to a new (though often disappointing) life in the urban shanties.

In answer to a direct question as to what most people in the village wanted, a total of 84 respondents replied by detailing a long list of precise needs. Most notably, these included: establishing clinics (54 respondents); founding a nearby school (27); developing shops and marketing facilities (41); upgrading the water supply (54); repairing roads (14); improving housing (15); organizing women's clubs and community centers (25); instituting agricultural schemes (16); creating homecraft centers (20); and erecting grinding mills for cassava and corn (13). Quite notably, the emphasis on education did not stand out as pre-eminently as was the case with the Kenya sample; rather, education shared a leading place with calls for improvements in health, water, and marketing facilities. It was apparent, moreover, that with the Kaunda government's emphasis upon administrative decentralization and rural development, the villagers have become increasingly conscious of their deprived situation. But this awareness of neglect has been more gradual than many realize -- particularly in the less advantaged areas. To make use of a hypothesis advanced in another context: the slower the rate of modern achievement, the smaller the feedback effect, the less thorough the "revolution of rising expectations."[21] In this instance, villagers in the Northern Province, above all those in the more isolated communities, frequently told interviewers of their basic contentment with their condition. They had no precise idea of what the village needed and in certain instances had never made any requests of the government. Some of those who had petitioned government were disappointed over what they felt to be a lack of response. They told of government help being promised but not materializing and felt neglected without knowing how to remedy the situation. To the extent that such village demands were presented by United National Independence Party officials to their superiors at district, provincial, and national levels, the villagers exerted some influence on government leaders. Yet by comparison with the influence exerted by the business elite and the urban workers, their political influence could not be anything but modest. Such a lack of political clout has the effect of reducing pressure on government, enabling it to pursue policies on corrective equity which "satisfice" (i.e., satisfy and suffice) rather than to maximize benefits.[22]

Finally, in Ghana, a focus upon the attitudes of the general public showed a variety of claims emphasizing the distributive side, with the peoples in the relatively advantaged subregions making heavier demands upon scarce resources than did their less advantaged neighbors. A "General Public Questionnaire," conducted by the Department of Political Science of the University of Ghana, Legon, in the summer of 1973, included a total of 2,122 interviews held in eleven local council areas. Although interviews were completed in all the subregions of the country, the comparison here will be restricted to two of the more relatively disadvantaged (Northern and Upper) on the one hand, and three of the more relatively advantaged subregions (Western, Ashanti, and Greater Accra) on the other. Such categorization by comparative advantage and disadvantage seems warranted at the outset by aggregate statistics on the key variable of school attendance. Whereas every area of Northern and Upper regions (except Tamale) showed less than one person in 25 who were over 15 years of age as having had any formal school experience, a similar sample in the administrative centers of Accra, Sekondi-Takoradi, and Obuasi revealed over one person in four having had some schooling.[23] Aggregate differences among subregions can also be shown with respect to medical and hospital facilities, access to telephones, consumption of electricity, small industry development, and other key variables; however, it should be kept in mind that behind these statistics on aggregate subregional disparities lie great internal differences within the subregions themselves.[24] This accounts in part for the different conclusions reached by scholars on the developmental ranking of the subregions; another critical factor, of course, is the variables they have used in coming to their conclusions. Classifying Ghana's subregions on the basis of most to least developed, K.B. Dickson puts forward the following list: Greater Accra, Ashanti, Eastern, Central, Western, Volta, Brong-Ahafo, Northern and Upper.[25] Kodwo Ewusi, however, has grouped the subregions as follows: more developed, Greater Accra; developed, Western and Central; developing, Eastern, Ashanti, Brong-Ahafo, and Volta; and less developed, Upper and Northern. Ewusi has also presented the following table setting out measurements of development for the subregions with Accra as a base and ranked from the most developed to the least developed: Greater Accra, 1.000; Central Region, 0.398; Western Region, 0.392; Eastern Region, 0.355; Ashanti Region, 0.340; Volta Region, 0.306; Brong-Ahafo Region, 0.265; Northern Region, 0.110; Upper Region, 0.071.[26]

As was the case with the Kenya and Zambian surveys, the Ghanaian respondents expressed discontent mainly with their immediate conditions. When respondents were asked to specify which services the local council should provide that it was not already furnishing, they gave considerable insight into the patterns of felt dissatisfactions on a subregional basis. In Northern and Upper Regions, these openly articulated discontents were concentrated upon a rather limited number of service needs (perhaps accounted for by a high percentage of "don't know," did not answer, and not applicable responses in these areas). Respondents in the Northern Region stressed sanitation (13 percent), market stalls (9 percent), and piped water (20 percent); those in Upper Region pointed to piped water (17 percent) and roads (9 percent). In the more advantaged areas, however, felt dissatisfactions were more diversified and more inclined to point toward improvements of a costly nature. People interviewed in Ashanti asserted the need for such services as market stalls (8 percent), latrines (19 percent), piped water (22 percent), electricity (8 percent), and hospitals (11 percent); those in Western Region called for the provision of sanitation (8 percent), piped water (11 percent), and entertainment facilities or services (12 percent); in Greater Accra, a desire was expressed by substantial numbers for market stalls (26 percent), latrines (13 percent), piped water (13 percent), and hospitals (19 percent). The significance of the Accra dissatisfaction over market stalls was pointed up even more sharply by another question which specifically raised the issue of how well the local council does in providing this service. Whereas 43 percent of respondents in Upper and 60 percent in Northern voiced some measure of discontent here, the percentages rose, in the more advantaged subregions, to 62 percent in Western, 79 percent in Ashanti, and 98 percent in Greater Accra.

In order to check the findings on the felt dissatisfactions of the Ghanaian general public, it is possible to make use of another question posed to the interviewees. When respondents were asked to specify which aspects of life they most disliked about living in their village or town, they again emphasized such factors as lack of sanitation, hospitals and clinics, roads, schools, and so forth. Subregional variations were most pronounced, however. Not only were those residing in relatively deprived Northern and Upper regions more reluctant to point to their major discontents, but their stated dissatisfactions were concentrated on a few items: in Northern Region they focused upon the physical environment (11 percent) and the high cost of living (7 percent), and in the Upper Region, upon lack of sanitation (32 percent), criminals and low morals (8 percent), and the high cost of living (7 percent). By contrast, respondents in more advantaged Western, Ashanti, and Greater Accra were less guarded about their dislikes and included a wider array of significant items of discontent among their answers. Interviewees in Western Region gave importance to such dislikes as the high cost of living (16 percent), and inadequate hospitals and clinics (12 percent); those in Ashanti Region set a high value on such aspects as high cost of living (31 percent), lack of sanitation (35 percent), and hospitals and clinics (8 percent); and, finally, those in Accra reckoned the following as significant: lack of sanitation (26 percent), hospitals and clinics (26 percent), roads (15 percent), and schools (17 percent). Clearly, the more advantaged the subregion, the more likely was the respondent of that subregion to demand improvements in a wide range of relatively expensive services or amenities.

In conclusion, the surveys of Kenya, Zambia and Ghana generally indicate public dissatisfactions in these middle African societies to be specific and distributional in nature. Some subregional variations were evident, however. Contrary to popular assertions on a "revolution of rising expectations," the collective dissatisfactions expressed and the claims put forward to governmental authorities by the peoples in the relatively disadvantaged subregions included a narrower range of relatively expensive services or amenities than was the case with the relatively advantaged.[27] Certainly, conditions in the relatively advantaged areas were anything but satisfactory for the great majority of their lower-class inhabitants. As was the case for our less advantaged subregions, Joan M. Nelson found the more recent, temporary migrants to urban centers to be less organized and less inclined than the more settled migrants to press a wide range of demands upon ethnic leaders.[28] Even so, the heightened political and social awareness of the members of the dominant political class from the relatively advantaged subregions as well as their proximity and access to decision-makers gave them an added advantage in the competition for the benefits of modernity.

Political Elite Demands for Improved Distributions

If the attitudes of the general public show the existence of felt dissatisfactions directed largely toward distributive issues, the statements and actions of ethnic group spokesmen suggest a greater variety and sophistication of demands, running all the way from the relatively low intensity politics of distribution to the relatively high intensity politics of political independence and territorial separatism. Thus a gap emerges between the public's "latent articulations"[29] and the dominant elite's selection and shaping of these dissatisfactions prior to channeling them into the political process. It is the dispersion of rural interests which accounts in part at least for government's ability to get by with a "satisficing" strategy. As Joel S. Migdal asserts, "The more divided the peasants, the less they can expect to have a class impact on state policy in return for their support."[30] Consequently, it is the collective brokers, the various formal and informal delegates from the constituencies at the center, who are in a position to make effective demands on the core decision elite. As members of the dominant class themselves, these political entrepreneurs can manipulate political symbols so as to secure resources for local development. Hence what the members of this class set as its primary values is critical to the manner in which scarce resources are distributed.

Certainly the constituency brokers (i.e., the parliamentarians, councilmen, traditional authorities, local notables, leaders of urban-based tribal unions, executive officials and administrators who rest their political base to some extent upon continued constituency support) are limited in the extent that they can diverge from the sentiments of the general public. Even in hegemonic systems which seek most energetically to enclose conflict and to block off discordant public messages, these intermediaries are not likely to achieve their goals or to survive for long without demonstrating minimal responsiveness to public wishes and expectations. In General I.K. Acheampong's Ghana, for example, government's unawareness of and inaction on the Northern rice growers' demands for higher prices and improved agricultural inputs led in 1975-76 to protests, lower production, hoarding and smuggling. The attendant effects of this Northern Ghanaian resistance upon the performance of the political system regarding food availability and the country's foreign exchange position were significant indeed.[31]

Although no regime, not even a military "command" regime, can in practice afford to close its avenues of access to the public, the dominant class (at the core and the periphery) nonetheless has considerable flexibility in managing demands and in setting policies designed to cope with these demands. The skill of this decision elite in arriving at a balance between state needs and legitimate public expectations is critical to achieving systemic growth and self-determination. The challenge is particularly exacting for the constituency brokers. At one and the same time their survival depends on their sustained linkage to dominant class interests as well as their ability to act (or be seen to act) as effective champions of geoethnic or racial claims. They must assure the growth of state capability even while demonstrating an ability to secure the benefits of modernity from central treasuries.[32] This involves a double personality role -- one which is likely to be misunderstood by constituents or by dominant class associates at any time.

In making their claims on the center, the collective brokers inevitably advance many of the same distributive claims put forth by their general public supporters. In classic interest group style, they utilize informal as well as formal channels, seizing upon whatever fleeting opportunities avail -- contacts with the bureaucracy, encounters with high-ranking officials touring the outlying areas, debates in parliament, possibly even threats of noncooperation -- to make their appeals for central largesse. The experiences over time of two closely linked subregions, the relatively disadvantaged Northern and Upper regions in Ghana, are instructive of the process of articulating political elite demands.

Ghana's North has not lacked in recent times for spokesmen to communicate its wishes to those in authority. Prior to independence, at least four means were used to extract concessions from the secretariat regarding amenities and services. Community leaders in the North, responding to pressures from their constituents, variously pursued tactics of bargaining, collective nonpayment of income taxes, local disturbances, and insistence upon distributive benefits at the territorial council. Northern brokers showed ingenuity in their use of bargaining opportunities. When officials (for example, the regional or district commissioners) requested the right to build roads or dams or to alter traditional land tenure practices, these local spokesmen could use the opportunity to secure a government commitment to a medical facility, school or post office. Within the subregions themselves, Northern representatives to the territorial council at Tamale presented numerous claims: they wanted the right to select representatives to the national assembly in a manner which accorded with local preferences, the allocation of subregional scholarships on the basis of individual merit rather than proportionality among ethnic peoples, and the extension of a priority to Northern candidates applying for public service posts in the subregions.[33] On the issue of social amenities, the spokesmen for Northern interests were equally assertive in pressing their demands upon government officials. In 1953, council members questioned the acting assistant regional officer keenly as to progress on the construction of feeder roads, the creation of a modern educational system, and the securing of adequate water supplies for their constituents. Members were also sufficiently distressed over the North's relative deprivation to doubt the wisdom of rapid movement toward the country's self-government and independence. Thus Bawumia Mumuni (Mamprusi) presented a motion to the effect that "this Council do request the Government to make available adequate grants to facilitate rapid development in the Northern Territories during the transitional period which the Gold Coast will pass before attaining full Dominion status."[34] Mumuni sought to close the gap between the relatively advantaged and disadvantaged prior to the transfer of power. His colleagues agreed with him and unanimously passed an amended motion urging government to make a concerted effort to enable their area to move rapidly toward equal status. Subsequently, as Ghanaian independence became imminent, a Northern delegation went to see Her Majesty's government in 1956 and urged the outgoing colonial power to make a lump sum grant for the development of this area prior to the transfer of power. J.A. Braimah, a leading Northern spokesman at the time, describes the circumstances around this plea: "There wasn't a promise from any quarter, but it was a demand by the representatives of the Northern Territories which they made to the British Government blaming them for the backwardness of the North. And if at that particular time, they had decided that the North should join the South in achieving independence then they should give the North thirty [million] pounds (£ 30,000 [m.]), to be able to carry on with the development of the area so they would be able to catch up."[35] Nothing ever came of this proposal. Although the British secretary of state for the colonies, A.T. Lennox-Boyd, stated that he would have given consideration to such a request from the Ghana government, President Kwame Nkrumah never made an explicit appeal for such funds, explaining that he did not want any strings attached to the grant of independence.

Northern assertiveness on allocation issues remained much in evidence after Ghana's independence as well. The Northern People's Party, which secured a majority of the area's seats in the National Assembly in the 1954 and 1956 elections (but less than a popular majority), championed the political and economic concerns of the North. Along with other opposition parties, it entered actively into the struggle for effective regional assemblies. Moreover, carrying out its mandate to secure "a progressively increasing share in the administrative and other services of the country," it focused national attention upon the area's relative deprivation and its demands for remedial action.[36] In the long run, however, this stress upon public goods and services did not prove sufficient to guarantee party survival. As Dennis Austin observes: "It was only by competing with the N.P.P. on its own terms, and by asserting its authority as a government to back its appeal as a party, that the C.P.P. later succeeded in splintering -- and eventually breaking -- the N.P.P."[37] Even so, as an agency for raising the general public's consciousness as to conditions in the North and for the communication of demands, the N.P.P. had an importance which must not be minimized.

Following the fall of the Nkrumah government and the handing over of power by the National Liberation Council in 1969, Northern spokesmen again used the parliamentary forum to make claims to equitable treatment. Northern MPs described the general neglect of their area and contended that it needed "something drastic in order to ensure that this country remains a stable country."[38] C.K. Tedam (Progress Party -- Chaina-Paga), for example, observed that by 1954 the North could boast only 0.02 percent of its people as formally educated, and he noted that of the allocations "earmarked" for secondary schools, less than 10 percent of the total was to be spent in Northern and Upper regions. "I feel," he concluded, "we have to face the realities of the day. The Government should step up aid for education in the North."[39]

These claims for corrective equity, interspersed through the parliamentary debates of 1969, were given coherence by an important motion introduced into the House on June 8, 1970 by Ibrahim Mahama (National Alliance of Liberals -- Tamale). Mahama's motion read as follows: "That since economically, educationally, and socially the Northern and Upper Regions are less developed than the other Regions of Ghana, this House requests the Government to draw up a special accelerated development programme for the North (beginning from the 1970-71 financial year) in order to ensure a rapid bridging of the gaps."[40] After describing the disparities among subregions in a detailed and comprehensive manner, Mahama and his colleagues stressed the duty of the central government to bridge the economic gaps among subregions. In particular, Mahama called upon the government to set up a "Northern Development Commission" to study the economic and technical problems of the area and to make recommendations on viable projects or enterprises which could be established there. A "Northern Development Bank," with the sole aim of promoting enterprises in these two subregions, was also urged upon governmental authorities. Despite pledges in the Progress Party manifesto to pay specific attention "to those regions of Ghana, especially the Northern, Upper and Volta Regions and those parts of the other regions which have been neglected by previous governments,"[41] government spokesmen opposed the motion, depicting it as "mischievous," "propagandistic," and "intended to gain political advantage." The opposition tried in the course of the debate to soften its original motion to take account of other relatively disadvantaged subregions (notably, Brong-Ahafo and Volta) and other areas within the remaining subregions. Nevertheless, the majority of members succeeded in pushing through an amendment still more general in its terms: "That this House urges the Government to draw up a special accelerated programme to offset the imbalance in regional development with particular reference to those regions where progress has been slow and comparatively negligible."[42] The demands of the relatively disadvantaged North were thus partially deflected by the powerful counterdemands articulated by the spokesmen for the other parts of the country.

With the eclipse of party government following the military coup led by Colonel Acheampong in January 1972, the opportunities for articulating demands through the electioneering and parliamentary processes no longer availed. Hence the style of the input process altered. Yet even in the face of these political changes, Northern spokesmen continued to find ways to press demands on government for favorable resource distributions. In less dramatic fashion, spokesmen (often civil servants, highly placed military officers, or traditional leaders) made known their constituents' wishes by means of face-to-face contacts with decision-makers. Not only did these intermediaries confirm such practices in interviews with this author in April 1976 (speaking of transmitting subregional interests directly to appropriate central ministries in Accra), but members of the Acheampong government have publicly referred to such communications.[43] Thus during his tour of the Northern and Upper regions in 1975, Colonel Acheampong took note of the appeals for amenities put forth by local notables. During his stopover at Damongo in the Northern Region on March 24, 1975, the _yabumwuta_, Ewuntomah Mahama, paramount chief of the Gonja Traditional Area, called upon the government to restore electricity to the Damongo Hospital and to macadamize the Yapei-Damongo-Sawla road. Two days later, while at Tumu in Upper Region, Acheampong entertained requests from the _tumukro_, Luri Kanton IV, for the provision of electricity, improved roads, a farmhouse, a variety of livestock and poultry, and some machinery for the Tumu Secondary School. That same day, in a brief visit to Chiana, Upper Region, the _chiana-pio_, R.A. Ayagitam (who was also the president of the Upper Regional House of Chiefs) "pleaded" with the head of state to construct dams and feeder roads, promote livestock breeding, facilitate the transportation of foodstuffs to market centers, and establish a groundnut processing plant.[44] In brief, praise for government was invariably mixed with demands on central officials for modern goods and services (some of which broadened noticeably as these local notables gathered together in such bodies as the Upper Regional House of Chiefs to appeal for the establishment of a university or a brewery in their area).[45] The military regime, desperate to expand its base of support, went a long way toward tacitly exchanging public resources for a degree of legitimization.[46] Under such circumstances, military government did not signify a total blockage of geoethnic messages from the periphery so much as an altering of the process by which these claims were communicated to the decision elite.

In 1979, with the return to civilian government and parliamentary elections, Ghana's demand process largely reverted to that prevailing prior to the coup. In addition to the normal pressures for favorable distributive practices on the part of civil servants and local notables,[47] parliament again became a vehicle for direct appeals to central decision-makers. A new thrust in this respect was the increased role of Northerners in key executive and party positions (most importantly that of the president, Hilla Limann). In seeking partisan victory in the elections, parties had pursued a tactic of wooing Northern support by offering top party positions to peoples from this area -- what one observer has described as "the northern factor in Ghanaian politics."[48] This approach had indeed worked out well for the victorious People's National Party, for Northerners threw considerable backing to its candidates, electing twenty-two MPs of the total of seventy-one that the party secured. In supporting the PNP so strongly, the North displayed a keen appreciation for the mechanisms of interest articulation. No doubt, should the society's structure for articulating demands shift once again, the political elite in these two relatively disadvantaged subregions would make an effort to adapt to the new circumstances. The form is less significant for them than the message and the outcome. As Ibrahim Mahama succinctly put the matter: "We want progress in action, not in words."[49]

The elite demands exemplified here by the case of Northern Ghana are fairly typical of middle African situations generally. In these societies, the elite possesses the skill, knowledge, initiative, and resources to enable it to exercise a disproportionate influence on outcomes. As members of the country's dominant class, those acting as ethnic articulators and/or brokers can bring an important influence to bear on the process of politics -- mainly in the way that they shape and communicate the general public's dissatisfactions to governmental decision-makers. What separates them from their less powerful constituents is largely their capacity for making selections as to which demand messages to channel to government and how intensively to push these particular demands. Certainly the geoethnic intermediary's fate is tied up with the type of linkages he or she effects both with the dominant class and with his or her constituent support base. As the community broker conforms to the wishes and expectations of the dominant class, that person tends increasingly to become a part of the political system, sharing in its rewards and benefits.

From the standpoint of constituents, the personally beneficial nature of the broker's inclusion in the dominant class's reward system would seem to be less important than that spokesman's loyalty to the group and effectiveness in securing benefits for his people. The survival of Ukambani leader Paul Ngei as one of the longest-serving of Kenya's cabinet ministers, despite the alleged maize scandal of 1966 and a court's finding of guilty in 1975 to an election offense, is testimony to the political entrepreneur's key role in the eyes of his constituents. In the 1975 by-election, Ngei made effective use of an intervention by Kenyatta on a bill allowing the president to permit leaders found guilty of election offenses to stand again for parliamentary election as a sign that the head of state wanted him to return to the inner circle of government. Under these circumstances, the people fully understood the message and sent him back to another term in office.[50] "There is a clear sense," declares John S. Saul, "in which politicized ethnicity is merely clientelism writ large, local/regional political barons [Kenya being, in fact, a case in point] rallying support for their own undertakings at the centre of the system by means of tribalistic demagogy and/or the promise to deliver some portion of the national pie to their constituents on a regional-cum-ethnic basis."[51] Politicized ethnicity, then, can be seen as part of a larger structure of relationships linking geoethnic and racial groups to the policy process.

The Critical Role of Brokerage

This discussion on the brokers' distinct and critical role as a shaper and communicator of messages points to a divergence between the general public's demands and those more effectively pressed by the community's spokesman upon government decision-makers. Certainly it is unwise to overstate the attitudinal differences between the general public and the political elite (as shown by the opinions expressed in separate surveys of chiefs, local councilors, members of development committees, and the general public in Ghana in 1973). Nevertheless, it is in their political roles and relationships at the center that political entrepreneurs alter their behavior and possibly even their priorities. This may be explained by two factors. First, in central level polities, the political entrepreneur, intent on furthering collective (and personal) interests broadly conceived, finds himself subjected to a wider range of pressures from within the dominant class -- both at the center and in the subregions -- than experienced by the public at large. As one author writing on group relations in Mexico notes, the intermediary must therefore "shape his behavior to fit these new expectations."[52] And second, the claims advanced upon central decision-makers must, if they are to prove effective, be presented in a more generalized, coherent, and sophisticated manner than is necessary in the case of the constituents. It is the requirements of the political process at the different levels which largely explain the variances in the types of distributive appeals made to government.

To illustrate the kinds of pressures emergent within the dominant class, it will be useful to turn again to the politics of Northern Ghana (as well as to the politics of the center and the international environment as it affects the politics of these subregions). In the Northern Ghana case, internal class pressures led to a decided overemphasis upon the needs and priorities of a single stratum of the dominant class, i.e., the politically powerful large-farming elements. Certainly if the political system has been responsive to the claims (or counterclaims) of any particular stratum, it has shown itself to be highly sensitive to the demands of large-scale commercial and farming interests. Indications of such a tendency were pointed out to this author in interviews conducted with Northern opinion-formers in April 1976. Respondents drew a sharp contrast between large and small farmers with respect to the former's available capital, productivity, and access to decision-makers. Other advantages they could boast included the ability to secure extensive tracts of land, gain favorable prices for produce, obtain access to feeder roads, acquire loans and financial backing, and get necessary supplies of fertilizer and seeds at subsidized prices. In fact, large-scale farmers, who own over 20 percent of Ghana's rice lands and have access to governmental subsidies for the purchase of tractors and for credits from the Agricultural Development Board, are a privileged socioeconomic class by any standard. As Charles Elliot noted in 1975: "With a profit of between N¢34 and N¢77 per acre, it is clear that these large rice farmers are earning high incomes by comparison with even the wealthiest cocoa farmers. By contrast their labourers, on an optimistic calculation, receive N¢90 per annum -- enough to buy no more than one and a half pounds of rice per day for the whole family."[53] Paradoxically, it is the most disadvantaged subregions (Northern, Upper and Brong-Ahafo) which have witnessed the heaviest reliance upon tractor ploughing and harrowing, with all the attendant social effects that such heavy mechanization is likely to have upon relatively poor areas."[54]

The large farmers of the North are certainly advantageously positioned to make effective demands upon public resources. A stratum of the larger dominant class, they can influence whatever political elite wields power at the time, i.e., the subregional military and civilian administrators, the central bureaucracy, the executive branch, or the parliamentarians. The informal manner in which they express their demands enables them to gain access without unnecessary fanfare or embarrassment; nevertheless, these approaches can be telling in terms of policy outcomes. The ruling political elite has generally responded to these influences by espousing goals of rapid economic growth and by showing itself to be particularly receptive to the claims of those best positioned to further these objectives. In encouraging Ghanaians (frequently civil servants and military officers) and foreigners to undertake commercial farming activities, the Acheampong government held out a wide variety of incentives: financial support; land availability on a leasehold basis; minimum guaranteed prices for all commodities grown under this scheme; tax concessions; prompt granting of import licenses; guaranteed immigrant quotas; waiver of the Selective Alien Tax; accelerated depreciation for plant, building, equipment, dams, access motorable roads, and other capital works; and a plan to accelerate the transfer of dividends of foreign companies participating in the Operation Feed Yourself Program. In 1972, some 50 percent of the Agricultural Development Bank's loanable funds (N¢14.57 million) went to large state agricultural corporations.[55] And with respect to the National Investment Bank, an even more decided support for "directly productive investments" in the private sector was a matter of the stated policy governing the bank's operations.[56] Clearly, the community brokers acted in a highly selective manner in determining which claims to push most strongly in their communications with those in control of public resources. Thus the various political entrepreneurs, pressed by powerful claims made from within the dominant class, tended to modify their preferences to adjust to the social environment at the central level.

The second feature differentiating the political broker from his constituents is his need to organize and present collective claims in a generalized, coherent, and sophisticated manner. This requirement arises, of course, from the political environment in which the intermediary operates. Effective competition with other political entrepreneurs for the scarce goods and services of the state necessarily entails the presentation of demands in a systematized and organized fashion.

To be sure, parliamentary debates in middle Africa are replete with calls for designated projects (new or improved roads, schools, universities, hospitals, water systems, markets, storage facilities, irrigation, electricity, postal services), for proportionality among ethnic peoples and subregions in matters of recruitment, and for increased fiscal allocations to the various subregions. Nigerian state spokesmen repeatedly urged the federal government to increase revenues allocated to their areas as well as to provide support for federal universities, educational systems, and basic services.[57] In Kenya, Liberia, Cameroun, Sudan and other African countries as well, political entrepreneurs strove by every means possible to secure from government the goods and services their subregions needed.

More general demands, such as that exemplified by Ibrahim Mahama's motion presented to the Ghana parliament, were also advanced from time to time. For example, an MP from Zambia's North-Western Province spoke on one occasion of the neglect of the "cinderella provinces" by the former colonial power, the lack of project implementation under the First National Development Plan, the subregions' needs for improved facilities, the government's policy on correcting past anomalies, and the inequitable allocations under the Second National Development Plan.[58] Here we see the dissatisfactions and needs of the general public in the subregion shaped and argued before the country's leading deliberative body in a logical and, given the time constraints under which he operated, a comprehensive manner.

It is the community broker, then, who gives direction both inside and outside of parliament to the diffuse claims of his constituents. His appeals, both for specific goods and services as well as the more general calls for the accelerated development of his area, often prove an effective catalyst for increased governmental attention and action. Inevitably, the political entrepreneur will be selective in the way that he articulates the call for more equitable distributions. He cannot possibly convey all of the general public's wants simultaneously -- and expect a favorable response. Consequently, his political judgment as to the choice and timing of demands as well as his ability to cope with counterclaims coming from within his own dominant class become critical to his performance as a constituency delegate.

Finally, an evaluation of the community broker's performance cannot be based solely upon his ability to extract the benefits of modernity from the center. To the extent that the community brokers are prepared to accept prevailing norms and values, their role as patron can be compatible with that of strengthener of "soft state" networks.[59] An adherence on the part of these intermediaries to generally accepted rules of political behavior has the effect of reinforcing the processes by which elites can engage in constructive conflict (i.e., the search for mutually beneficial outcomes). Provided that these political entrepreneurs "make their demands properly," as Nigeria's General Yakubu Gowon once stated it,[60] they can be said to be contributing to the political "development" of their country. An acceptance on their part of the restraints of recognized and accepted rules, even while engaging in anxious interplay over the extraction of material resources, is an act of statesmanship which fosters a sentiment of transcending state-national interests vital to intergroup negotiation and compromise.

Policy Implications

Two major policy implications follow from the propensity of the relatively disadvantaged and their spokesmen to make only limited claims on government. First, these minimal demands allow government to put "satisficing" policies into effect. Lower demand levels in the relatively disadvantaged areas afford the central government greater latitude in ranking priorities as to resource allocation. Not only do the counterclaims of the relatively advantaged place constraints on determined policies of corrective equity, but the lack of forcefulness with which the relatively disadvantaged press their demands on central patrons contributes to the minimizing strategy adopted by government decision-makers. These decision elites reduce their sights precisely because the costs of such a strategy remain low. Obviously, central decision-makers must be sure that every significant group "wins a satisficing quantity of outcomes."[61]

Nevertheless, even within this broad limitation, government retains much greater maneuverability in meeting the demands of the relatively disadvantaged than is the case with their relatively advantaged counterparts. Such flexibility enables government to maintain countrywide stability by distributing development resources according to the principle of demand intensity rather than need as such.

Second, although the intensity of demand and, as a consequence, bargaining strength, may differ as between the relatively disadvantaged and advantaged, the low intensity politics prevailing on these distributive issues are nonetheless conducive to exchange relationships. Certainly the "horizontal" cleavages between African geoethnic groups are narrower than the "vertical" cleavages dividing racial groupings from one another. Because common nationalist experiences and aspirations foster a sense of solidarity within the majority African community, the interethnic boundaries may be reduced and the demands for corrective equity may be enmeshed with the overriding goal of development. Moreover, as the survey results indicate, a political culture might be emerging in certain settings which is conducive to bargaining encounters (direct or tacit) between the relatively advantaged and disadvantaged on distributional issues. The very fact that such a high proportion of respondents among the relatively advantaged supported extending government help of a redistributional nature to Kenya's smaller ethnic peoples may augur well for the possibilities of constructive conflict management.

The foregoing reinforces the general conclusion that, as far as the general public is concerned, distributive claims remain negotiable. This is not to say, however, that a regime strategy of "satisficing" is not without its definite limitations. As the peoples in the relatively disadvantaged subregions come to make their demands more forcefully and effectively (the likely consequence over time of the feedback process), their bargaining capacity can be expected to increase. In that case government would seem to need to keep abreast of changing demand patterns and intensities in order to assure the achievement of system goals on equity, stability, and capacity over the long term.

Yet there is certainly nothing automatic about this tendency to present geoethnic demands more forcefully over time. As we have seen, the policy process, as it relates to interregional allocations and opportunities, is affected by the manner in which geoethnic brokers fashion the demands they communicate to decision elites at the center. As members of the dominant political class themselves, these group spokesmen are subject at times to conflicting class and ethnic pressures. In the way they shape their behavior to conform to the expectations of what may be different class-ethnic publics, such brokers are likely to have a significant impact upon policy outcomes, in this case, the nature of the mix between distributive and redistributive policies.

Footnotes

1. Government of the Mid-West State, "Understanding the Nigerian Crisis," West Africa -- Supplement, July 20, 1968, p. 6. (Italics in text.)
2. On the process of conflict escalation, see Milton M. Gordon, Human Nature, Class, and Ethnicity (New York: Oxford University Press, 1978), p. 80; Leo Kuper, The Pity of It All (Minneapolis: University of Minnesota Press, 1977); Quincy Wright, "The Escalation of International Conflicts," Journal of Conflict Resolution, 9, 4 (December 1965): 434; Richard Smoke, War: Controlling Escalation (Cambridge: Harvard University Press, 1977); and Lincoln Bloomfield and Amelia C. Leiss, Controlling Small Wars (New York: Alfred A. Knopf, 1969).
3. See, for example, Dov Ronen, The Quest for Self-Determination (New Haven: Yale University Press, 1979), p. 14.
4. On the externalist-internalist distinction as used here, see Donald Rothchild, "The Two Senses of Ethnonational Self-Determination," Africa Report, 26, 6 (November-December 1981): 56-58.
5. Jonathan S. Barker, "The Paradox of Development: Reflections on a Study of Local-Central Political Relations in Senegal," in Michael F. Lofchie, ed., The State of the Nations (Berkeley and Los Angeles: University of California Press, 1971), p. 56.
6. Richard Rose and Guy Peters, Can Government Go Bankrupt? (New York: Basic Books, 1978), pp. 237-238.
7. Onyeonoro S. Kamanu, "Secession and the Right of Self-Determination: An O.A.U. Dilemma," Journal of Modern African Studies, 12, 3 (September 1974): 361; also see Ved P. Nanda, "Self-Determination Outside the Colonial Context: The Birth of Bangladesh in Retrospect," in Yonah Alexander and Robert A. Friedlander, eds., Self-Determination: National, Regional, and Global Dimensions (Boulder: Westview Press, 1980), pp. 203-205.
8. See Samuel P. Huntington's "Foreword," in Eric A. Nordlinger, Conflict Regulation in Divided Societies (Cambridge: Harvard University Center for International Affairs, 1972). Also see the comments on this in Arend Lijphart, Democracy in Plural Societies (New Haven: Yale University Press, 1977), p. 46. For an eloquent attack upon secession in Third World countries, see Ali A. Mazrui, The Trial of Christopher Okigbo (London: Heinemann, 1971), especially p. 118.
9. T. Dunbar Moodie, The Rise of Afrikanerdom (Berkeley and Los Angeles: University of California Press, 1975), p. 263.
10. Ernst B. Haas, Robert L. Butterworth, and Joseph S. Nye, Conflict Management by International Organizations (Morristown: General Learning Press, 1972), p. 54. Also see West Africa, May 18, 1968, p. 567, and December 20, 1969, p. 1565.
11. K.J. Binns, Report of the Fiscal Review Commission (Lagos: Federal Ministry of Information, 1965), p. 13.
12. Donald Rothchild, Toward Unity in Africa (Washington D.C.: Public Affairs Press, 1960), pp. 42-44.

13. See Gabriel A. Almond's "Foreword," in Myron Weiner, *The Politics of Scarcity* (Chicago: University of Chicago Press, 1962), p. ix. Also see Gabriel A. Almond and G. Bingham Powell, Jr., *Comparative Politics: A Developmental Approach* (Boston: Little, Brown and Co., 1966), p. 61.
14. Robert H. Bates, "Modernization, Ethnic Competition and the Rationality of Politics in Contemporary Africa," in this volume.
15. Richard L. Sklar, "The Nature of Class Domination in Africa," *Journal of Modern African Studies*, 17, 4 (December 1979), p. 550. Also see Michael A. Cohen, *Urban Policy and Political Conflict in Africa: A Study of the Ivory Coast* (Chicago: University of Chicago Press, 1974), p. 194.
16. Goran Hyden and Colin Leys, "Elections and Politics in Single-Party Systems: The Case of Kenya and Tanzania," *British Journal of Political Science*, 2, 4 (October 1972): 415; and Joel D. Barkan, "Legislators, Elections, and Political Linkage," in Joel D. Barkan with John J. Okumu, eds., *Politics and Public Policy in Kenya and Tanzania* (New York: Praeger, 1979), pp. 67-77.
17. For further information on the background and methodology used in these surveys as well as the data, see Donald Rothchild, "Ethnic Inequalities in Kenya," *Journal of Modern African Studies*, 7, 4 (December 1969): 689-711; Donald Rothchild, *Racial Bargaining in Independent Kenya* (London: Oxford University Press, 1973), pp. 453-461; Donald Rothchild,"Rural-Urban Inequities and Resource Allocation in Zambia," *Journal of Commonwealth Political Studies*, 10, 3 (November 1972): 223, 226, 240; Donald Rothchild, "Comparative Public Demand and Expectation Patterns: The Ghana Experience," *African Studies Review*, 22, 1 (April 1979): 127-147.
18. Letter to *The Reporter* (Nairobi), May 19, 1967. On charges of Kikuyu domination of certain ministries, see *House of Representatives Debates*, 10, 4th session (October 14, 1966), cols. 883-884, 1385-1386.
19. See International Labour Office, *Employment, Incomes and Equality* (Geneva, 1972), p. 301.
20. The manifesto is reprinted in *Standard* (Nairobi), October 13, 1979.
21. Iro K. Feierabend, Rosalind L. Feierabend, and Betty Nesvold, "Social Change and Political Violence: Cross National Patterns," in Hugh Davis Graham and Ted Robert Gurr, eds., *Violence in America: Historical and Comparative Perspectives* (New York: New American Library, 1969), p. 620.
22. Sidney R. Waldman, *Foundations of Political Action: An Exchange Theory of Politics* (Boston: Little, Brown and Co., 1972), p. 45; Herbert Simon, *Models of Man* (New York: John Wiley, 1957), pp. 247-254; Aaron Wildavsky, *The Politics of the Budgetary Process* (Boston: Little, Brown and Co., 1964), pp. 12-13; and Warren F. Ilchman, "Decision Rules and Decision Roles," *African Review*, 2 (1972): 239.
23. Walter Birmingham, I. Neustadt, and F.N. Omaboe, *A Study of Contemporary Ghana*, vol. 2 (London: George Allen and Unwin, 1967), p. 53.

24. Central Region, described by most observers as a developed subregion, appears to fit into the relatively deprived category if its main urban centers are removed from it.
25. Kwamina B. Dickson, "Development Planning and National Integration in Ghana," in David R. Smock and Kwamena Bentsi-Enchill, eds., The Search for National Integration in Africa (New York: Free Press, 1975), p. 106.
26. Kodwo Ewusi, "Disparities in Levels of Regional Development in Ghana," Social Indicators Research, 3 (1976): 88-89.
27. See Ted Robert Gurr, Why Men Rebel (Princeton: Princeton University Press, 1970), p. 71.
28 Joan M. Nelson, Access to Power (Princeton: Princeton University Press, 1979), p. 241; also see Wayne A. Cornelius, Politics and the Migrant Poor in Mexico City (Stanford: Stanford University Press, 1975), pp. 184-185, 196-197.
29. Almond and Powell, Comparative Politics, p. 86.
30. Joel S. Migdal, Peasants, Politics, and Revolution (Princeton: Princeton University Press, 1974), p. 222.
31. On the inflated price of rice smuggled into Upper Volta, see Daily Graphic (Accra), November 7, 1975, p. 12. The economic crisis in Ghana in the late 1970s is discussed in Donald Rothchild, "An African Test Case for Political Democracy: President Limann's Economic Alternatives," in Colin Legum, ed., African Contemporary Record 1979-80 (London: Rex Collins, 1981), pp. A137-145.
32. On the role of rural traditional intermediaries who "serve to balance the demands of the populace and the bureaucratic groups," see Norman Miller, "The Political Survival of Traditional Leadership," Journal of Modern African Studies, 6, 2 (August 1968): 187.
33. Record of the Seventh Session of the Territorial Council Held at Tamale on 18th, 19th, 20th and 21st July, 1950 (Tamale: Government Press, 1950), pp. 23, 25, 35, 45, 47.
34. Record of the Sixteenth Session of the Northern Territories Council Held at Tamale on 3rd and 4th September, 1953 (Tamale: Government Press, 1953), p. 20.
35. Interview with J.A. Braimah conducted by Dr. Kofi Frimpong (1973), typescript copy (Accra: Ghana Academy of Arts and Science, 1975), p. 15. Although the figure 30,000 pounds was recorded on this typescript copy, it seems apparent that 30 million pounds was intended. See Paul Andre Ladouceur, Chiefs and Politicians: The Politics of Regionalism in Northern Ghana (London: Longman, 1979), p. 161.
36. Northern People's Party, The Constitution, Aims and Objects, 1954, as quoted in Dennis Austin, Politics in Ghana, 1954-1960 (London: Oxford University Press, 1964), p. 184.
37. Austin, Politics in Ghana, p. 228.
38. Republic of Ghana, Parliamentary Debates, Official Report, 2nd series, vol. 1, no. 8 (December 1, 1969), col. 207. Statement by Ibrahim Mahama (N.A.L. -- Tamale).
39. Ibid., vol. 1, no. 6 (November 27, 1969), col. 120. Of course, to the extent that funds are allocated to educate northerners outside of these two subregions, such data are likely to prove somewhat misleading. See the remarks of Dr. Jones Ofori-Atta,

ibid., vol. 3 (June 9, 1970), col. 565.

40. *Ibid*., (June 8, 1970), col. 520.
41. *Ibid*., (June 9, 1970), col. 572.
42. *Ibid*., (June 12, 1970), col. 687. In support of a balanced approach to subregional allocations, see the editorial, "National Development and Regional Claims," *Legon Observer*, 5, 13 (June 19-July 21, 1970): 1-2.
43. In this vein, Colonel Acheampong went to great pains to deny that any Ghanaian subregion was neglected in the provision of social amenities on the occasion of a visit to Castle Osu by a seventeen-member delegation from the Volta Region House of Chiefs. *Daily Graphic*, December 27, 1975, pp. 8-9. Similarly in Nigeria, it was the military governors of the less advantaged areas who represented their constituency's interests on such issues as resource allocation, appointments, and revenue allocation. See, for example, Omo Oye, "The Politics of Revenue Allocation," *Daily Times* (Lagos), April 29, 1974, p. 7.
44. Republic of Ghana, *Fourth Year in Office of Colonel Ignatius Kutu Acheampong* (Accra: Office of the Press Secretary to the Supreme Military Council, 1976), pp. 74, 79, 80. Also see *Ghanaian Times*, November 24, 1975, p. 12.
45. *Daily Graphic*, March 18, 1976, p. 7; March 22, 1976 p. 12; and April 14, 1976, p. 5.
46. On Acheampong's achievements (albeit mixed) in the area of subregional reallocation in Ghana, see Donald Rothchild, "Military Regime Performance: An Appraisal of the Ghana Experience, 1972-78," *Comparative Politics*, 12, 4 (July 1980): 472-476.
47. See the report of the appeal made by the Yabumwura Abudu Mahamah Kurabaso, the president of the Northern Region House of Chiefs, for central government support in repairing bridges during a *durbar* in Tamale attended by President Limann. *Ghanaian Times*, November 12, 1979, p. 1.
48. *Legon Observer*, 11, 6 (April 6, 1979): 126.
49. Republic of Ghana. *Parliamentary Debates, Official Report*, 2nd series, vol. 3 (June 8, 1970), col. 534.
50. See Republic of Kenya, *Report of the Maize Commission of Inquiry, 1966* (Nairobi: Government Printer, 1966), pp. 166-168; and *Weekly Review* (Nairobi), June 22, 1979, pp. 7-8.
51. John S. Saul, "The Dialectic of Class and Tribe," *Race and Class*, 20, 4 (Spring 1979): 351.
52. Eric R. Wolf, "Aspects of Group Relations in a Complex Society: Mexico," *American Anthropologist*, 58 (1956): 1072.
53. Charles Elliott, *Patterns of Poverty in the Third World* (New York: Praeger, 1975), p. 55.
54. Data on a limited survey of tractor service distribution by subdistricts shows the Northern Region, with its numerous large-scale rice farms, to be the heaviest user of tractor services. See B.E. Rourke, *Wages and Incomes of Agricultural Workers in Ghana*, Technical Publication Series, no. 13 (Legon: Institute of Statistical, Social and Economic Research, 1971), p. 75.
55. Agricultural Development Bank, *Annual Report for the Year Ended 31st December, 1973* (Accra, 1974), p. 6. It should be noted, however, that by 1974, 82.9 percent of ADB loans were allocated to small-scale farmers. *Annual Report for the Year Ended 31st*

December, 1974 (Accra, 1975), p. 12.

56. The Bank's lending policy emphasizes commercial banking practices and places a lower limit of N¢10,000 on amounts normally loaned. The National Investment Bank: Objectives and Functions (Accra, n.d.), p. 6.

57. See, for example, Daily Times (Lagos), May 23, 1973, p. 32, November 22, 1973, p. 7, and May 16, 1974, p. 3.

58. Republic of Zambia, Parliamentary Debates, Second National Assembly, 4th session (February 11, 1972), cols. 733-739. Statement by Mr. Wisamba (Solwezi).

59. Gunnar Myrdal, Asian Drama -- An Inquiry into the Poverty of Nations, vol. 2 (New York: Pantheon, 1968), pp. 895-900. See also Daniel Bell, The Cultural Contradictions of Capitalism (New York: Basic Books, 1976), pp. 154-155.

60 West Africa, March 7, 1970, p. 273.

61. Waldman, Foundations of Political Action, p. 177.

11
Comparative Claims to Political Sovereignty: Biafra, Katanga, Eritrea

Crawford Young

> [The Member States affirm]...respect for the sovereignty and territorial integrity of each State and for its inalienable right to independent existence. -- Charter of the Organization of African Unity (OAU), Article III, Paragraph 3

Territorial Integrity: The Master Principle

In the postcolonial state system in Africa, no principle has been more fundamental than the sanctity of the existing sovereign units, within their current frontiers. This view, which crystallized only with the advent of the independence era, was already strong by the time a consort of African nations was institutionalized through the OAU in 1963, as evidenced by the important place accorded to an affirmation of this principle in the charter. Since that time, it has become more firmly entrenched as a basic postulate, upon which internal stability of individual states, and harmonious relations within the African state system, must rest.

Useful insights into the operation of this principle, and its problematic elements, may be gained by examination of three instances where particular groups in different ways challenged this premise, claiming to override it with a higher principle: the right to self-determination. The cases discussed, Katanga, Biafra, and Eritrea, are not the only such instances; the clash between territorial integrity and self-determination is also encountered in the greater Somalia movement and the Western Sahara conflict. Under the rubric of secessionism, one may note such important examples as the southern Sudan, Buganda, and a sizable number of less consequential claimants. While a comprehensive exploration of these issues would require consideration of the full universe of cases, the three examples chosen here do permit an examination of many of the dilemmas of state definition and nationality formation in contemporary Africa.

Self-Determination: Genesis and Evolution

While these issues are quite complex, their origin is simple: the reception of the doctrine of self-determination by African nationalism, and its translation into an ideological weapon for the defeat of colonialism. The idea of self-determination received conscious elaboration in the eighteenth and nineteenth centuries; in its first version, it was advanced in vindication of claims to political independence by Western hemisphere colonies. "We the people" as then understood essentially meant the immigrant population from the mother country, and usually had as territorial referent the administrative or political units which the colonizer had established. Then, in nineteenth century Europe, the polyglot empires of Ottoman Turkey, Austro-Hungary, and Russia came under growing challenge by linguistically defined groupings; their claims drew great inspiration from the examples of Germany and Italy in creating new, large and powerful states on the basis of the common linguistic heritage. The Irish case then showed that self-determination could be grounded on a cultural identity and historic consciousness of oppression, even without a solid linguistic base. The apotheosis of this period was the Wilsonian version of self-determination, which gained partial recognition at the Versailles Peace Conference. This connoted a right for collectivities which could establish common cultural claims to seek political independence. With colonialism at flood tide, there was no question of applying such a doctrine to Africa or Asia; the international state system, assembled in Versailles, heard only European claims for self-determination.

However, as self-determination won gradual acceptance as international jurisprudence, it proved an invaluable weapon in the emergent struggle of colonized peoples in Asia and Africa to assert the right to independence. With a small number of exceptions, however, anticolonial nationalism found inconvenient the notion that cultural affinities were a necessary basis for exercise of this right. A shared condition of oppression and alien rule was the essential cause of revolt. This could be remedied only by the pooled efforts of all who participated in a given framework of subordination. Thus the particular colonial territory was the necessary framework for challenging foreign hegemony. Nationalists, in seeking united support of all inhabitants of a given territory to sanction the independence demand, embraced the colonial entity itself as the defining basis for the "people" to whom self-determination should apply.

The remarkable triumph, in this form, of the anticolonial movement worldwide brought 100-odd new sovereign units into the world comity of nation-states, most of them multicultural. The nationalist leadership was acutely aware of the ambiguities of the self-determination doctrine, and the risks that segments of the new states, whether regionally or culturally defined, would invoke the principle in their turn. Thus it is no surprise that the principle of sanctity of existing sovereign states and their boundaries commanded early and general assent in Asia and Africa. There was quickly widespread acceptance of the territorial integrity norm as having precedence over self-determination. Were it not for the power of the doctrine that self-determination could apply only to the territorial states of the decolonization settlement, there is good reason to believe that all three of the entities in question would today be independent.

To pursue our inquiry, we will first set forth the basic outline of each case. We will then successively give comparative examination to the conceptualization of the unit laying claim to independent status, the justifications for this claim, the ways in which the sovereignty demand was pursued, and the responses of the international system. Finally, we will consider what instructive lessons may be gleaned from these three cases.

Katanga: Origins of Secession

We begin with the Katanga case,[1] because it was first to acquire international visibility.[2] The roots of the separatist idea extend back to the early years of the colony, the particularity which grew up around the strong concentration of European enterprise, and the remoteness from the headquarters of colonial administration at Kinshasa (until 1925, Boma). The giant mining corporation, Union Minière du Haut-Katanga (UMHK), was a world of its own; around its activities arose a network of subsidiary activities, and a small but clamorous European settler population (31,887 whites in 1960, mostly in the mining towns). Until the centralizing administrative reforms of 1933, Katanga was ruled by a vice-governor general, who was in practice largely autonomous. Much more attuned to southern African patterns than the rest of the Belgian Congo, the Europeans had longed for the kind of dominant governing role they saw in Rhodesia and South Africa, and complained of diversion of wealth generated in the province to other regions of the colony. As ealy as 1920, leading company managers and Catholic mission officials called for complete separation for Katanga.

"Katanga" as an African concept was much more recent, and emerged as a rallying cry in the later 1950s. It arose out of the crystallization in the Copperbelt cities of new forms of ethnoregional solidarities. Population densities in southern Katanga, where the principal mines were located, were low; the voracious labor demands were met in good part by recruitment in adjoining provinces. Kasaiens were particularly numerous; many came from areas where schools were established at an early date, and by the 1950s the Kasaiens tended to predominate in the white collar strata of African society. After World War II, as growing numbers of Katangans flocked to town, they became aware of a social pre-eminence of those they viewed as "strangers." When many Kasaien candidates generally triumphed in the first municipal elections in 1957, resentment at social advantage fused with fears of political domination. The slogan began to be heard that expanding opportunities for African social promotion should accrue in the first instance to "authentic Katangans."

In the tumultuous two years which preceded independence, many of the Kasaien intellectuals found the unitarian ideals and vaguely radical nationalism associated at the national level with Patrice Lumumba particularly appealing. This stimulated a marriage of circumstance between the settlers and corporate interests, fearful of the aggressive brand of Lumumbist nationalism, and the "authentic Katangans," apprehensive over "stranger" domination. As spokesman for this alliance emerged Moise Tshombe, son of one of the rare Africans to have achieved wealth through business dealings, and linked through his mother to a Lunda royal lineage. His political movement was the _Confédération des Associations Tribales du Katanga_ (CONAKAT).

At a pace that no one had foreseen, the colony by January 1960 found itself careening toward an independence date of June 30. National and provincial elections were scheduled for May 1960, to establish assemblies which in turn would select ministerial teams at both levels. CONAKAT, with a circmscribed and sparsely inhabited regional base, counted for little at the national level, with only eight of 137 seats. However, in the Katanga provincial assembly it emerged as the largest party, with twenty-five of sixty seats. The rise of CONAKAT had provoked in turn a hostile response from an ethnic coalition representing the much poorer northern part of the province, the _Association des Baluba du Katanga_ (BALUBAKAT). The BALUBAKAT alliance, fearful of marginalization by the southern Katanga groups, affiliated in national politics with the Lumumbist coalition; in the provincial elections, it won twenty-two seats.

Aided by a last-minute Belgian amendment to the provisional constitution,[3] the CONAKAT was able to form, by winning over a number of independents, a homogeneous CONAKAT government, led by Tshombe. Meanwhile, at the national level, after tortuous negotiations, Lumumba was finally able to form a government including a wide range of parties. Though CONAKAT received two minor posts, Tshombe believed his party had been excluded from significant influence. The stage was now set for secession, which had been under serious consideration by CONAKAT and its European allies since late in 1959, as one possible option.

The Rise and Demise of the Katanga Secession

A scheme to declare the secession two days before independence was squelched by the Belgian administration, unprepared to thus accept the full blame for an immediate breakdown in the new state at the moment of the power transfer festivities. However, a new occasion quickly offered itself: the precarious decolonization settlement was unhinged when on July 5, 1960, a mutiny broke out in the army against the exclusively European officer corps, and quickly spread throughout the country. The sudden loss of the central government's control over its coercive instruments made it possible for Tshombe to declare, on July 11, that Katanga was a sovereign and independent state, and to "appeal to the whole free world, and ask all to recognize in us the right of every people to self-determination."[4] He also asked Belgium for immediate technical, financial, and military aid.

He already had at his disposal the very considerable resorces of the mining company and the European community. Belgium declined to accord recognition, but did provide swift and efficacious assistance. King Baudouin declared on July 21 that "entire ethnic groups headed by men of honesty and worth...ask us to help them construct their independence.... It is our duty to respond favorably."[5] Belgian functionaries, who for the most part fled the growing disorder provoked by the mutineers in the other parts of the country, were ordered to remain at their posts in Katanga. The day preceding the secession, Belgian troops intervened in several cities, on the grounds that the European population required protection. Those in Katanga disarmed and expelled the leaderless soldiery, and hastily began creating a Katanga gendarmery. This Belgian support for the consolidation of the secession was crucial to its initial survival.

Zaire at once became the focal point for a world crisis. There were already enough African independent states for their reactions to weigh in international forums; they were uniformly outraged at what appeared to them as a conspiracy to destroy not just Zairian independence, but also radical African nationalism. The Lumumba government appealed for UN intervention, envisaged as an international force under the authority of his government, to restore its writ throughout the country, including Katanga, and to secure the removal of Belgian troops. However, the actual mandate given by the Security Council was much more ambiguous: to maintain order in the country; to protect human lives; and to bring about the withdrawal of Belgian troops.

While an international peacekeeping force was quickly assembled, and deployed in most major towns, its entry into Katanga was delayed. Tshombe declared he was willing to have a UN garrison, but not to have Lumumba's authority re-established under UN auspices; unless guarantees were provided, Katanga would resist. The UN, unwilling to risk casualties, went back to the Security Council for a new mandate; on August 8 the interpretation of General Secretary Dag Hammarskjold that the UN should remain apart from internal conflict was upheld. Thus, although UN detachments were present in Katanga from August 13, 1960, they had little leverage in the provincial situation; the UN thus coexisted with a secession which it had theoretically opposed.

In desperation, Lumumba then ordered his own ill-disciplined troops to invade Katanga, a campaign which at once degenerated into murderous assaults on civilian populations in neighboring Kasai. By late August, the Lumumba regime was unravelling; on September 5, in an American-sponsored move, President Joseph Kasavubu removed him from office. Though Lumumba at once challenged the legality of this deposition, he was never able to recapture power, and was finally imprisoned, then transferred to Katanga, where he was assassinated in January 1961. From September 1960 to August 1961, Kinshasa lacked a widely accepted government, permitting the Katanga secession to consolidate itself. During this period, Katanga also engaged in dilatory negotiations offering to rejoin a loosely confederal Zaire.

Katanga, however, also had its difficulties. Despite arduous efforts, not a single country in the world could be found to accord diplomatic recognition. Despite Katangan affirmations of Western solidarity, the bitter opposition by Afro-Asian states made the price of embracing the Tshombe regime much too high for the West. Also, by September 1960, violent revolt had broken out in the BALUBAKAT zones of the north. Brutal pacification campaigns carried out by the Katanga gendarmery and white mercernary auxiliaries could not contain this spreading rebellion.

The formation of a compromise national government under Cyrille Adoula with full parliamentary sanction in August 1961 altered the situation. The UN was now under growing pressure to bring the Katanga secession to an end, while at the same time it was denied a clear mandate to use force for this purpose. The Soviet bloc, the Afro-Asian states, and most Western members were by now united -- for very different reasons -- in the conviction that Zaire had to be restored to its 1960 boundaries. A series of efforts was made to bring this about by negotiations; Tshombe, however, proved an elusive bargainer. The impasse was exasperating for many UN officials. Western powers refused to allow the Security Council to authorize direct force to terminate the secession; a number of states (not only Western), as well as UN officials, were uneasy about the implications of this precedent. At the same time, it became apparent that Tshombe could spin out the negotiation process almost indefinitely, while Katangan independence might over time become irreversible. One way out of the impasse might be for UN forces to find a pretext for a swift and decisive strike which in turn would make the end of the secession a _fait accompli_. This indeed transpired -- abortively in September 1962, when the operation miscarried, and then successfully in late December 1962. On the latter occasion, on the initiative of local UN officials (though possibly with the tacit knowledge of their superiors), full military occupation of the Katanga strongholds was achieved before the operation could be halted by pressure on New York, because of the lack of Security Council or Secretariat authorization. On January 13, 1963, Tshombe and his ministers, from a Northern Rhodesian exile, conceded that the secession was over.

Roots of Secession in Nigeria

In a quite different way, the genesis of the Biafran secession also finds its roots in the colonial period.[6] Northern and Southern Nigeria, amalgamated only in 1914, had quite different administrative traditions. The subsequent division of the southern provinces into two regions in 1939 completed an administrative structure hinged upon the three major cultural groupings (Ibo/Eastern, Yoruba/Western, Hausa-Fulani/Northern), each of which numbered roughly two-thirds of the respective regional populations. This tripolarity acquired political content when, in the decolonization process, responsible government was introduced first at the regional level, with a federal framework utilized for the national government. By a dialectic process, there emerged in each of the three regions a separate dominant party identified with the most numerous cultural community, though they each disclaimed ethnic exclusiveness. Beneath the facade of Westminster parliamentary democracy, the remorseless power struggle of First Republic Nigeria closely resembled a three-player ethnic game, with control of the regional governments providing each actor with ample institutional and material resources.

Ethnicity as a crucial political determinant emerged earlier and more forcefully in Nigeria than anywhere else in tropical Africa. Fears of domination appeared at a precocious stage in the crystallization of Nigerian nationalism, among the three major groups with relation to each other, and among the smaller groups -- "minorities," in the parlance of the day -- vis-à-vis the three large agglomerations. Thus we find uncertainties expressed about Nigeria as a political entity which were uncharacteristic of African nationalism in the terminal colonial period. At different points in time, in pre-civil-war Nigeria, all three regions had occasion to threaten secession.

Thus Nigeria began the venture of independence with three highly self-conscious ethnic groupings, and a multitude of smaller ones chaffing at the domination -- nationally and regionally -- of the big three. As events transpired, it was the Eastern Region which finally embarked upon the adventure of separation, although it is perfectly possible to construct scenarios in which either of the other two might have proclaimed their withdrawal from the Nigerian federation. Not the least of the ironies of history is that, in all probability, a secession of either the Western or Northern Region could not have been militarily subdued; conversely, through its defeated separation, the Eastern Region was unintended midwife to the birth of a new and far more securely consolidated single Nigeria.

The Ibo Background

The birth of Biafra revolves around the Ibo cultural community. Not all Ibo supported the secession; some continued in federal service throughout the civil war, and the large Ibo community in present Bendel state remained ambivalently neutral. As well, some non-Ibo elements initially supported the secession, particularly some among the "minorities" of the Eastern Region, but also some other southern Nigerians who at the time felt it would hasten the breakup of Nigeria as a whole. Yet the seeds of separation are primarily found in collective Ibo perceptions of their situation within the Federation in 1966.

Iboland was the last major part of Nigeria to be thoroughly subdued by the colonial administration; initially the colonizer held a quite negative stereotypical view of Ibo culture and capacities. In the 1920s, Iboland was swept with a highly competitive "getting up" mood, pitting one village against another to seek relative advance by securing such symbolic appurtenances of progress as a local school. Young Ibos began migrating to towns throughout Nigeria; by the 1930s, they were locked in intense rivalry in Lagos for social ascendancy. The strong Yoruba-Ibo animosities, which played such a crucial role in First Republic politics, have their origins in this Lagos competition.

At that time, virtually all politically articulate Ibo were ardently devoted to a Nigeria-centered nationalism. The first major nationalist party, the National Convention of Nigeria and the Cameroons (NCNC, later National Convention of Nigerian Citizens) included many prominent Ibo, and was led by the ebullient Nnamdi Azikiwe. Though the NCNC sought to organize on a national basis, and had no conscious intent of serving as instrument for an ethnic clientele, the visibility of Ibo amongst its leadership created the perception, especially among Yoruba, that it was an Ibo movement. Thus emerged, in 1948, the Action Group (AG), as a Yoruba-centered movement. Not long afterward, the Northern Peoples' Congress (NPC) came into being, to protect against what seemed a growing threat of hegemony by southern Nigerians.

By this time, the early colonial images of Ibo as a backward, bush group had long since given way to the stereotype of the dynamic, aggressive, upwardly mobile Ibo. However satisfying such a representation was to the Ibo themselves, it was a source of fears and anxieties on the part of others. The passionate nationalism of Ibo leaders of the 1950s was a logical reflection of a social situation in which they stood ready -- in terms of their relatively large representation in the ranks of the educated -- to play a leading role in the emergent national institutions, and in which their exceptional geographic dispersal throughout the Federation gave them a strong stake in its preservation.

Toward Secession: Decline of the First Republic

In the formation of a national government in 1959-60, the corrosive logic of the three-player game began to operate. The calculus of relative incompatibilities and animosities produced, in essence, a coalition of East and North against West. An effort was then made to splinter the Western Region, with each coaltion partner hoping to secure new allies for itself; thus the non-Yoruba areas were hived off as the Midwest Region, and an opportune moment chosen to nuture a split in the AG. Within the Federal political establishment, the North proved the stronger partner; within various national institutions, a grinding struggle began to reduce the proportions of Ibo personnel, judged excessive by other groups.

By the time of the 1964 national elections, a differently defined set of battle lines emerged, with two multiparty electoral alliances (the Nigerian National Alliance, and the United Grand Progressive Alliance) anchored respectively in North and South. The elections demonstrated not only the intense passions which could be generated by ethnically defined electoral competition, but also that fraud, intimidation, and violence would be used at the regional level to permit the dominant party to keep others out. As the Northern Region contained roughly half the total population, this situation assured a permanent dominance of the national institutions by the NPC and its allies. Nigeria, thus constituted, began increasingly to appear as a cultural threat to the Ibo.

However, disenchantment with the corrupt and violent patterns of First Republic politics was found in all regions. When a group of young majors mounted a partly aborted coup on January 15, 1966, the initial reaction throughout Nigeria was relief and rejoicing at the demise of an _ancien régime_ which had rendered itself contemptible. The majors succeeded in assassinating the Federal prime minister, two of the four region premiers, and a number of senior officers. However, they did not succeed in winning power for themselves; forces still loyal to the Federal government rallied to prevent their power seizure in Lagos. Instead, a demoralized rump cabinet turned power over to the commanding general of the Nigerian armed forces, J.T. Aguiyi Ironsi.

There can be little doubt that the majors acted out of an idealistic commitment to the Nigerian nation. However, popular perceptions of their motivations began to change when further details about the coup became generally known. There was an undeniable ethnic selectivity both in the victims of the coup attempt, and in its leadership. Six of the seven majors involved were Ibo; of the seven senior officers killed who outranked the majors, only one was Ibo. The two regional governors who were murdered were not Ibo; the two who survived were.[7]

General Ironsi, who succeeded to power, was also an Ibo. To boot, he proved singularly insensitive to the inflamed cultural fears which now permeated the country. His closest civil service advisors were coethnics. In May 1966, promotions to replace those eliminated in the coup casualties, though made on the basis of seniority, saw nine of twelve top army vacancies go to Ibo. The final straw was Ironsi's sudden proclamation on May 24, 1966 that Nigeria was to be a unitary state. A wave of anti-Ibo riots at once swept the Northern cities.

On July 29, a second coup occurred, now directed by northern officers. This time, twenty-seven of the thirty-nine officers (including General Ironsi), and most of the 191 of other ranks who were murdered, were Ibo. The fate of Nigeria hung by the merest thread. The ranking military officer declined to serve as president; next in line was Yakubu Gowon, a Christian northerner from a small "minority" group, who reluctantly assumed power.

As Gowon maneuvered uncertainly to find some basis for a restored Nigerian government, Ibo troops regrouped in the Eastern Region, now under the control of Ibo Lieutenant Colonel Chukwuemeka Ojukwu. Many ranking Ibo professionals and intellectuals also took refuge in the east; by late August, most were committed to the goal of separation. Their determination was confirmed by a series of rampages by soldiers and civilians in northern cities during September and October, directed at the mercantile and professional southern Nigerian communities, with the Ibo the most visible target group. Estimates of the deaths range from 5,000 to 50,000, with between 700,000 and 2 million fleeing southward.[8] Now the issue of separation began to be posed as a question of cultural survival.

However, the official declaration of secession was delayed until May 30, 1967, as desultory negotiations were pursued. During this period, efforts were begun to acquire arms, while the option of a confederation so loose that de facto sovereignty was achieved was also pursued. Hopes were held that the complete breakup of Nigeria was imminent. Finally, Ojukwu proclaimed: "The territory and region known as and called Eastern Nigeria together with her continental shelf and territorial water shall henceforward be an independent sovereign state of the name and title of 'The Republic of Biafra.'"[9]

Initially, the Federal position was quite precarious. The commitment of its remaining components to a civil war was by no means certain; earlier that month, Awolowo and Yoruba "leaders of thought" had both issued statements that if any region seceded, the Federation was automatically defunct, and that the Western Region thereby became independent as well.[10] The remaining Federal army numbered only 7,000, without tanks or combat aircraft, and there was an inconsequential navy. Britain had made clear that it would not militarily intervene to save Nigeria, and neither London nor Washington would make a clear statement of support. Relations with the Soviet Union had never been very warm, and until this point Soviet ideological analysis of Nigeria had identified the north as "feudal," the west as irremediably capitalist, and only the east as having progressive social composition.[11] Even other African states, though they pledged not to support the secession, were initially muted in their response.

Nigerian and Biafran Objectives: The Diplomatic Struggle

For the 959 days of the secession, the struggle proceeded both on the international and the military front. For the Nigerians, the critical diplomatic goals were to secure an arms supply which would ultimately assure military supremacy, to build a solid base of support in the African state system, especially in the OAU, and to head off any external military backing for Biafra. For the Biafrans, crucial objectives were to obtain the legitimization which would come from official diplomatic recognition from some states, to find mechanisms for arms purchases on the private international market or through clandestine governmental supply, and to mobilize the sympathies of Western public opinion. In fact, both sides were able to achieve these objectives.

While Britain would not supply combat aircraft, Nigeria was assured access to light arms and ammunition through London, despite growing public hostility and the opposition of a significant number of Labour Party MPs. By July 1967, the Soviet Union had re-evaluated its perceptions of class struggle in Nigeria, and indicated its willingness to sell reconditioned MIG 17s; the availability of Soviet arms was crucial leverage in keeping the British supply channel open. While the OAU gave Nigeria some anxious moments, the three OAU summit conferences held during the civil war gave unambiguous backing to Nigeria, and condemned the secession. The relatively solid African support imposed high risks for any government supplying arms to Biafra; the most careful study of Biafran military acquisitions concludes that only France, from the latter half of 1968 on, sent significant amounts of weapons (about $5 million worth, funneled through Gabon).[12]

Biafra did have some success on the diplomatic recognition front, denied to Katanga and Eritrea. Five states acknowledged Biafran sovereignty: Tanzania, Ivory Coast, Zambia, Gabon (April-May 1968), and Haiti (May 1969). The Haiti case is idiosyncratic, and François "Papa Doc" Duvalier took his reasons with him to the grave. But the justifications for the African cases bear examination.

The crucial Tanzanian decision was partly attributable to skillful Biafran representation in Dar es Salaam, matched by equally inept Nigerian diplomacy. But Nyerere became persuaded that moral considerations intervened which overrode the OAU commitment to the integrity of existing states. In contrast to Katanga, the secession had not been instigated by sordid imperial interests. The Federal government had set terms for surrender which left Biafrans no option but to continue the struggle. As Nyerere put it:

> Leaders of Tanzania have probably talked more about the need for African Unity than those of any other country.... Unity can only be based on the general consent of the people involved.... For ten months we have accepted the federal government's legal right to our support in a "police action to defend the integrity of the State."... We watched the federal government reject the advice of Africa

> to talk instead of demanding surrender before talks could begin.... It seemed to us that by refusing the existence of Biafra we were tacitly supporting a war against the people of Eastern Nigeria.... We could not continue doing this any longer.[13]

President Félix Houphouet-Boigny of Ivory Coast was perhaps less firmly attached to OAU doctrine than many other African leaders. Further, he retained from the final colonial era a hostility to federations which were imposed by the metropolitan power; by this reasoning, Nigeria was not necessarily entitled to coverage in the principle of integrity of states. Biafran emissaries also pushed hard on a religious argument: eastern Nigeria, and especially the Ibo heartland, had been the major center of Catholic mission endeavor; 68 percent of Nigerian Catholics were in the east.[14] Houphouet-Boigny, as a devout Catholic, was not insensitive to such an appeal, particularly when the implication was added that Islamic domination threatened. Perhaps the thought also crossed his mind that a shrunken Nigeria was a less potent West African rival. Once he did act, he also brought along Gabon. It is worth noting that no other Francophonic state followed, though Biafrans had high hopes for Senegal. Worse, from the Biafran standpoint, Cameroun remained steadfastly pro-Nigerian, precluding its use as a supply channel.

Kenneth Kaunda, like Nyerere, was especially prone to viewing issues in moralizing terms. Even more strongly than Houphouet-Boigny, he carried a continuing animus toward colonially constructed federations, born of the struggle to extricate Zambia from the settler-dominanted Federation of Rhodesia and Nyasaland. In all four cases, recognition of Biafra was seen as providing pressure toward negotiation, and not necessarily as a commitment to see Biafra survive as an outcome of a peaceful settlement.

Biafra also was remarkably successful in securing enough arms, even including a few aging aircraft, to continue its military resistance. The weapons flow, as Stremlau persuasively demonstrates, came mainly through the foreign exchange spin-off of the estimated $250 million of humanitarian relief supplied through the International Committee for the Red Cross, Caritas (Catholic relief agency), and the World Council of Churches.[15] Even more effective was Biafran propaganda aimed at Western public opinion, already influenced by the Vietnam war. On this front, Biafrans placed particular stress on the genocide claims; the Ibo people, it was asserted, faced the prospect of extermination at the hands of the Nigerian army. The September-October 1967 pogroms in northern cities were cited as eloquent testimony to this risk. In the final stages of the civil war, a second theme was added, suggesting that even if the Biafran army were defeated, the imminence of genocide would leave the population no choice but continuing guerrilla war. The concurrent ability of the Vietnamese to sustain a guerrilla struggle against the 500,000-man American army lent credibility to the claim of unending insurgency. The mobilization in most Western countries of militant and vocal pro-Biafra groups was not enough to force any government into actual support for the secession. It did, however, mute their public backing for Nigeria, make arms acquisition more difficult for the Federal government, and lead to strong official backing for the relief effort as a device for deflecting popular sentiments.

The Civil War

On the military front, the Federal government had overwhelming advantages: recognized national sovereignty, which made possible official arms purchases; an enormous manpower reserve, permitting expansion of the armed forces from the initial 7,000 to 250,000; and ample fiscal and foreign exchange resources to finance the war. Within a few weeks all major figures in Nigeria had thrown their support behind the war effort; the slogan "To keep Nigeria One is a task that must be done" won increasing acceptance in public opinion as well. The Federal war effort thus faced little internal dissent; its problems all lay in the international arena.

The only possibility for Biafra to actually triumph came in the very first weeks of the war. Federal troops invaded on July 5, 1967, and quickly seized some important towns in northern Biafra, as well as the coastal oil terminal at Bonny. However, on August 9 a surprise Biafran attack across the Niger River into the Midwest Region, coordinated with a coup by Midwest Ibo officers in the Benin garrison, permitted the uncontested capture of this regional capital. Had the Biafran column at once continued toward Ibadan and Lagos, rather than waiting a week to sort out internal problems, it is quite possible that it might have occupied the Nigerian capital and triggered a final breakup of the Federation. But Federal forces were able to regroup, and to halt the Biafran strike force at the gates of Ibadan; from that point forward, it was impossible for the Federal army to lose.

What was still uncertain was whether it could win. Biafra still had, at the outset, defensible boundaries on three sides (Niger River, the sea, and Cameroun). The hope persisted that, if Biafra could hold out long enough, world opinion would force Nigeria into negotiations which would result in de facto independence, or that the pressures of the war would bring about the disintegration of Nigeria. Neither of these hopes was fulfilled; further, the Biafran leadership had underestimated one fatal flaw in their internal situation -- the alienation of the non-Ibo minorities within the Eastern Region. For these groups, strategically located along the coast, the Cameroun frontier, and the oil-bearing region of the Niger delta, membership in Nigeria, whatever its shortcomings, was more attractive than the threat of Ibo domination within a Biafran framework. Thus, outside the Ibo heartland, the Federal army could count on local support. Within a year, Biafra was sealed off from the sea and the Cameroun border area. By the end of 1968, only about 5,000 square miles of the initial 30,000 remained.

Even with the inflated Nigerian army, the final stages were agonizingly slow. The Federal forces were never able to make effective use of air power, nor could the vast new levies be quickly formed into a potent combat instrument. Finally, on January 12, 1970, the airstrip which was the final link to the outside was captured, and the secession collapsed. Contrary to the threats of unending guerrilla struggle, the dissolution of the Biafran state apparatus brought with it immediate cessation of resistance. So also was the haunting fear of genocide at once laid to rest; for the most part, Federal troops remained under control, and did not exact vengeance upon the Ibo populace.

Eritrea: The Background

Eritrea contrasts with the preceding two cases in a number of respects. Above all, it remains an unfinished chapter in the history of African state formation. The issue of its status has a much longer history, beginning with British occupation in 1941. The international dimensions of its conflict with Ethiopia have been far more important than in Katanga and Biafra, both in the origins of the dispute and its unfolding. Further, the campaign for realization of an independent Eritrea has been carried out by guerrilla struggle, rather than armed conflict conducted by an organized though unrecognized state.[16]

Eritrea originated as an imperial gleam in the eye of newly united Italy. By 1890, the colony was proclaimed as a territorial entity in approximately its present boundaries. The territory has three major regions: 1) a coastal strip, arid and torrid, where settled agriculture is impossible; 2) a highland plateau and montane region, sufficiently watered for rain-fed cultivation; 3) a dry but stream-fed plain in the west, opening onto Sudan. The coastal zone has two major ports, Assab and Massawa. In the southern part, around Assab, Danakil (Afar) predominate, most spending part of the year in Ethiopia. In the central area, Tigrai is the most widespread language amongst the pastoral populations. In the northwest, the large Bani Amer confederacy includes Beja- and Tigrai-speaking clans, with significant Arabic bilingualism. Arabic is also an important language in both Massawa and Assab; the coast is entirely Muslim. In the highlands, with the capital of Asmara at the center, the major group is the Tigre, Tigrinya-speaking, and Coptic Christian.[17] They are an extension of the Tigre community in northern Ethiopia, which historically was one of the regional principalities in the usually loose-knit Ethiopian empire. In the western plain, in addition to elements of the Bani Amer confederacy, one finds, on the edge of the highlands, some Tigre, as well as small numbers from Nilotic groups who in language, religion, and origin diverge sharply from all other Eritrean groups.

British Occupation

In 1940, the 300,000-man Italian East African army stood poised in conquered Ethiopia to strike at Sudan and Egypt; the neutralization of this force became a key British objective. Extensive leafletting was done, with appeals to the numerous Eritrean soldiers to defect; the British, psychological warfare services promised, came to bring the Eritreans liberation and their own flag.[18] The Italian army proved far less formidable than its numbers suggested, and defections by Eritrean soldiers soon became numerous, partly induced by the British promises. Eritrea was swiftly overrun at the beginning of 1941, and a temporary wartime administration was installed.

To Eritrean disappointment, the British maintained that they were installing only a caretaker military occupation, essentially retaining the Italian colonial administration under British direction.

Not only had Eritrean expectations been raised, but so also had Ethiopian. Haile Selassie, at the time of his triumphant return in April 1941, had expected that he would be able to lay claim to the former Italian colonies -- not only Eritrea, but also Somalia. But the disposition of the Italian possessions, he was told, would have to wait until the end of World War II.

The Postwar Debate: "Disposal" of the Italian Colonies

At the conclusion of the war, it was decided that the Big Four would determine the future of the occupied Italian colonies. Very diverse perspectives and interests had emerged during the war; the three powers with the most direct interests and most clearly formulated objectives were Italy, Britain, and Ethiopia. The Italian view was quite simple: Italy should be afforded the opportunity, through trusteeship if necessary, to complete its "civilizing mission." As a constituency for this claim, there was a still numerous (perhaps 40,000) Italian community, and at first some Eritrean clientele (veterans and pensioners, employees of Italian enterprises).

British officials in Eritrea had in large part been drawn from the cadres of African colonial administrations, with views characteristic of that milieu; Ethiopia, though a wartime ally, was seen as a primitive and backward land, quite incapable of providing progressive tutelage to new territories. As Stephen Longrigg, chief administrator of Eritrea during much of the war, put it: "The reasons must be sought in the isolation which, for mistaken motives, [Ethiopia] imposed upon itself; in the unreceptive pride or arrogance of its people; and perhaps in some defect or essential limitation in the African character."[19]

Eritrean independence, from the British perspective, was precluded by the lack of "viability" of the territory. "Viability," a particularly intriguing concept, appeared to mean primarily the ability to sustain through local revenues the cost of colonial administration. In London, the impulses for imperial expansion had grown cold; there was not much interest in adding this economically uninteresting real estate to the embattled empire. Thus the logical solution was dismemberment. The Danakil areas might go to Ethiopia, as these pastorals were transborder seasonal migrants anyway, and the majority of their groups resided in Ethiopia. The Tigrean highlands were admittedly linked historically to the kingdom, and their sentiment for union was at that time strong; however distasteful it was to surrender this territory to Ethiopia, there seemed no choice. The port of Massawa was irrevocably tied to its highland hinterland, and would have to share the same destiny. This left western Eritrea, which might conveniently be added to Sudan; its water courses were crucial to the potential expansion of irrigated agriculture in the eastern Sudan. The "viability" concept precluded separate independence for this zone; thus Sudanese annexation was the only possibility.

The Ethiopians, outraged to find themselves largely excluded from the diplomacy of the postwar settlement, adduced a number of arguments in support of annexation. The Ethiopian brief began with the historical claim that these lands had always been part of the millenial kingdom. While these arguments had a solid foundation for the Tigre highlands, they were more controversial with respect to the Muslim lowlands; here the historical connection had only been in the form of loose tributary relationships at moments of peak effectiveness of the empire. Commonalities of cultural heritage were next invoked, a theme more recently given elegant scientific synthesis in the Levine vision of Ethiopia as a historically ordained ingathering of peoples, rather than a beleaguered Amharic-Tigre Christian state encircled by a hostile Muslim or pastoral periphery.[20] Thirdly, Ethiopia had an inherent right to an outlet to the sea. Interestingly, in the late 1940s, nearly everyone accepted this argument, which has long disappeared from African international law, with the advent of fourteen landlocked states. However, those opposing annexation suggested this need could be satisfied by a "Polish corridor" to Assab. Fourthly, Ethiopian security needs required annexation of these lands, which had been used as invasion springboards in recent centuries by the Turks, Egyptians, British (the Napier expedition in 1868), and Italians (1889, 1896 and 1935). Finally, the economies were held to be interdependent.

Other powers had their own interests to pursue as well. The USSR at first hoped to acquire trusteeship over one of the Italian colonies. The United States, which had developed a communications facility near Asmara in 1942, became interested in retaining this base as the Cold War gathered force; initially, Washington backed trusteeship leading to independence, but, as the possibility of a military relationship with Ethiopia emerged, by 1950 swung to some formula for attachment to Addis Ababa. The French were concerned with securing their Red Sea base at Djibouti.

The UN as Arbiter of Eritrean Self-Determination

The Big Four, however, proved unable to agree upon a formula, and in 1948 turned the problem over to the UN General Assembly. Here new external participants entered the picture. A number of Latin American delegations were receptive to Italian solicitations. Pakistan and the Arab states saw themselves as protectors of the interests of the Muslim population against encroachment by a Christian state. Liberia was a strong supporter of the Ethiopian cause; South Africa believed its interests at stake in the disposition of any African territory.

The outlines of a compromise finally emerged by 1950, through an internationally guaranteed federal link to Ethiopia. The Italians had by now given up on their trusteeship hopes. The British were unable to find any backing for their partition scheme, which was fatally weakened by the inability of their administration to discover any Eritreans who supported the idea of attachment of the western region to Sudan.

The role of Eritreans themselves in this process was quite secondary. Two international commissions were dispatched to ascertain their preferences, the first by the Big Four in 1948, and the second by the UN General Assembly in 1950. Both commissions came back with badly divided reports (Americans and British versus French and Soviets in the first, and Norwegians, Burmese and South Africans versus Guatamalans and Pakistanis in the second). In truth, there was no single Eritrean view, nor a very clear sense of what the alternatives were. By 1948, there was widespread hostility to any restoration of Italian authority, in trusteeship or any other form. A substantial and aggressive Unionist Party had emerged, based on the Tigre highlands, though it included some Muslims. According to Longrigg, who was unsympathetic to their cause, the Unionist supporters were "the young race-conscious and usually Mission-educated intelligentsia of Asmara; the Coptic priesthood, who favour the Emperor in the hope that he will favour them; a small proportion of the chiefs and village heads; and a very few of the merchant class."[21] The Unionists, with obvious backing from Ethiopia, were increasingly militant in their action; by early 1950, there was a wave of terrorist activity, aimed at both Italians and leading opponents of unification. Most Muslim spokesmen wanted separate statehood, though one western faction wanted secession and independence for its region. The prolonged uncertainty and repeated external canvassing served to politicize the religious cleavage and provoked not only the Unionist terrorist campaign, but an upsurge in _shifta_ ("bandit") activity in many coastal zones.

The Failure of Success: The Federal Compromise

The federal solution was finally accepted, with varying degrees of reluctance and apprehension, by all parties. Ethiopia from the outset considered the formula dangerous and unworkable, but acceded only because it could bring Eritrea within its sovereignty. While there were enthusiasts for the federation in the Christian highlands, some intellectuals were beginning to look uneasily at the authoritarian style of the Ethiopian state. Muslims were not wholly reassured by Ethiopian claims that there was full religious tolerance in a state whose inner core appeared to them Christian, nor were they persuaded by such Ethiopian explanations of the smaller numbers of Muslims in top positions as that offered in rebuttal to Pakistani charges at the United Nations: "If more Ethiopian Muslims did not take part in the administration of their country, it was because they preferred to devote themselves to commerce."[22]

A constitution embodying the federal compromise was drawn up under UN auspices and presented to a newly elected Eritrean assembly and the Ethiopian government for acceptance. Both bowed to the inevitable and duly ratified the document. The scheme adopted had a number of anomalous features, which would quickly lead to difficulties. An Eritrean chief minister was to be elected by the Assembly; he would appoint a team of ministers and a judiciary. Eritrea, with its own budget, would assume all governmental functions except foreign affairs, defense, currency, communications and trade. The federal government, which was to be identical with the Ethiopian government, would have a resident representative in Asmara. However, Eritrea would have its own flag, police, and languages (Tigrinya and

Arabic).

At the beginning, a mood of amity and relative optimism prevailed.[23] However, Ethiopia found a number of the provisions of the constitution anathema. A full-blown parliamentary democracy in Eritrea, when political parties were forbidden in Ethiopia, was constraining. The lack of direct police and taxing powers was frustrating. Tigrinya and Arabic as official languages ran counter to nation-building concepts then in vogue in the capital, which saw Amharic as the linguistic instrument of unification and integration of Ethiopian peoples.

With the UN effectively removed from the picture, Ethiopia set about undoing the imposed formula. In 1956, the Assembly was suspended; earlier, the constitution itself had been put in abeyance. Amharic was declared the sole national language. New Assembly elections were held without political parties, and returned a docile chamber. This new body voted in 1958 to discard the Eritrean flag, in 1959 to accept the Ethiopian penal code, in 1960 to redesignate the territory as the "Eritrean Administration" rather than "Government," and -- apparently -- in 1962 to dissolve the Federal Act and make Eritrea a simple province of Ethiopia.[24] Neither the UN nor any other external power took official note of this change.

The Beginnings of Liberation Struggle

By the mid-1950s, an exodus of middle-class Eritreans associated with opposition to union had begun. In 1958, several well-known figures, mostly Muslim, formed the Eritrean Liberation Movement (ELM) in Cairo. In 1961, setting a pattern which has remained constant in Eritrean liberation politics, one faction of the ELM broke away to found the Eritrean Liberation Front (ELF). With a £ 6,500 purchase of antique Italian rifles, an ELF band launched its first military attack in September 1961.[25] The ELM in 1965 attempted to launch its own armed campaign, but its handful of insurgents were liquidated by the ELF. Then, in July 1970, a more ideologically self-conscious and radical faction in the ELF broke away in turn to form the Eritrean Peoples' Liberation Front (EPLF). At first, the EPLF was mainly composed of Christian Tigre, but it soon enlarged its clientele to include many Muslims. From 1971-74, a ferocious civil war pitted the ELF and EPLF guerrillas against each other. In 1976, a further split occurred when the EPLF severed connections with Osman Saleh Sabbe, who had headed their "foreign mission." Osman, a Muslim spokesman since colonial times, opposed the socialist orientation of the EPLF, but was also at odds with the ELF; he accordingly formed his own splinter, the ELF/PLF (Eritrean Liberation Front/Popular Liberation Forces).

In the two decades of armed insurrection, there has been a short-term pattern of cyclical fluctuations in the nature and intensity of conflict, and a long-term trend toward increasingly massive confrontation between the Ethiopian state and the Eritrean people. The ebb and flow of action reflected changing degrees of arms availability, fluctuations in the external conjuncture, and varying strategies in Addis Ababa for dealing with the insurrection. The long-term trend was determined by the incorporation into the struggle of widening segments of the Eritrean population, transformation of the insurgency from quasi-*shifta* skirmishes to a people's war, a progressive politization and ideological radicalization at all levels of the Eritrean forces, and increasing violence.

On the Eritrean side, there was a progressive build-up of an exile and refugee population, which provided the external support base. A handful of those leaders who had opposed union took refuge abroad beginning in the 1950s. Growing numbers of Eritrean students pursued their educations in the Middle East, Europe, and America. Significant numbers began to find employment in the Middle Eastern oil states. Then, beginning with the first major anti-insurgency sweep by the Ethiopian army through Muslim zones in 1967, a refugee population of rural people took form in Sudan. The first punitive military measures produced about 30,000 refugees; by 1979, the figure was over 500,000, and today may total nearly a third of the population.[26] With the overthrow of the emperor by the military Dergue in 1974, and the bitter and bloody power struggle in Ethiopia which began with the wave of executions in November of that year, many of the substantial Eritrean population of professionals, civil servants, and students in Ethiopian cities fled either abroad or into the guerrilla zones. This influx into the resistance and refugee ranks of well-educated, often ideologically radical, and mainly Christian elements tended to dilute the earlier Muslim predominance in liberation organization ranks, and hastened the doctrinal evolution along radical and Marxist-oriented lines.

Guerrilla Insurgency: From Shifta Skirmishes to People's War

In tandem with these changes went a gradual metamorphosis of the military struggle. Through much of the 1960s, the links between the exile leadership and internal insurgents were weak. The first skirmishes appear to have been episodic ambushes carried out by *shifta* bands induced to strike by the shipment of arms across the Red Sea.[27] By degrees, the thin political cloak to endemic banditry by pastoral clans became a guerrilla *maquis*, operating in the Muslim lowlands where settled administration had always been thin. From 1975-77, extensive liberated zones opened, with free access to external leadership in neighboring Sudan, and the whole infrastructure of a modern revolutionary movement swiftly crystallized: a countergovernment in gestation, with an active social program of schools, medical outposts, and intensive ideological education.[28] In March 1977, Eritrean forces captured the district capital of Nafca, and by the end of that year controlled 90 percent of the territory and nearly all towns except Asmara, Massawa, and Assab; this made possible the establishment of an embryonic state. Ethiopian counterattacks in 1978, logistically backed by Cubans and Soviets, and benefiting from the huge inflow of Soviet weaponry during 1977, pushed the Eritreans back into their rural redoubts, and into refugee camps. At their high water mark in 1977, the various Eritrean guerrilla forces totalled nearly 40,000, well-equipped with light weapons.

Initially, Ethiopia responded to the insurrection by seeking to isolate, contain, and conceal it. The occasional skirmishes were dismissed as mere *shifta* activity, an interpretation which initially had a kernel of truth. As the Ethiopian administration was mainly concerned with the port cities and Christian highlands, where most economic activity was centered, a degree of disorder in peripheral lowland zones could be ignored. By the middle 1960s, some officials and military circles began to fear that the maquis was consolidating itself, and to push for more vigorous countermeasures. The 1967 sweep, with destruction of villages accused of harboring insurgents and aerial attacks against suspected guerrilla units, was the first major response. Guerrilla action died down for a time, but flared up again in 1969, and was now extended to external targets: plane hijackings and attacks against Ethiopian Airways in Frankfurt and Karachi. A second and more determined set of military operations was undertaken in 1970-71, again momentarily reducing the scale of insurgency. In 1972-74, the liberation movements were preoccupied with decimating each other, permitting the Ethiopian army to stand aside. After a brief interlude when it appeared that the postimperial regime might seek accommodation, a new and far more lethal set of repressive operations was undertaken in 1975, including a denial of drought relief supplies to rural Eritrea -- afflicted like other northern regions of Ethiopia with a devastating rainfall failure.[29] A period of internal conflict and weakness in the Dergue, the preoccupations of the revolutionary land reform, and other insurgent challenges around the periphery opened the way for the Eritrean triumphs of 1977. However, with the Ogaadeen campaign won with the aid of Cuban forces and Soviet arms, the Ethiopian army was able to launch its counteroffensive in 1978 with even greater ferocity.

The International Dimension

Finally, swiftly fluctuating external contexts played a critical role. From the Eritrean side, at the beginning hopes were placed in the UN, to which a number of petitions were addressed in the 1950s. Within the African state system, the OAU Charter firmly excluded Eritrean claims. Notwithstanding, several African states -- Libya, Egypt, Sudan and Somalia -- did at some point provide help. The politics of pan-Arabism provided openings to the Arab world; Syria gave relatively consistent support, and Iraq, Kuwait, Saudi Arabia and South Yemen provided conjunctural backing. However, heavy reliance upon Arab support involved claiming an Arab vocation and stressing the Islamic dimension, which alienated the Christian Tigre highlands. Ideological radicalization, from the late 1960s on, opened the door to Cuba, Libya (after 1969), China and South Yemen.

Arms supply could reach the combatants either through Sudan, or across the Red Sea; this made the attitudes of Khartoum and the coastal states particularly crucial. Ethiopian diplomacy was able to close the Sudanese conduit at several moments (1967, 1970 and 1980) with a crucial impact on the scale of guerrilla operations. Egyptian military operations in North Yemen provided an abundant source of weapons in the early 1960s. The Qaddafi coup in 1969 provided a generous benefactor from then until the Soviet-Cuban entry in 1977, which also sealed off the South Yemen pipeline.

On the Ethiopian side, diplomatic strategies under both emperor and Dergue have been constant: seal off the sanctuaries, deny any external recognition, and maintain a military link with a major arms supplier. The last two have always been assured. The first has been pursued with considerable skill, although with varying success. Under Haile Selassie, Ethiopia had the leverage of possible supply to the Anya-Nya insurgents in southern Sudan; more recently the burgeoning Eritrean refugee population has constituted a heavy burden for Sudan.

The final outcome in Eritrea remains indeterminate. Twenty years of armed struggle, while it has not brought unity to the insurgents, has produced a far-reaching politization of the Eritrean population; veteran student of Third World revolutionary movements Gerard Chaliand has called the EPLF, now the stronger of the liberation organizations, "by far the most impressive revolutionary movement produced in Africa during the last two decades" (thus ranking it ahead of the PAIGC and FRELIMO).[30] It now confronts a revolutionary state with a powerful army, formidable equipment, and potent external allies.

In 1976, the Dergue offered a vaguely defined autonomy, formulated in the not very reassuring phraseology of Soviet nationality policy:

> The program of the Ethiopian National Democratic Revolution has affirmed that the right of self-determination of nationalities can be guaranteed through regional autonomy which takes due account of objective realities prevailing in Ethiopia, her surroundings, and in the world at large. To translate this into deeds, the government will study each of the regions of the country, the history and interactions of the nationalities inhabiting them, their geographic positions, economic structures and their suitability for development and administration. After taking this into consideration, the government will, at an appropriate time, present to the people the format of the regions that can exist in the future. The entire Ethiopian people will then democratically discuss the issue on various levels and decide upon it themselves.[31]

By 1981, a new strategy was becoming visible, by which language and ethnicity might be played against the Eritrean national idea. There were hints that different linguistic zones of Eritrea might be given separate administrative status and nominal autonomy; preparations were well advanced to implement such a design for the Danakil (Afar) zone in southern Eritrea and contiguous areas of Ethiopia. Autonomous zones might also be created for the Tigre, Tigrai, and Kunam linguistic areas.[32] While such a scheme would be strongly resisted in the Tigre and Tigrai areas, where the Eritrean liberation movements are strongly implanted, it might well work with the Kunam as well as the Danakil, who have remained on the sidelines of the struggle.

The Dergue in 1981 was in a far stronger military position than in 1977. The new entente with Sudan, closing that critical supply route, would necessarily circumscribe the scale of guerrilla actions by the Eritrean movements. But a _maquis_ on a more modest level can probably sputter on for a very long time, in the hope that the external conjuncture may again turn favorable. While negotiations are again under discussion, a real basis for settlement is not yet visible. The Dergue, like its imperial predecessor, would be fearful of an autonomy which permitted Eritreans to acquire real control of regional institutions, and possibly make another lunge for independence. The Eritreans remember all too well the fate of the autonomy constructed for them by the UN in 1952. While a settlement is not inconceivable, the barriers of mutual distrust are very great.

Laying Claim to Self-Determination: The Unit

We turn now to a comparative examination of our three cases, beginning with the issue of the conceptualization of the unit laying claim to political sovereignty. Here the first striking fact is that, in contrast to many separatist movements in other parts of the world (Basques, Bangladesh, Scotland, the Tamil regions of Sri Lanka, Kurdistan, Nagas in India, Karens and others in Burma, to mention but a few), none of these three are culturally defined. The right of self-determination is asserted for territorial segments which themselves originate in arbitrary colonial divisions; Katanga and Biafra were provinces, and Eritrea a separate colony. Until the 1950s, while "Congolese" was a widely employed synonym for Africa under Belgian rule, "Katangan" had only trivial importance. "Biafra" is a designation borrowed for the purposes of secession; its predecessor, "Easterner," lacked social resonance. Eritrea is a Greek word for the Red Sea; all observers agree that little national consciousness existed in the territory before the uncertainties as to its disposition, and above all the guerrilla struggle, created one.

In fact, the self-determination claims were critically weakened -- fatally for Katanga and Biafra -- by the ambivalence or rejection of these concepts as a basis for sovereignty by important elements of the population. The rebellion in the Luba zones, which could never be eliminated, compromised the Katangan claims from the outset. Had the secessionist province been able to demonstrate that the entire population supported its initiative, its chances of winning some international recognition would have been greatly improved. For Biafra, the hostility of the Ijaw, Ibibio, Efik, and other non-Ibo groups to prospects for an Ibo-dominated independent state had a major impact on the military campaign, permitting the Federal forces to entirely surround the Ibo redoubt.

In the Eritrean case, the support in the Christian highlands for unification in the 1940s was crucial to the Ethiopian success in winning UN approval for the federation compromise. Both the visiting commissions recorded most Muslim opinion as hostile to union; had they encountered the same response in the Tigre areas, the Ethiopian claim would have been rejected at that point. Eritrea, like Somalia, would have been placed under some form of trusteeship, which would have led to independence as a single state. During the early guerrilla struggle period, the independence movement was mainly Muslim. The persistent disunity amongst the liberation movements is not simply a reflection of religious and ethnic cleavage, but these are undeniably major factors. In the 1970s, the rise of ideology and influx of Tigre Christians into both the ELF and the EPLF has blurred the religious factor; however, each movement has a pronounced regional clientele and orbit of operation. (EPLF is stronger in the Christian highlands, with its Muslim support from Massawa urbanites and the Sanhar and Saho ethnic groups in the central coastal zones, while the ELF predominates in western Eritrea and is mainly Muslim.)[33]

The other side of this point is that the underlying basis for the secession claim propounded in territorial terms may be found in grievances held by particular cultural communities. In Katanga, the resentments over social predominance by Kasaiens were felt most strongly by the several southern Katangan ethnic communities who formed a political alliance which they termed a "confederation of tribal associations." The Biafran movement essentially stemmed from Ibo frustrations and fears. The Eritrean liberation idea originates primarily in the anger of coastal Tigrai Muslims at being coerced into an essentially Amharic Christian unitary state over time, with Tigre Christians becoming progressively disaffected by Ethiopian policies. The Danakil (Afar), and culturally distinct populations of southeastern Eritrea remained outside the conflict. The crucial point is that, while owing their genesis to cultural pluralism, secessionist claims, through articulation in territorial terms, alter their nature.[34]

As the separating units, like the national states, originated in the colonial partition, the secessionists could not easily claim an historical sanction. Some efforts were made in this direction; however, Tshombe maintained that Katanga was no mere European administrative entity, but the contemporary embodiment of prestigious precolonial kingdoms:

> To serve certain political designs, people have pretended that Katanga did not exist, that it was a construction of the colonizers. This is to deny that, when the first white explorers discovered the part of Africa called Katanga, they found three monarchies which were not only bound by family, economic and social links, but, and this is by far the most important, their historic destiny had been linked for centuries.... These monarchies constituted in the heart of Africa an entity apart, matured slowly over a long historic period, with July 11, 1960, only the manifestation of an awakening self-consciousness.... When the Belgians and the English...tried to lay their hands on Katanga, the Baluba, Lunda, and Bayeke chiefs were united in the face of the new danger which threatened their sovereignty.... It was for the first time the common resistance to a foreign effort to impose its will on Katanga.[35]

Such claims had little credibility within or without Katanga. Historically the arguments were problematic, as the kingdoms concerned did not coincide with contemporary Katanga; while there were links, their relationships were primarily conflictual. More importantly, everyone knew that the contending sets of political and economic interests in 1960 had nothing to do with precolonial history.

In the Biafran case, there were occasional references to the "mistake of 1914," and the unreasonable yoking of southern Nigeria to what was labeled the backward, feudal, Muslim north. The Eastern Region per se was not put forward as a historically predestined ensemble. The possibility remained open that Nigeria might entirely break up, and that other parts of the south might become linked on some new basis.

In Eritrea, historical arguments did play somewhat more of a role, above all negative. As the Ethiopian brief for annexation rested in part upon claims that Eritrea was an historical part of the traditional Ethiopian state, Eritreans needed to disprove these links. This in itself was troublesome, because the Ethiopian arguments had solid foundation with respect to the Christian highlands. The clinching point, however, was a legalistic one: the Ethiopian state, as a free and sovereign entity, had voluntarily signed with Italy successive treaties establishing the frontiers of kingdom and colony. Eritrea existed not only by Italian creation, but by Ethiopian renunciation.

In the Biafran and Eritrean campaigns for external sympathy and support, tacit use of a cultural self-definition was employed. The quite efficacious Biafran propaganda operations in Western countries did employ imagery of the Ibo as an embattled and persecuted people, threatened with extermination. The echoes of holocaust themes were by no means coincidental, and subtly linked elements of the Ibo stereotype of success-oriented upward mobiles to those of Jews -- to whom they had not infrequently been compared. However, it was never suggested that the large Midwest Ibo community should automatically be incorporated; to do so would have conceded that Biafra was an ethnic, and not a territorial state. Eritrean external representatives, in their ceaseless search for arms and money as well as support, found Arab states the major source of backing. The ELF, and Osman Sabbe Saleh, were willing in return to identify themselves with Arab causes, and even to imply that Eritrea formed part of the Arab nation, broadly defined; similar reasoning led Somalia to join the Arab League. As one ELF pamphlet put it: "Our relation with the Arab nation is not an emotional or superficial, but militant, organic, historical, and cultural one based on bonds of the joint destiny, mutual and common interests, and solidarity in face of menace and aggression."[36] Such declarations were not well received outside of Muslim milieux, and were one source of friction with the EPLF.

Within the African state system, grounding the sovereignty claim on a territorial and not a cultural basis was indispensable to gaining any support at all. Biafra was able to make some headway with Zambia and Ivory Coast on the argument that the territory was coerced into a colonially designed federation which it could not accept. Eritrea, however, had a stronger juridical claim, in African terms; it had been a distinct colonial territory. Indeed, its situation is closely analogous to that of the Western Sahara, whose claim to separate and independent statehood has been accepted by nearly half the African states. The crucial difference is historical timing. The federation and then annexation of Eritrea occurred before the OAU Charter was drafted; by a kind of "grandfather clause," its potential claims were extinguished. Ethiopia, which as host to the founding OAU conference played a major part in the adoption of the charter, was second to none in its anxiety to see the state integrity clause formulated so as to cover its own current boundaries.

Justifying Separation to the External Audience

The presentation of full justification for separation aimed at the external audience is of paramount importance for a would-be secessionist. Separation has virtually no chance of success -- assuming, as is invariably the case, that the existing state resists dismemberment -- unless some international backing can be obtained. It is striking that, since World War II, Bangladesh is the only state to succeed in an opposed secession, with the Indian army playing a critical role.[37] Sovereignty must be recognized to be effective. Internal resources will never be enough to withstand the military efforts of the existing state to preserve itself; various forms of assistance will be required from at least some other existing states. Accordingly, the briefs for secession presented externally by the three territories are worth scrutiny.

Katanga had the potential advantage of making its case before the sanctity of existing state boundaries doctrine was fully institutionalized. Had African support been unanimous within Katanga, and had the separation not served so transparently the interests of the mining corporations and European settlers, the case might have been justifiable to at least some African states; the separation of Rwanda and Burundi, governed as a single unit by Belgium, had reluctantly been accepted by African opinion when close inspection revealed the total incompatibility of these two entities, especially after the 1959 Hutu revolution. But Katanga stood convicted in African opinion from the outset; if this secession were to succeed, then, to many African leaders, nothing could stop capitalist, imperialist, and southern African racialist interests from destroying African independence through dismemberment everywhere. Only the Congo-Brazzaville of Abbé Fulbert Youlou openly espoused the Tshombe cause.[38]

Katanga arguments could thus only be addressed to Western powers. Tshombe offered a warm reception to Western capital, and portrayed himself as an implacable foe of communism. The radical nationalism of Lumumba could only bring disorder and chaos, which in turn would lead to communist domination. Katanga offered an oasis of peace, desiring to preserve its Belgian ties while opening the doors to all other Western suitors. Not even the Belgian government could be won over fully by this appeal; too much was at risk in the remainder of the country to permit total commitment to Katanga. The propaganda directed at Western opinion won support only in right-wing circles: Suez Tories, Senator Thomas Dodd, some of the *algèrie française* "ultras."

Biafra, which excelled in its diplomacy, focused upon a quite different audience. Moral mobilization, particularly in universities, was exceedingly strong over the Vietnam war. The Biafran case was presented in terms calculated to divert some of the emotional energies of the antiwar movement to the Biafran cause. Biafra was thus not an historically sanctioned entity, but a martyred people desperately seeking protection against annihilation. Only the shield of sovereignty could bar this tragic outcome. In Catholic milieux, quiet use was made of the heavy concentration of Catholics in

Iboland, an argument which, as we have seen, was influential with the Vatican. For church audiences generally, note was made of the Muslim hordes at the gates. The spectre of genocide, and the implicit Vietnam analogy of interminable guerrilla war, both pointed to the conclusion that only negotiated settlement could end the tragedy and avert the holocaust.

The key claim advanced by the Eritreans is that they are engaged in a national liberation struggle against a colonial occupation. Their struggle is a true people's war against a foreign oppressor; it deserves support by all progressive people precisely because it is not a simple separatist movement, but a revolutionary combat of the same category as the Vietnamese, Algerian, Mozambiquan, and Guinea-Bissau liberation wars. While Eritrea was initially admittedly only a colonial territory, it has become through struggle a true nation, whose oppressed masses have in the course of two decades of privation and sacrifice developed a revolutionary consciousness.

Since unification with Ethiopia, they argue, all the classical elements in colonial subjugation may be observed. An alien language, Amharic, has been imposed; the Ethiopian state is an instrument of Amhara domination. Eritrea is economically exploited for Ethiopian benefit; no major industrial or agricultural development projects have been sited in Eritrea, and the main public investment has been in Assab port, which serves Addis Ababa.

The initial federation, Eritreans maintain, was imposed by the UN, without their full consent. Criteria were used to refuse them independence (the "viability" concept, the right of a landlocked country to access to the sea) which were not applied to other African states. Because free choice was denied, the right to self-determination had never been exercised.

Eritrean appeals have never enjoyed the external audience which the Biafrans achieved. In the first years of the Eritrean struggle, they appealed mainly to the UN, on the grounds that Ethiopia had violated the terms of the original constitution. However, the 1952 settlement had made the issue an internal affair of a sovereign member state, and Eritrea found little sympathy. More generally, Ethiopia in the 1950s still seemed a progressive symbol of African independence. The flaws in its political system were not so manifest as they became by the last years of the monarchy. From 1953 on, it had close military links with the United States, which gave American official milieux a strong interest in propagating the Ethiopian cause.

Until 1969, Eritreans made no major effort to secure the attention of the international media. At that point, they advertised their presence by a few terrorist spectaculars, like the plane hijackings, and also began to seek out Western journalists of the left.[39] They began to stress, at this juncture, the "feudal" character of the Selassie regime, and the role of American imperial interests in sustaining it.

However, as their self-portrayal became more radical, they were overtaken by the unanticipated advent of the Ethiopian Revolution. The ideological schema of the Dergue, by 1976, was essentially the same as that of the more radical Eritrean movement, the EPLF -- a coincidence which gave rise to the abortive Soviet-Cuban scheme, put forward by Fidel Castro in spring 1977, to form a broader socialist commonwealth in the Horn, including South Yemen and Somalia as well as Eritrea and Ethiopia. These confusing events cut off the Eritrean movements from the kind of backing they might otherwise have received from activist groups on the left.

For a brief moment in 1976-77, the Eritreans could invoke the most compelling argument of all: success. When the guerrillas could show visiting Western observers impressively functioning liberated zones with improvised but no less inspiring social services, there was an image of inevitability to their triumph. This, however, evaporated in the massive Ethiopian counteroffensive of 1978.

The Role of Force

Whatever the importance of international norms at the state level, or moral values which may activate segments of public opinion, secessionists must almost always defend their claims by simple force. International law has as one defining feature the absence of either authoritative organs whose writ is universally accepted, or machinery for enforcing compliance. Thus the military capabilities of separators are crucial.

Katanga and Biafra fundamentally differ from Eritrea in that the proponents of separation in the former two cases possessed an existing apparatus -- a major subdivision of the repudiated state. Paradoxically this both facilitated their initial steps in seeking international legitimacy and made them fatally vulnerable to complete defeat through conquest. A state-based separation is not convertible into a guerrilla liberation movement.

In the Katanga case, the pre-existing provincial institutions did not include a significant coercive vehicle. There was a small police force under provincial command which had been locally recruited in colonial times, plus unarmed local authority constabularies. The initial consolidation of the secession was made possible only by the center's total loss of control over its army (which included a large Lubumbashi garrison), the intervention of Belgian troops to disarm and expel the mutineers, and the availability of a small but armed and trained volunteer corps of local Europeans.

At top speed, a Katanga gendarmery was brought into being, mainly composed of newly recruited soldiers, with the leadership and logistical support mainly supplied by Belgian and a few French officers. This force, which may have numbered 20,000 at its peak, was mainly recruited in the southwestern parts of the province where the core support for the Tshombe regime was concentrated. It was soon buttressed by several hundred white mercenaries, organized into separate units. The workshops and large vehicle stocks of the mining companies, in addition to their financial resources, also played a key part.

The combat effectiveness of this force was open to doubt, as demonstrated by its inability to cope with the BALUBAKAT dissidence in the north, an insurrection carried out without modern firearms. However, during the period of the secession, the Katangans were not seriously threatened by the national army, which recovered very slowly from the trauma of the mutinies. The decisive military fact throughout this period was the superiority of the UN peacekeeping force to any Zairian units. Kinshasa could not in practice invade Shaba without UN consent and support, which was never given.

Thus the Katanga secession could only survive through diplomatic action to forestall the use of force by the UN to end the separation; for thirty months, this worked. UN detachments were posted in the two leading Shaba towns (but not elsewhere in the province) from August 1960. However, as we have seen, the ambiguities of their formal mandate precluded firm military action to end the secession. The impasse was only ended by an unauthorized initiative from the field.

In the Biafran case, the first consolidation of secession was also made possible by the absence of national troops in the Eastern Region. The unravelling of the army which occurred as a consequence of the January and July 1966 assassinations left it unable to immediately respond. In August 1966 there was still a detachment of northern troops in the Eastern Region, but Ojukwu was able to persuade Gowon to remove them. Unlike in Katanga, Ojukwu had a solid military nucleus available, in the form of Eastern men and officers from the Nigerian army. Ibo had been quite numerous in the officer corps, and also were strongly represented in the supply and technical services. At the very beginning of the civil war, Biafran forces were at near parity with the Federal army. As noted, Biafran forces very nearly brought the war to a swift end in August 1967.

Thereafter, Biafra had no hope of matching the capacity of the Federal forces for expansion. Its military strategy was simply defensive: prolong the war long enough so that international pressure was brought to bear on Nigeria, or the rest of the federation fell apart. As Nigerian military preponderance grew, the threat of the guerrilla war became more important as a component in external propaganda; the international arena needed to be persuaded that a Nigerian army triumph through conventional means was not enough to extinguish Biafra.

However, though the guerrilla menace seemed credible to many at the time, it was an empty threat. A regular army does not easily undergo metamorphosis into guerrilla units; its routines, habits, and mentalities do not permit such adaptation. The rural infrastructure of partisan warfare must be painfully constructed over time; the sea is not automatically ready to receive its fish. An existing state bureaucratic structure is singularly maladapted to such a task. The major guerrilla movements of modern times -- for example in occupied Europe during World War II -- never emerged from a conversion of existing state structures and forces into a _maquis_.

The Eritrean liberation struggle rests upon guerrilla action -- hampered, as we have seen, by its permanent duality. Such a campaign can triumph only if the central state itself decomposes -- as it nearly did in 1977. However, once well established and institutionalized, the guerrilla movement is correspondingly difficult to totally defeat.

Another particularity of the Eritrean situation is that the combat zones, for much of the war, have been in the pastoral regions, where in the best of times, under Ethiopian sovereignty, administration was a low-density, town-based, indirect rule structure operating through clan headmen. The state did not ask very much from the lowland clans; taxation was unusually low during the Italian period (partly because Rome, unlike virtually all other colonial powers, did not insist that the colony pay for its own administration; only briefly at the end of the 1920s was the Eritrean budget ever self-supporting).[40] Neither did the semi-settled or nomadic groups ever receive very much from the state; they were only very loosely incorporated into the state domain. The local politics of agnatic segmentary societies, with a continuing swirl of rivalry and conflict, and a predatory tradition toward intruders, provided a supply of fighting men who could be sporadically mobilized with an influx of arms and money. Over time, however, there was a gradual construction of a revolutionary consciousness, and of a true guerrilla force by the 1970s. Once well rooted, a guerrilla movement is a formidable adversary.

However, even a well-established *maquis* requires a flow of arms and other resources from abroad, and external sanctuary to which units can retreat if necessary. As conduit and sanctuary, Sudan was particularly crucial. So also is the position of riparian Red Sea states. The periodic loss of the Sudan base was invariably attended by a sharp reduction in the scale of guerrilla action.

Finally, the guerrilla phase of the struggle is unlikely to be enough; a strong national army cannot be actually defeated by guerrilla forces. The prolongation of the *maquis* provides the base for an external diplomatic campaign, and the ultimate erosion of the desire of the national state to maintain this territory. There is, of course, no guarantee this will ever come, particularly if insurgency can be contained in isolated rural areas at modest levels of intensity.

The International Arena and Secession Claims

This brings us to the international arena, which in such contests plays an important role. While the giant powers in the world cannot be easily swayed by international pressures when their vital interests are at stake, smaller states cannot ignore external factors. In the three cases examined, not only the insurgents, but also the national states, required military succor from the outside, in one form or another. While there are many sources of arms supply in a divided world, and there is a strong probability that at least some major bloc or power can be persuaded that its interests lie in providing weapons to the national states involved, dependency for arms does give leverage to foreign actors.

In dealing with the world arena, two major sets of variables are involved. One is the shadowy, contradictory, but no less real complex of norms, treaties, and precedents which make up international law. The other, and ultimately more powerful, lies in the *raison d'état* of the major external actors, separately and in the clusters into which the world order resolves (Western, Communist, African, Arab, and the like). To a lesser extent, the concerns of major transnational forces in the economic realm (multinational corporations) may also figure.

At the global level, both law and interest confer a powerful advantage to the existing states. Such institutional framework of world order that exists -- primarily the United Nations and allied agencies -- is founded upon national states as constitutive units.

The UN must then be assumed to have a structural hostility toward dismemberment of states which are full members.

Since 1960, and in particular since the creation of the OAU in 1963, an African system of states has become a reality, whatever its weaknesses.[41] It has successfully asserted the right to define an international code of African orientation, and place it before the world as a pre-emptive definition of state integrity. The self-determination concept has been rewoven to preclude legitimate claims for segments of existing states, whether territorially or above all culturally defined. The dominant African view was well-stated in a 1966 Nigerian document:

> The principle of self-determination in its purely theoretical context may be at variance with the other important principle of territorial integrity. This means that some sections of existing states may claim self-rule on the principle of self-determination. This, however, is not what is meant by the principle in the context of our African policy.[42]

The general support which this territorial integrity doctrine has in Africa imposes considerable diplomatic costs on external actors supporting a secession. Only for compelling reasons would a state with significant stakes elsewhere in Africa wish to place them at risk in such a cause. None of the three separators in question have been able to overcome this handicap.

In those cases where governments can be placed under pressure by public opinion, primarily the Western system of states, aspirants to sovereignty could invoke broader moral appeals, which do not belong to the domain of international law. Our three separators each tried to trigger such public pressures, appealing to strikingly different sectors of opinion. Katanga offered order, anticommunism, and rich grazing for corporate interests. Biafra employed the genocide theme, while Eritrea in the 1970s portrayed itself as a popular liberation movement with socialist goals. Only Biafra -- perhaps because it aimed at the large middle ground -- had much success.

The Arab constellation of states has a distinctive perception of international norms, as they apply above all to the Arab world itself. The commitment to Palestinian rights, and the evanescent yet enduring dream of a broader Arab nation, partially conflict with the doctrine of the sanctity of existing states. This internal fluidity of the contemporary Arab state system is illustrated in the endless succession of ephemeral amalgamations of states in the Arab world, with Libya in recent years as the most tireless promoter of such schemes. This was of no use to Katanga or Biafra, but has played a critical role in the Eritrean struggle -- to the extent that Eritreans were willing to portray themselves as a part of the Arab world.

International economic actors have played a perceptible though less crucial role. There is, of course, no single "international capitalist" interest in such conflicts; however, particular corporate groups may see advantages for themselves in state dismemberment in given circumstances. This was most visible in the Katanga case, where at the beginning UMHK did play a crucial role. Once the Katanga affair had become a world crisis, however, UMHK was swiftly eclipsed by much more powerful forces; even its parent company, the *Société Générale de Belgique*, was at best ambivalent toward this adventure, and was soon persuaded its various interests were best served by a single Zaire. In the Biafra case, some effort was made to interest certain oil companies in new ventures that secession might open, and great pressure was placed on Shell Oil to make its revenue payments to Biafra. Very quickly, however, the oil-bearing zones were in Federal hands, and whatever flickering oil company interest there might have been in Biafra went out. In Eritrea, external financial forces had little involvement.

Concluding Reflections

By way of conclusion, let us put these three examples into a broader African framework, and ask what instruction they may have for the future. We need first remind ourselves that these claims to full sovereignty are the exception rather than the rule. However culturally plural most African states may be, demands for self-determination by internal segments -- whether territorial or cultural -- have not been especially frequent. Are there then some rules which may be deduced from these exceptions? Five tentative propositions come to mind, on the basis of these instances.

Firstly, an African system of states has become consolidated, with a growing code of international norms of its own. The now-powerful consensus behind state integrity has a potency which should not be underestimated, however divided the OAU may seem to be in dealing with particular crises within the African system, and despite the rise of interstate conflict within Africa. This African code places enormous obstacles to securing external support.

Secondly, claims to sovereignty are perhaps only possible on the basis of the territorial code, if they can be made at all.[43] In the Western Sahara case, which does deeply divide Africa, even though most of the population belongs to a single confederacy of clans this was never put forward as a nationality principle. Western Saharan claims to sovereignty rest on its identity as a former, distinct colonial territory. In Africa, grounding of self-determination on the ethnic principle would encounter innumerable obstacles: the very fluidity and complexity of ethnicity itself; the lack of congruence with extant territorial divisions either internally or across states; the growing importance of polyethnic cities; as well as the total illegitimacy of ethnicity as a basis for sovereignty within the African state system.

Thirdly, the hostility of the world arena to state dismemberment makes it extremely unlikely that a sovereignty claim by a state segment can succeed. Without recognition and succor from abroad, secession cannot withstand the inevitable armed attack by the national state involved. The pre-emptive effect of the African state system code makes this almost impossible to obtain.

Fourthly, beyond simple repression, there are many strategies available to national states to reconcile themselves with cultural pluralism. The new Nigerian constitution is one example; the regionalization of politics in Sudan is another. Experience throughout the world in this age of more highly politicized cultural solidarity movements shows that co-optation of these forces is well within the range of calculated statecraft.

Fifthly, only the extraordinary conjuncture of a virtual decomposition of the state really opens the way for successful separation. This paved the road for both the Katanga and Biafran secessions, and provided the one moment of near-triumph for Eritrean liberation movements. Several states have experienced such moments, and the dismal economic prospects in the 1980s for many nonmineral states suggest that others as well may face grave crises. But what is remarkable is that despite the energetic policies of state destruction pursued by such unlamented former leaders as Idi Amin, Pierre Tombalbaye, Francias Macias Nguema, Jean-Bedel Bokassa, or Ignatius Acheampong, no African nation-state has yet broken apart. Indeed, the persistence of Chad as a theoretically single entity (if we set aside as a mere hypothesis its announced merger with Libya), when its state appartus virtually ceased to exist, is miraculous.

Thus the future does not necessarily portend a multiplication of Katangas, Biafras, and Eritreas. Nor does dismemberment of states normally commend itself as an inspirational solution to the many problems confronting contemporary Africa. Whatever else may lie ahead, "respect for the sovereignty and territorial integrity of each State" appears one of the safer political forecasts.

Footnotes

1. In 1971, the Katanga region (province) was renamed Shaba by President Mobutu Sese Seko as part of a far-reaching change of nomenclature. We will retain the designation "Katanga" in this essay because the events considered were anterior to the name change, and, more importantly, because the affair acquired international notoriety under that label. For all other place names, we will use the current designations.
2. The best study of the Katanga episode is Jules Gérard-Libois, _Katanga Secession_ (Madison: University of Wisconsin Press, 1966), pp. 167-208; see also Crawford Young, "The Politics of Separatism: Katanga 1960-63," in Gwendolen M. Carter, ed., _Politics in Africa_ (New York: Harcourt, Brace & World, 1966); René Lemarchand, "The Limits of Self-Determination," _American Political Science Review_, 56, 2 (June 1962): 404-416.
3. The initial version of the _loi fondamentale_ required a two-thirds majority for a quorum, making it possible for formation of provincial governments to be blocked in Katanga as well as former Leopoldville province. The amendment permitted a simple majority to form a quorum.
4. Young, "The Politics of Separatism," p. 167.
5. _Ibid_., p. 182.
6. Of special value on the Biafra question are Anthony Kirk-Greene, _Crisis and Conflict in Nigeria_, 2 vols. (London: Oxford University Press, 1971); John J. Stremlau, _The International Politics_

of the Nigerian Civil War 1967-1970 (Princeton: Princeton University Press, 1977); Victor Olorunsola, _The Politics of Cultural Sub-Nationalism in Africa_ (Garden City, N.Y.: Doubleday, 1972); B.J. Dudley, _Instability and Political Order_ (Ibadan: Ibadan University Press, 1973); Robin Luckham, _The Nigerian Military_ (Cambridge: Cambridge University Press, 1971); and N.V. Akpan, _The Struggle for Secession 1966-1970_ (London: Frank Cass, 1971).

7. Luckham, _The Nigerian Military_, pp. 43-50, has elaborate calculations on this arithmetic.
8. These figures are taken from Stremlau, _International Politics of the Nigerian Civil War_, p. 38.
9. Kirk-Greene, _Crisis and Conflict in Nigeria_, vol. 1, pp. 451-452.
10. Stremlau, _International Politics of the Nigerian Civil War_, pp. 45-54. Some argue that the Awolowo statement was intended as a warning to Gowon not to let the East go; I am indebted to J.I. Elaigwu for this insight.
11. Oye Ogunbadijo, "Ideology and Pragmatism: The Soviet Role in Nigeria, 1960-1977," _Orbis_, 21, 4 (Winter 1978): 803-830.
12. Stremlau, _International Politics of the Nigerian Civil War_, pp. 224-252. Other rumored governmental sources -- Portugal, Israel, South Africa, Rhodesia, China -- gave little if any, Stremlau concludes.
13. _Ibid_., pp. 129-141.
14. _Ibid_., p 120. This argument was also employed with the Vatican. On October 31, 1967, Pope Paul declared, "What are We to say when violence reaches such proportions that it becomes almost equivalent to genocide and pits tribe against tribe within the borders of a single nation?"
15. _Ibid_., pp. 238-252. Part of the relief came in the form of monetary gifts; this was converted into Biafran currency for local purchase of food and goods, while the foreign exchange was available for arms purchase. The foreign currency stipends of relief personnel in Biafra, which they converted to meet their local living costs, were another significant source.
16. Eritrea differs as well in being less documented; there are no solidly based monographs on the actual operation of Ethiopian administration in the territory, nor on the real character of the internal guerrilla struggle before the 1970s. Particularly useful are Richard F. Sherman, "Eritrea in Revolution" (doctoral dissertation, Brandeis University, 1979); G.K.N. Trevaskis, _Eritrea, A Colony in Transition_ (London: Oxford University Press, 1960); Donald N. Levine, _Greater Ethiopia: The Evolution of a Multiethnic Society_ (Chicago: University of Chicago Press, 1974); Stephen H. Longrigg, _A Short History of Eritrea_ (Westport, Conn.: Greenwood Publishers, 1974) (first published 1945); E. Sylvia Pankhurst and Richard K.P. Pankhurst, _Ethiopia and Eritrea_ (Woodford Green, Essex: Lalibela House, 1963); Tom Farer, _War Clouds on the Horn of Africa: The Widening Storm_, second edition (New York: Carnegie Endowment for International Peace, 1979); and J. Bowyer Bell, "Endemic Insurgency and International Order: The Eritrean Experience," _Orbis_, 18, 2 (Summer 1974): 427-450. Longrigg and Trevaskis are former British

administrators; Sherman is a fervent partisan of Eritrean liberation, while Levine and the Pankhursts are greater Ethiopia enthusiasts.

17. Tigrinya and Tigrai are closely related but mutually unintelligible Semitic languages, with ancient Ge'ez as their common root. Considerable confusion arises because the ethnonym Tigre has an identical pronunciation with Tigrai, a designation commonly used both for the language and for the speech community. To make matters worse, both have various spellings. We will use "Tigre" to refer to the highland Tigrinya speakers, and "Tigrai" for the lowland, Muslim pastoralists.
18. Pankhurst and Pankhurst reproduce specimens of these documents in _Ethiopia and Eritrea_, pp. 22-23.
19. Longrigg, _A Short History of Eritrea_, p. 18. Sylvia Pankhurst recounts pungent conversations with British officials during her 1944 visit to Eritrea, finding a "serene, unquestioning confidence in [their] own efficiency and humanity...bitter dislike for the Eritrean Unionists [favoring Ethiopian rule] and obstinate refusal to admit the possibility of any solution for the territory, save a prolonged quasi-permanent extension of British rule." Pankhurst and Pankhurst, _Ethiopia and Eritrea_, pp. 96-98.
20. Levine, _Greater Ethiopia_.
21. Longrigg, _A Short History of Eritrea_, p. 160.
22. Pankhurst and Pankhurst, _Ethiopia and Eritrea_, pp. 191-225.
23. See the extremely positive appraisals of the immediate post-federation period by Duncan Cameron Cunning, "The Disposal of Eritrea," _Middle East Journal_, 7, 1 (Winter 1953): 18-32; "Eritrea: A Successful Beginning," _External Affairs_, 5, 6 (June 1953): 191-195.
24. Sherman, _Eritrea in Revolution_, pp. 53-58. There is some controversy as to whether the packed assembly in 1962 ever did actually vote the liquidation of Eritrea.
25. Bell, "Endemic Insurgency and International Order," p. 432. This account of the guerrilla phase draws primarily on Sherman, _Eritrea in Revolution_, pp. 77-161. See also John F. Campbell, "Rumblings along the Red Sea: The Eritrean Question," _Foreign Affairs_, 48, 3 (April 1970): 537-548.
26. Sherman, _Eritrea in Revolution_, p. 174.
27. This is the interpretation of Bell, "Endemic Insurgency and International Order."
28. Sherman, _Eritrea in Revolution_, pp. 173-224. Also, at this time foreign journalists were able to travel freely into liberated zones, bringing back usually laudatory reports.
29. Farer, _War Clouds on the Horn of Africa_, p. 44.
30. Cited in _ibid._, p. 159.
31. Clause two of the Nine Point Program of the Dergue, reproduced in Sherman, _Eritrea in Revolution_, pp. 292-293.
32. The non-Muslim, non-Tigre-Tigrai Sudanic populations on the Sudan frontier of southeastern Eritrea.
33. Farer, _War Clouds on the Horn of Africa_, p. 46.
34. I am indebted to Ali Mazrui for this point.

35. Cited in Young, "The Politics of Separatism," p. 171.
36. Sherman, Eritrea in Revolution, p. 97.
37. The Syrian withdrawal from the UAR in 1961, the Senegal withdrawal from Mali in 1960, and the Singapore withdrawal from Malaysia in 1965 were not opposed; all three were very short-lived amalgamations.
38. For details on Youlou's motivations, see René Gauze, The Politics of Congo-Brazzaville (Stanford: Hoover Institution Press, 1973), pp. 125-134.
39. As examples of this impact, see Jack Kramer, "Hidden War in Ethiopia," Venture, 21, 5 (May 1969): 18-23; Georgie Anne Geyer, "Eritrea: A Name to Remember," Progressive, 34, 6 (June 1970): 24-27; Fred Halliday, "The Fight in Eritrea," New Left Review, 85 (May-June 1974): 57-67.
40. Longrigg, A Short History of Eritrea, p. 132.
41. For a valuable development of this concept, see Martin Wight, Systems of States (Leicester: Leicester University Press, 1977).
42. Stremlau, International Politics of the Nigerian Civil War, p. 12.
43. For intriguing confirmation of this point, see David Brown, "Borderline Politics in Ghana: The National Liberation Movement of Western Togoland," Journal of Modern African Studies, 18, 4 (December 1980): 575-610. While this movement expressed the ethnic interests of the Ewe, their concern has always been discovering which existing territorial state best served their needs, not claiming an Ewe independent state.

12 African Public Policies on Ethnic Autonomy and State Control

Donald Rothchild
Victor A. Olorunsola

As ethnic groups press their demands for autonomy upon central decision-makers, some form of governmental response is required. This entails a setting of priorities by governments on the allocation of scarce political and economic resources. In relatively poor African societies, the determination of priorities on the nature of social relations proves a particularly trying task, for public authorities lack the capacity to satisfy the claims of all contending parties at one and the same time. Hence scarcity necessitates choice among alternative courses of action. "When dealing with sectors," Warren F. Ilchman and Norman T. Uphoff have written, the statesman "must often make choices in the face of incompatible or mutually exclusive demands.... Choices will usually favor one sector and not another because in most cases resources are too scarce to satisfy all demands."[1]

The decision as to a desired course of action is not simply a "rational choice" based upon mechanical calculations of costs and benefits; policy-makers are also influenced by a number of factors emanating from the environment, i.e., information, values, belief systems, analytic capacity, and bureaucratic organization.[2] In addition, the decision model used in arriving at public choices can have a significant impact upon the policies adopted. A state's utilization of hegemonial or bargaining models to regulate social relations is critical, for each implies the acceptance of different values, norms, and interactional processes among groups. These models emphasize varying priorities and objectives, and are distinguishable in their investment of resources in the decision process as well as in their tolerance of openly displayed group conflict. Broadly speaking, hegemonial models seek to control conflict from the top downward; bargaining systems are based upon a mutual adjustment of conflicting interests. The former strengthens authoritative institutions at the center, with some risk of system overload and low responsiveness; the latter decentralizes control, increasing autonomous participation at some cost in coordinated decision-making.

The Arena of Choice

Allowing considerable scope for overlaps between our general decision models, we associate four conflict regulating strategies with both the hegemonial and bargaining approaches (see Table 1). Whereas subjection, avoidance, isolation, and displacement are generally connected with a hegemonic decision model, buffering, protection, redistribution, and sharing are linked with bargaining. There is an obvious tendency here to group strategies displaying relatively low levels of political interaction under hegemony, and to put those strategies tolerating a high level of political exchange under the bargaining rubric. Certainly this table does not include a universe of available choice strategies. Such alternatives as genocide and assimilation are not dealt with, as they would transform the conflict through an elimination of the boundaries between groups within the state.

With respect to the degree of reliance on mechanisms of coercion, choice strategies on subjection and sharing are considered polar opposites, with six alternative strategies interposed between them. Subjection assumes a social stratification system highly inequitable in its effects. It proceeds to cope with this relationship of inequality by maintaining an outward appearance of order, utilizing the coercive power of the state to perpetuate existing patterns of ethnic dominance. Such dominance may succeed in giving a superficial impression of political stability; in reality, however, beneath the apparent calm, interethnic conflict remains intense.

By contrast, a strategy of power sharing -- which embraces such widely disparate institutional arrangements as federalism, confederalism, regional autonomy, devolution, parliamentary coalition, executive power sharing, mutual vetoes, and so forth -- assumes the moral equality of rival groups. It makes use of the institutions of state to assure a limited access on the part of ethnic spokesmen to decision-makers as well as to provide for a minimal participation by ethnic representatives in the exchange process. This ongoing bargaining encounter between ethnic collectivities is by no means a static one. As shown by Kenya's experience before and after the transfer of power to majority African hands, transactions among the main ethnic peoples at the center did indeed take place, shifting over time from a process of direct to tacit bargaining.[3] The consequence of such interactions was to reduce the intensity of interethnic tensions. Norms, values, and procedures gained in credibility and effectiveness; a process of social learning enabled groups to work a "semicodified set of procedures" linking racial sections and geoethnic groups to the political process.[4] Hence even in deeply divided societies such as preindependence Kenya, a sharing strategy can exhibit a constructive potential for moderating interethnic conflict.

Quite clearly, the choice of strategy can significantly affect outcome -- intensifying or moderating conflict. A strategy of isolation, taking the particular form of partition (Cyprus, Northern Ireland, possibly South Africa) or of de facto secession (Angola, Chad), may keep the contending groups at arms' length, but the anticipated costs of delinkage are high and the subsequent economic and social dislocations are likely to be extensive. Avoidance mechanisms may

Table 1
Strategies for Regulating Conflict

Strategy	Central Thrust of Strategy	Intensity of conflict	Longer-term potential for moderating conflict
A. Hegemonic Decision Model			
1. Subjection	Mechanisms of coercion applied throughout political system to maintain a struc- of inequality	high	low
2. Isolation	Conflict regulated by separating contending groups into distinct political systems	high	low
3. Avoidance	Decision-makers insulated from ethnic demands in order to restrain direct conflict	high/ medium	low/ medium
4. Displacement	Population transferred to alter the nature of the group encounter	high/ medium	low/ medium
B. Bargaining Decision Model			
1. Buffering	Organization of rules for social interaction by third parties	medium	medium
2. Protection	Concessions of legal and constitutional guarantees to ethnic minorities	medium	medium/ high
3. Redistribution	Reallocation of interethnic opportunities by means of redistribution programs	low/ medium	medium/ high
4. Sharing	Regularized reciprocity through coordinated participation in the decisional process	low/ medium	high

Source: This table is drawn from a forthcoming volume by Donald Rothchild tentatively entitled Politics of Ethnicity and Regionalism in Middle African Societies.

insulate decision-makers from some pressing geoethnic demands (Afghanistan in the late 1960s, Ghana under General I.K. Acheampong, Upper Volta in 1980), but at a high price in terms of establishing intergroup linkages critically needed for the long-term amelioration of conflict. Fiscal redistribution (Zambia, Ghana, Nigeria) may facilitate the integration of the hinterland peoples into the political system, but at a possible trade-off cost in terms of rapid economic growth in the core areas. Sharing mechanisms, often viewed somewhat skeptically by African political elites, may, through their legitimization of pluralistic tendencies in party and state affairs, complicate the decision process and heighten interethnic tensions. Thus difficult trade-offs between objectives are well-nigh inevitable, especially in situations where limited public resources coincide with unlimited interest group demands for resources.

If choices among strategies are unavoidable, what of the choices within strategies? Disparate emphases are also present here. We will have more to say on this topic in the next section on altering choice, but it is important to note now the significant differences that exist among policy mechanisms within each type of conflict-regulating strategy. A brief examination of the diversity of forms applying to subjection and sharing will be used to make this point.

Subjection systems are distinguishable among themselves in various ways: subsystem autonomy, the degree of central control and intervention, popular participation, ideological commitment and innovation, and adaptability and responsiveness to the claims of interest groups.[5] In the African context, it is necessary to distinguish the no-party from the one-party system and then to differentiate between reforming pluralistic and "revolutionary-centralizing" patterns.[6] A further distinction, and one with immediate implications for interethnic relations, can be made with respect to cleavages evident within a country's social structure. Thus Samuel P. Huntington writes that "successful one-party systems have their origins in bifurcation [class as well as ethnic]; the party is the means by which the leaders of one social force dominate the other social force." Huntington then proceeds to describe two policies pursued by different one-party regimes in regulating cleavage relationships: 1) an exclusionary policy which restricts political participation by subordinate sections (the Americo-Liberian approach in the precoup period); or 2) a revolutionary policy which attempts to eliminate the cleavages in the society by means of assimilation or genocide (Nazi Germany's "final solution").[7] Obviously, then, a wide variety of dominating mechanisms exists for regulating conflict in divided societies. If these systems share features on hierarchical control, they differ markedly as to the nature of dominating decision elites, constituency bases, social cleavages, control techniques, participation, and so forth.[8]

Variety is also marked within a sharing strategy. To be sure, all sharing systems are similar in their basic commitment to reciprocity, exchange, and compromise on the part of morally equal sections; beyond this fundamental value agreement, however, it is easy to detect a sharp contrast in sharing mechanisms in terms of operating styles and outcomes. Centrally coordinated polities differ markedly from mixed centrally and noncentrally coordinated systems in their definitions of responsibilities to be dealt with at each authority level.[9] And within these categories, further distinctions are evident. Systems based on areas (i.e., federal devolution, regional autonomy, and confederation) could be arranged on a continuum with respect to powers reserved for regional action; moreover, contrasts emerge in the process of mutual adjustment at the center as competing group actors establish norms on adversarial and consensual principles, majoritarian as against proportional decision rules, cabinet versus presidential forms of government. Despite the fact that all of these sharing mechanisms may prove insufficient to cope with the strains of economic scarcity and geoethnic conflict, the proponents of these various institutional arrangements are little discouraged from putting forth their respective merits with apparent ardor.[10]

Changing the Choices

It is when a strategy imposes unacceptable costs for the state that decision-makers may decide to alter existing arrangements. There is nothing certain about this. Political leaders can find themselves locked into structures and relationships which are hard to transform, leading to a process of ethnic polarization culminating in violent encounter (Algeria, Zanzibar, Rwanda, Burundi).[11] Such encounters are the very essence of highly destructive conflict and give rise to the extreme solutions of subjection, displacement, and isolation.

Those high-cost strategies still allowing for a change of choices tend to be limited, then, to situations where domestic interethnic relations are coherent (i.e., where linkages between individuals and groups within the state are maintained),[12] or external actors are no longer prepared to pay the political costs of a status quo policy (i.e., continued French military repression and torture in Algeria).[13] So long as some local connections remain intact and minimal reciprocal interests can be recognized on all sides, opportunities are present for the revising of basic strategies and structural relations.[14] Since these circumstances have availed in the contemporary African experience, it is not surprising that there has been evidence of changes in choice strategy and in the rules for regulating interethnic conflict.

Several instances of changed choice strategies in the face of unacceptable costs deserve mention at this point. Where the costs of maintaining a system of dominance comes to be seen as prohibitive (colonial Kenya and Rhodesia-Zimbabwe), the decision elite, domestic and/or foreign, has decided at times to back off from an intensifying confrontation and to enter into negotiations with its former adversaries. In these encounters, some room for maneuver remained in evidence; thus emergent elites can formulate new rules for the structuring of intergroup competition. For example, the costs of the Sudan's continuing war with the southern-based Anya-Nya forces came to be perceived as prohibitive after sixteen years of fighting, causing President Jaafar el Nimeiry to change from a confrontational to a reconciliatory approach. Not only did the 1972 Addis Ababa Agreement and accompanying policy statements and regulations shift the relationship from unitary control to regional autonomy in the South, but provisions were put into effect which included a general amnesty for those who had participated in the civil war; the recruitment of Anya-Nya soldiers into the army, police and prison services; the inclusion of southern leaders in the central decision-making process; and the allocation of increased revenues from the center to the southern region.

But military governments in Uganda, Benin, Upper Volta and elsewhere have gone the other way and retreated from mechanisms with a sharing potential, such as open parliamentary elections; it is contended that the adversarial politics implicit in such institutions have been shown to be costly in terms of heightened interethnic tensions. By contrast again, other military rulers, such as those in Nigeria and (somewhat reluctantly) Ghana, have presided over a return to open, adversarial politics, and precisely on the grounds that continued military rule worked against the corporate interests of the military and made difficult the building of accepted political processes and linkages in the society at large. In Nigeria, Lieutenant General Olusegun Obasanjo and his military colleagues spoke in the formative 1978-79 period of the need to establish "laid-down rules" to promote healthy competition under the new constitution.[15] Toward this end, the Supreme Military Council invoked a variety of measures, including a decree to the effect that party executives must reflect the federal character of the country, a nineteen-state federal structure, a complex system of voting designed to "de-ethnicize" the presidential election process, and, following difficulties with the first presidential election under the new constitution, amendments to the constitution altering the run-off and appeals procedures. The thrust of all these initiatives was to facilitate Nigeria's return to competitive politics, but this time to channel the inevitable conflict among groups along constructive lines.[16] Thus if some leaders have been unswerving in their commitments to an initial strategy on the regulation of intergroup relations (colonial Algeria, Amin's Uganda, Rwanda, Burundi), others, quite pragmatically, have shown themselves prepared to shift strategy -- once they have concluded that the costs of existing patterns of relations are excessive.

Although observers have taken note of shifts in general strategy on regulating intergroup relations, they have shown less concern with respect to substrategies, i.e., the lesser choice options offering creative possibilities for redirecting conflict along positive lines. The state principle verges on being so fundamental and lacking in divisibility that policy-makers tend too easily to lose sight of the scope for choice open to them at the margins of policy. As Yehezkel Dror comments on the important role of rational techniques regarding these lower level alternative policies:

> In those many cases where incremental change is optimal, a correct way to formulate alternatives is to ask whether one should add or subtract from current policies. Rational methods can be used both to identify such alternatives and to estimate their probable benefits and costs, and are also of some use in synthesizing new alternatives from parts of past policies.[17]

To gain an understanding of the full range of choice, it is necessary to separate basic strategies into smaller components. Hopefully such a process of breaking up general strategies into smaller parts will enable policy analysts to find new structures and values in which mutual state-ethnic interests overlap. What is critical to successful regulation of conflict is to pull back from absolute positions and to search out the area of common ground between principles.[18] Suboptimization may contribute to this by creating additional opportunities for reconciling ethnic autonomy and state control, making those statesmen who are actually reaching for new ideas aware of the varied and intricate points of agreement appropriate in each context.

In fact, an examination of our eight strategies reveals considerable scope within each of them for altering the pattern of group interactions. As shown in Table 2, some of the available suboptimized choice strategies have themselves been the subject of protracted and heated debate. The proponents of revenue allocation principles on need and derivation, decision rules on proportionality and merit in recruitment practices, provisions on rigid amendment clauses, second chambers, or reserved seats to safeguard minorities, or the economic viability of subregions in a federal system have all assumed critical stances on issues with great meaning for the outcomes of intergroup encounters. Thus in postindependence Nigeria, where a disparity existed between the North's greater size and population and the South's more rapid modernization, formulas for representation in the federal legislature proved extremely complicated and contentious. At the 1950 general conference on the review of the constitution, spokesmen for North and South came into direct conflict on this issue. The East and West called for equal subregional representation; the North demanded 50 percent of the seats for itself, with the remainder being divided equally between the other subregions. Even after the drafting committee had proposed a compromise ratio of 30:22:22, the North remained steadfast in its demands. It was unable to sway the general conference to its view; however, in the ensuing months, Southern members of the legislative council did reluctantly come to accept Northern calls for parity in the interests of Nigerian unity.[19] Further significant changes in parliamentary representation followed in later years, in particular the decision to divide the country into 320 single-member electoral districts -- an action which facilitated the Northern People's Congress rise to power at the center. Suboptimized strategies are definitely genuine choices in themselves, even if the gradations between them may prove on occasion to be quite narrow.

At the same time that one calls attention to the role of suboptimized strategies in expanding choice and thereby enhancing the prospects for successful negotiations, it is important also to take note of the limits of such an approach, i.e., that leaders must accept some values in common which are associated with "rational" objectives. Rational choice options may have little meaning for irrational tyrants bent upon destructive purposes or for political ideologues unchanging in their dedication to absolute principles, irrespective of human costs. However, for the remaining leaders, including those who come to view the costs of subjection as no longer tolerable, an awareness of the scope for change may be helpful in

Table 2
Some Available Suboptimized Strategies for Conflict Regulation*

Strategy	Suboptimized Strategies
1. Subjection	Different degrees of control -- Majority versus minority domination -- Denial of principle of proportionality in recruitment and resource allocation -- Inequitable tax policies -- Laws against the writing of local languages -- Different degrees of responsiveness to group demands -- Channeling of political participation by subordinate sections -- Disenfranchisement -- Expulsion -- Undue restrictions on citizenship -- Ethnic restrictions on land sales and occupation -- Ideological inflexibility -- Transformations of ethnic relations versus maintenance of status quo --
2. Isolation	Full regional autonomy -- Limited administrative autonomy -- Granting of personal or territorial autonomy -- Autonomous sites -- Autonomous boroughs -- Boundary rectification -- Millet system --
5. Buffering	Arbitration -- Mediation -- Conciliation -- Good offices -- Judicial review and interpretation -- Third-party administration --
6. Protection	Application of principles of equality, proportionality, and/or nondiscrimination in recruitment and resource allocation -- Antidiscrimination laws -- Blocking minorities -- Mutual vetoes -- Quotas -- Controls on land sales -- Ombudsman -- Bilingualism -- Minority councils and boards -- Parliamentary committees on differentiating legislation -- Bills of rights -- Rigid amendment procedures -- Second chambers -- Reserved seats and special voting qualifications -- Proportional representation -- "Loading" of parliamentary representation to favor disadvantaged areas or peoples --
7. Redistribution	Emphasis on principle of need versus on derivation --

	Special communal legislative bodies -- Decentralization by delegation -- Deconcentration -- Legislative devolution -- Executive devolution -- Delegation to parallel organizations -- Regional councils -- Centralized federalism -- Decentralized federalism -- Confederation -- Partition --
3. Avoidance	Values supportive of depoliticization -- Single-party system -- No-party system -- Ban on the formation and operation of ethnic parties -- Political socialization -- Insulation of decision-makers from group demands -- Restraints on participation by subordinate sections -- Limitations on open partisan contestation --
4. Displacement	Population ban -- Population expulsion -- Population transfer -- Exchange of populations -- Relative increase of one section's population -- "Regroupment" -- Nationalization policies with special impact on one section --

	Revised goals on subregional allocations -- Recruitment quotas reflective of changed priorities -- Affirmative action -- Africanization or localization programs -- Changed priorities on siting industries -- Special scholarship and training programs -- Capital assistance schemes -- Revised regulations on land tenure and landholding -- Land redistribution --
8. Sharing	Majoritarian versus consociational politics -- Presidential as opposed to parliamentary-type government -- Tensional vis-à-vis cooperative federalism -- Varying distributions of federal and subregional responsibilities -- Judicial review and interpretation -- Various patterns of executive-legislative relations -- Formal executive power-sharing -- Grand coalition -- Proportionality in recruitment and resource allocation -- Joint administration -- Internationalization of contested areas --

*Source: This table is drawn from a forthcoming volume by Donald Rothchild tentatively entitled <u>Politics of Ethnicity and Regionalism in Middle African Societies</u>.

striking a balance between conflict and stability.

Table 2 indicates just how wide a range of lesser choices is accessible to decision-makers. Since isolation, with its emphasis upon the distancing of contending geoethnic groups within the same polity, is particularly relevant to the reconciling of state control and ethnic autonomy, it makes sense for us to discuss suboptimization by reference to the refinement of this strategy. Such a focus conforms to the "federalizing influence" that ethnic groups have upon the political process.[20] Certainly the effect of placing isolation's various combinations along a continuum is to highlight both the extent of choice and the nuances of difference between these lesser options.

But a caveat is necessary here. These differences are only meaningful when made precise, and this requires mutual acceptance on all sides as to how these arrangements are expected to operate. The possibilities for misinterpretation abound. What John Foncha and the West Camerounian elite interpreted as a loose federal system at the time of reunification in 1961, Ahmadou Ahidjo and the powerful Francophone interests around him regarded as a temporary accommodation which would be transformed into a unitary system as soon as possible.[21] If the French talked of applying federalism to Chad in 1980, this was viewed by the Nigerians and others as a step toward de facto autonomy and partition, not as an attempt to put the Dicey-Wheare model into practice in that country. No doubt, the Nigerians were influenced in this regard by their own disintegrative experiences in the past.[22] Similar misunderstandings over the implications of federalism were apparent in the transactions, under UN auspices, leading to the integration of Eritrea with Ethiopia in the 1950s; what the Eritreans considered a protection for their autonomy the Ethiopians under Emperor Haile Selassie regarded as a brittle and transitory stage on the way to full unification.[23] Hence it is critically important that major actors come to an agreement on what their transaction entails for all parties. In working out such joint interpretations, protracted negotiations are themselves of vital importance as an instrument for social learning. Not only do actors work out precise meanings as to their institutional arrangements, but they establish for themselves rules of procedure which may be carried over into the postnegotiation stage. If this process of clarification and social learning is put off until after the exchange has occurred, it will only have to take place later, under the more difficult circumstances of day-to-day governance.

Assuming sensitivity, knowledge about options, and a spirit of accommodation on the part of the decision elite, a considerable range of choices on institutional arrangements is available to leaders under the general strategy of isolation. To start with, it is useful to distinguish generally between systems of territorial and personal (cultural) autonomy. The principle of group personality applies to the special collective status and rights of nondominant ethnic peoples "irrespective of their location within the State."[24] Ethnic, linguistic, and religious groups are granted a limited jurisdiction over communal affairs with respect to such distinctive issues as religion, education, language, monuments and holy places, and cultural affairs, permitting them to preserve their group identity in a manner little threatening to the society at large. Past experience with such arrangements is minimal but not insignificant. It includes the special regimes under which Jews lived in medieval European cities, the Millet system of the Ottoman Empire, the minority protections put into effect in eastern Europe after World War I, the autonomous rights exercised by the Greek and Turkish Cypriot communal legislative bodies over cultural and religious matters under the Cyprus constitution of 1960, and the personal law systems in the African states affecting religious, family and land tenure matters.[25] In modern Canada the support given by a number of provincial authorities to a two-denominational educational system, Catholic and non-Catholic, is cited as evidence of a partial application of the group personality principle.[26] In addition, possible uses of this principle to solve knotty questions of group rights in other countries have not gone unnoticed. The call by the 1973 Political Commission of the Study Project on Christianity in Apartheid Society for the establishment of representative councils for the major non-European races, with legislative and executive powers over matters mainly affecting the communal group (subject to some central government control), is an indication of cautious consideration among "verligte" opinion-makers in South Africa of a limited concession of personal autonomy.[27] Similarly, observers have concluded that the use of personal rather than territorial autonomy might work to ensure the future of Jerusalem as a united city. By allowing the Arab and Jewish inhabitants of the city to participate in the political life of their respective neighboring countries, it opens up an alternative to state sovereignty in the strictest sense.[28]

The principle of territorial autonomy is relevant to interethnic relations when circumstances permit a distribution of the powers of the state, temporarily or permanently, among various levels of administration or government. Because such a distribution of powers includes numerous gradations and refinements, it gives decision-makers a broad scope for reconciling the demands made by state and geoethnic spokesmen for unity and diversity. In situations where subregional appeals for autonomy are perceived as relatively weak, decision-makers may opt for formulas of administrative decentralization and deconcentration. Such arrangements offer the prospect of increased central efficiency, enabling civil servants to learn first hand about the special needs and requirements of peoples far removed from the capital city and therefore to be more effective in applying regulations and monitoring results. However, this central control and penetration of the periphery is not likely to satisfy local groups' aspirations for control over their own affairs. Hence stronger demands for subregional self-management may be made, and may give rise to further concessions of central administrative responsibility. These concessions may assume a variety of forms, i.e., subregional councils, planning boards, and other administrative bodies with a limited capacity for initiative and action.

Geoethnic dissatisfaction with such circumscribed concessions to local governance may lead, at times, to more forceful claims to autonomous jurisdiction. Here one sees a move, possibly for a limited period only, into the area of tensional governmental systems. The new political authorities no longer act as agents of the central administration in the districts and provinces, but begin to assume responsibility for a wider array of functions; their entrenchment acts as a symbolic expression of separate ethnic influence and power on the country's political scene. Such tensional governmental arrangements raise some troublesome questions, however, about subnational participation in political affairs at the central and subregional levels. In particular, can statesmen decentralize authority while at the same time ensuring effective subregional participation in central decision-making activities? And can governments which place limitations on their own exercise of authority at the periphery guarantee the full and equal participation of minority ethnic peoples in the political and economic affairs of the subregional units? The different constitutional arrangements which have taken shape in contemporary Africa attest to the variances in state-subnational power configurations as well as the distinct policy preferences of leaders as to formulas for isolation.

The forms making possible a stratification of authority along horizontal lines are many and varied, running across a continuum from devolution to centralized federal schemes (Kenya's majimbo; Uganda's semi-federalism), to de jure regional autonomy (Sudan), to classical federalism (Nigeria, 1960-66), to de facto regional autonomy (Angola; Southern Chad, 1980-81), to confederation (proposed as a solution to Nigerian-Biafran relations in 1968-69 and to South Africa, 1981),[29] to leagues (OAU). These, and other political arrangements as well, differ in the proportions of power assigned to the state and its subregions, but they share in common the objective of reducing conflict by separating groups into distinct political systems, each of which possesses a decisional capacity over certain specified activities in its own right. Because subregional authorities can bring a considerable measure of influence to bear upon central decision elites, and most convincingly where the ethnoregional unit is itself so integrated and commanding of resources as to be a candidate for statehood, the problem of maintaining the stability of the existing state can never be far removed from the consciousness of central leaders. In the real world, it is not always possible for these leaders to increase central capacity by accommodating autonomy claims. Not only are connections between center and periphery brittle at times, but unpleasant memories of past interethnic encounters may complicate relations, and elite preferences for unitary control and an extended functional role may prove strongly fixed. Ethiopia's Mengistu Haile Mariam, despite counsels to the contrary,[30] has thus far rejected a mutual adjustment approach on the autonomy question. Although such self-imposed limitations on his sovereignty (in practice at least) might mollify armed resistance from the Eritreans, the Somalis, the Tigreans, the Afar, and the Oromo in Bale Province, he may view it as an overly costly concession which entails a worrisome secession potential. This possibility is made more ominous in such scarcity-prone regimes as Mengistu's by the weak institutional resources currently at his disposal. And experiences in other countries show his fears to have substance. When external actors have inserted themselves into center-periphery relationships, as in the cases of Angola and Chad where de facto secessions have occurred, incoherence has become apparent and conflict has appeared to be intense as well as difficult to regulate. In light of the fragility of Ethiopia and other African states, it would seem imprudent to discount the possibility of foreign military resources playing a role in other de facto secessions in the future.

The list of suboptimized strategies noted above by no means exhausts the options available under the territorial principle of autonomy. To start with, variations seem virtually endless among the institutional forms already mentioned. For example, a federal-type state may involve units unequal in size, population, and resources (i.e., Nigeria, 1960-66). Subregional units in a federation may also lack uniformity with respect to the powers granted them under the basic law (Uganda, 1962-66). The subregions may be large and partially viable (Nigeria, 1960-66), or subdivided in such a way as to accentuate central hegemony (Nigeria, 1976; Sudan, 1980). At the time the Sudan was divided into five self-governing subregions (in addition to the autonomous Southern Region, which had already been established), President Nimeiry is reported to have said that a federal system was necessary "to reconcile the country with itself."[31] Federalism and the acceptance of a limited subregional autonomy were Nimeiry's means of contributing to state coherence and integration.

Experiments with the territorial principle of autonomy have also been proposed for other types of conflict situations in Africa. Leonard Doob and his associates brought together Somali, Kenyan and Ethiopian intellectuals who recommended the establishment of a joint administrative area for the Ogaadeen and the North-Eastern Province of Kenya; an agreement toward this end would include provisions for the military neutralization of the area and for freedom of movement across the borders by both people and animals.[32] A Commission of Privy Counselors inquiring into the dispute between the kingdoms of Buganda and Bunyoro in 1962 suggested that a site of patriotic and sentimental importance to the Banyoro, the Witch Tree on Mubende Hill, be given special protection by including Mubende on the list of central government towns, thereby denying its control to any one ethnic group.[33] And in South Africa, where a small number of "moderate" whites seem increasingly aware of the excessive costs of a subjection strategy, cautious consideration of alternative public policies, including such isolation strategies as federalism, regional autonomy, confederation, partition, and so forth, is currently taking place.[34] Although a substantial proportion of black South African opinion, and particularly that of the more militant African National Congress and the Pan Africanist Congress, prefers a multiracial, unitary state solution, it is also pertinent to note that there was a considerable acceptance among more conservative homeland leaders and supporters of Chief Gatsha Buthelezi for partition and consociational democracy.[35] Surely no form of territorial autonomy seems likely to advance the goals of ethnic self-determination and state coherence unless it provides for a great measure of autonomous power, economic opportunity, and moral equality for all groups.

What is true for isolation is equally the case for the remaining political strategies; that is, a significant range of choices, general as well as suboptimized, is at hand in ethnically based conflict situations, and knowledge of these options may prove critically important to decision-makers searching for agreement as to policy mechanisms. Quite clearly, a spirit of accommodation among leaders facilitates the process of working out a mutually satisfactory balance between the claims of state control and ethnic autonomy. What is often overlooked, however, is the important contribution detailed information on the gradations of possible choices can make to the deliberative process. Such knowledge can provide statesmen with maps of the world indispensable to their ability to cope with the environment.[36]

Conclusion

Positing the necessity of a multinational African reality, it seems not unlikely that incompatible values will at times arise regarding such abstract concepts as state control and ethnic autonomy. Each principle can, for legitimate and understandable reasons, boast its ardent supporters. But what public policies can be utilized to reconcile these apparently exclusive aims? Recognizing that a determination to press the absolute rights of state or subcommunity will lead to unacceptable choices, how can leaders be encouraged to place effective restraints on principle so as to promote jointly advantageous outcomes? The challenge here is exceedingly complex. Clearly, a number of attitudes are essential preconditions for successful accommodations among decision-makers: a minimum consensus on values, a spirit of compromise, sensitivity to the concerns of competing groups, and an adherence to rational objectives. In the exceptional circumstances where these prevail, a policy focus may contribute to the enhancing of choice and the reduction of intense and destructive conflict. It accomplishes this, in part at least, by providing decision elites with detailed information on the comparative costs and benefits of various forms of interaction. Provided that decision-makers are willing to limit their own powers, and thereby to qualify their objectives for the sake of common benefit, it will always be possible for people to cut loose from their enclosures and to shape more secure and positive social relationships. A transformation of group interactions necessarily requires a breaking of old values as well as a forging of new, substitute directions of common purpose.

The altering of choice involves an awareness of the full range of options present in each time-place context. By recognizing both the high cost of destructive conflict and the broad range of alternatives available, it becomes possible to restructure conflict along more constructive lines. Contradictions between principles clearly necessitate compromise. Alfred Cobban, commenting that rights must be qualified by circumstances, observes as follows: "The rights which we derive from the general conception of right are, however, not absolute, but are susceptible of, and indeed demand compromise, whenever they come into practical conflict with one another, as experience proves that they constantly do."[37] Compromise is indeed critical to assure reciprocity; however, once this point is accepted, the issue becomes largely one of determining the most appropriate means of structuring relations in order to emphasize the area of overlapping interests.

To probe the structuring of relations is to enter the delicate sphere of political management. As has been shown in this chapter, the process of working out an agreement suitable for application in a particular conflict situation is facilitated by a focus upon the refinements existing among and within strategies. Here an awareness of the range of choice options as well as precision in delineating the distance between these options is critical to the conflict-regulating effort. Well-drawn maps of intergroup relations can present a detailed scaling of political alternatives with respect to such factors as intensity of conflict and potential for moderating conflict. It is then up to the negotiators themselves to take from these maps such information as will help them to plan the basis for an effective journey ahead.

Footnotes

1. Warren F. Ilchman and Norman T. Uphoff, *The Political Economy of Change* (Berkeley and Los Angeles: University of California Press, 1969), p. 38.
2. These are discussed in Donald Rothchild and Robert L. Curry, Jr., *Scarcity, Choice, and Public Policy in Middle Africa* (Berkeley and Los Angeles: University of California Press, 1978), pp. 99-105. Bargaining and command models are examined on pp. 47-49.
3. Donald Rothchild, *Racial Bargaining in Independent Kenya* (London: Oxford University Press, 1973), chapters 1, 4, 13; and his "Changing Racial Stratifications and Bargaining Styles: The Kenya Experience," *Canadian Journal of African Studies*, 7, 3 (1973): 419-431.
4. Joel D. Barkan, "Legislators, Elections, and Political Linkage," in Joel D. Barkan with John J. Okumu, eds., *Politics and Public Policy in Kenya and Tanzania* (New York: Praeger, 1979), p. 71.
5. Carl J. Friedrich and Zbigniew K. Brzezinski, *Totalitarian Dictatorship and Autocracy*, second edition (New York: Praeger, 1972), pp. 21-27; Gabriel A. Almond and G. Bingham Powell, Jr., *Comparative Politics: A Developmental Approach* (Boston: Little, Brown and Co., 1966), pp. 312-314; and Alexander J. Groth, *Comparative Politics: A Distributive Approach* (New York: Macmillan, 1971), pp. 14-17.
6. See James S. Coleman and Carl G. Rosberg, Jr., eds., "Introduction," *Political Parties and National Integration in Tropical Africa* (Berkeley and Los Angeles: University of California Press, 1964), p. 5. Also see Rupert Emerson, "Parties and National Integration in Africa," in Joseph LaPalombara and Myron Weiner, eds., *Political Parties and Political Development* (Princeton: Princeton University Press, 1966), pp. 274, 281.
7. Samuel P. Huntington, "Social and Institutional Dynamics of One-Party Systems," in Samuel P. Huntington and Clement H. Moore, eds., *Authoritarian Politics in Modern Society* (New York: Basic Books, 1970), pp. 15-17.
8. Ian Lustick, "Stability in Deeply Divided Societies: Consociationalism versus Control," *World Politics*, 31, 3 (April 1979): 333-334; and Robert A. Dahl and Charles E. Lindblom, *Politics,*

Economics and Welfare (New York: Harper and Row, 1953), p. 229.
9. Charles E. Lindblom, The Intelligence of Democracy (New York: Free Press, 1965), pp. 25-28.
10. See, for example, Kweku Folson, "The Plea for a New Parliamentary Tradition," Legon Observer, 12, 3 (February 15, 1980): 53-57.
11. Leo Kuper, The Pity Of It All (Minneapolis: University of Minnesota Press, 1977), chapter 4.
12. Manfred Halpern, "Changing Connections to Multiple Worlds," in Helen Kitchen, ed., Africa: From Mystery to Maze (Lexington, Mass.: Lexington Books, 1976), pp. 11-13.
13. Robert Forster writes: "The French paratroopers actually 'won' the Battle of Algiers...but at the unacceptable price -- for the French public at home at least -- of secret military torture. The French press rose to the occasion. The 'pacification' of the countryside also appeared to succeed militarily, but at a political cost of sending a quarter-million draftees to North Africa and thereby threatening every French family, once again, with a youth 'mort pour le Patrie.'" Review of The War Without a Name, in Manchester Guardian Weekly, 123, 27 (December 28, 1980): 18.
14. Ralf Dahrendorf, Class and Class Conflict in Industrial Society (Stanford: Stanford University Press, 1959), p. 209.
15. West Africa, October 2, 1978, p. 1937. On the centripetal forces at play in Nigeria, see Victor A. Olorunsola, The Politics of Cultural Sub-Nationalism in Africa (New York: Doubleday, 1972), and his Societal Reconstruction in Two African States (Washington: University Press of America, 1977), pp. 32-33. For a charge that "political intemperance" has in fact prevailed in recent Nigerian party politics, see Claude Ake, "The Perils of Political Intemperance," Daily Times (Lagos), April 30, 1980, pp. 3, 11.
16. For the most recent experiences of Nigeria under the new constitution, see the chapter by Isawa Elaigwu and Victor A. Olorunsola, "Federalism and Politics of Compromise in Nigeria," in this volume.
17. Yehezkel Dror, Public Policymaking Reexamined (Scranton: Chandler Publishing Co., 1968), p. 179.
18. Leonard W. Doob, William J. Foltz, and Robert B. Stevens, "Toward a Solution?" in Leonard W. Doob, ed., Resolving Conflict in Africa (New Haven: Yale University Press, 1970), p. 163.
19. See Donald Rothchild, "African Federations and the Diplomacy of Decolonization," Journal of Developing Areas, 4, 4 (July 1970): 517-518.
20. J.A. Laponce, The Protection of Minorities (Berkeley and Los Angeles: University of California Press, 1960), p. 84. Also see Claire Palley, Constitutional Law and Minorities, Report No. 36 (London: Minority Rights Group, 1978), passim.
21. Frank M. Stark, "Federalism in Cameroon: The Shadow and the Reality," in Ndiva Kofele-Kale, ed., An African Experiment in Nation-Building: The Bilingual Cameroon Republic Since Reunification (Boulder: Westview Press, 1980), pp. 104-111.

22. _West Africa_, September 29, 1980, p. 1894.
23 Tom J. Farer, _War Clouds on the Horn of Africa_ (New York: Carnegie Endowment, 1976), pp. 26-27. Confusion over federalism as a form of government distributing power between center and subregions and as a guarantee of a "right" to national self-determination is widespread. Thus one writer notes: "The United Nations itself has, _de facto_, recognized the Eritrean nationality and its right to self-determination by agreeing to a _federal_ relationship with Ethiopia rather than total amalgamation in 1952." Richard Lobban, "The Eritrean War: Issues and Implications," _Canadian Journal of African Studies_, 10, 2 (1976): 340.
24. Benjamin Akzin, _States and Nations_ (New York: Doubleday, 1966), p. 157.
25. Akzin, _States and Nations_, pp. 157-160. For a 1957 plan to apply the principle of personal autonomy to the Algerian Moslems in a French-controlled society, see Laponce, _Protection of Minorities_, p. 95.
26. Kenneth D. McRae, "The Principle of Territoriality and the Principle of Personality in Multilingual States." Paper delivered at the Ninth World Congress of the International Political Science Association, Montreal, August 19-25, 1973, p. 6.
27. Peter Randall, ed., _South Africa's Political Alternatives_ SPRO-CAS Publication no. 10 (Johannesburg: Ravan Press, 1973), p. 233. Also see the comments in F. Van Zyl Slabbert and David Welsh, _South Africa's Options_ (New York: St. Martin's Press, 1979), pp. 110-112.
28. Benjamin Akzin, "The Future of Jerusalem," _Jerusalem Post_, April 25, 1980, p. 14.
29. On Nigeria, see _West Africa_, October 5, 1968, p. 1163; May 10, 1969, p. 541; and September 6, 1969, p. 1069; on South Africa, see _Christian Science Monitor_, February 23, 1981, pp. 1, 10.
30. Sudan's President Jaafar el Nimeiry, drawing on his own experience in resolving geoethnic conflict in his own country, has reportedly counseled his Ethiopian counterpart on the advantages of regional autonomy over a costly confrontationist stance. Acting as a mediator in Ethiopia's dispute with Eritrea, Nimeiry presented the Ethiopians with a seven-point proposal to end the conflict. This proposal reportedly includes a recommendation for a referendum supervised by an international commission which would allow the Eritreans to declare their preferences on such political options as federalism, regional autonomy, and independence. See Gregory Jaynes, "Sudan Offering to Mediate in Ethiopia's Eritrea War," _New York Times_, December 4, 1980, p. 3. Also see the interview with Sudanese First Vice President Abdel Majid H. Khalil in _Sudanow_ (Khartoum), 6, 4 (April 1980): 14.
31. Quoted in _Weekly Review_ (Nairobi), February 3, 1980, p. 3.
32. Doob, ed., _Resolving Conflict in Africa_, p. 168.
33. Colonial Office, Uganda, _Report of a Commission of Privy Counsellors on a Dispute between Buganda and Bunyoro_, Cmnd. 1717 (London: H.M.S.O., 1962), pp. 19-20.
34. One analyst reports that between 1974 and 1977, the proportion of Nationalist Party respondents supporting the idea of a qualified franchise on a common roll rose from 5 to 12 percent. Most

nationalists were reported as appearing to favor "separationist" policy solutions (i.e., homeland development). Lawrence Schlemmer, "Change in South Africa: Opportunities and Constraints," in Robert M. Price and Carl G. Rosberg, eds., The Apartheid Regime (Berkeley: Institute of International Studies, 1980), pp. 250-251, 253.

35. Using a baseline of 100, the values, by African support group calling for three main political strategies in the South African context are reported in one opinion survey as being the following:

	Followers of:				
	Homeland Leaders	Buthelezi	ANC	PAC	Black Consciousness
Partition into two separate racial states............	136	119	54	56	109
One-man one-vote majoritarianism...	60	101	115	124	129
Consociational democracy.........	123	98	81	84	86

Source: Theodor Hanf, Heribert Weiland, and Gerda Vierdag, Südafrika: Friedlicher Wandel? (Grunewald: Kaiser, 1978), p. 376.

36. See Walter Lippmann, Public Opinion (New York: Harcourt, Brace and Co., 1922), p. 16.

37. Alfred Cobban, National Self-Determination (Chicago: University of Chicago Press, 1944), p. 47.

13
The State, Public Policy and the Mediation of Ethnic Conflict in Africa

Edmond J. Keller

> I do not know how many of us would survive the test of self-determination, as expressed by our colleague from France. Are we to allow the principle of self-determination to be applied at the village level, at the district level, at the provincial level, let everybody decide what he wants to do? I do not know. Perhaps instead of having 146 States Members of the United Nations we would have up to 1,000.... I dare to predict that many members of this Council also would not survive that test. -- Salim Ahmed Salim, permanent representative to the United Nations, Republic of Tanzania (1973)

The State, Ethnic Conflict and the Self-Determination Dilemma

Among the numerous problems confronting the new states of Africa, none seems more critical than the periodic claims to self-determination lodged by disgruntled constituent ethnic communities. The genesis of the self-determination dilemma is well known. It can be traced to two important, interrelated facts of recent African political history. First, in most cases, contemporary African states, instead of being cohesive political entities with long traditions of unity and cooperation, are relatively recent artificial creations, the products of the partitioning of the continent among European imperialists at the Conference of Berlin (1884-85). For the most part, the partitioning took place without much attention to the ethnic boundaries which already existed in one region or another. Consequently, it was common for colonial states to be fashioned into unitary wholes more on the basis of administrative convenience than anything else, governed by a central colonial bureaucracy which superimposed itself over a multiplicity of distinct ethnic communities. Civil order was assured not by normative or utilitarian consensus, but by the coercive might of the colonial state. There was no effort on the part of the colonialists to encourage the development of a sense of national consciousness among the disparate groups which made up the colonial polity; nor was there any attempt to develop the concept of individual African rights.[1] In fact, both would have been anathema to the colonial regime given its primary objectives of political control and economic exploitation.

The second important fact contributing to the self-determination dilemma is the perception that the boundaries of independent nation-states are inviolable. Despite the fact that modern African states were largely arbitrarily constructed during the colonial era, at independence new African governments were unanimously disinclined toward the idea of balkanizing their respective polities in the interest of allowing constituent ethnic communities to exercise their right to self-determination. Instead, existing boundaries were accepted as permanent, and it was agreed that they should be respected as legitimate by all other states, both in Africa and the world community as a whole. This sentiment was confirmed in the Organization of African Unity Charter which was adopted at Addis Adaba, Ethiopia, in May 1963.[2] The signatories pledged themselves, among other things, to work for the eradication of all forms of colonialism in Africa and to defend the sovereignty and territorial integrity of independent African states. These twin tenets form the basis of the concept of self-determination as it is generally interpreted by the leaders of present-day African states. As such, self-determination is understood to mean no more than territorial self-determination for those states dominated by some recognized colonial authority. It is not perceived to mean the right of every distinct ethnic community in Africa to secede from an established polity either because of the alleged historic incompatibility of one's own group with another, or because one's group feels that it is the victim of systematic exploitation, inequality, or violations of human rights.

Self-determination in Africa, then, becomes a conditional principle. It seems to apply mainly when one racial community is dominated and exploited by another racial community;[3] it does not seem to apply when the acceptance of such claims would lead to the fragmentation of an established African state. Given this interpretation of the concept, the OAU has consistently avoided interjecting itself into internal disputes even when secession was a real possibility.

In contrast to the conditional nature of the concept of self-determination, the principles of territorial integrity and the continuity of existing boundaries appear to be seen by African states as absolute. This is reinforced by prevailing arguments in the scholarly literature on this issue.[4] For instance, O. Kamanu argues that "the right to self-determination cannot mean the freedom of every self-determining ethnocultural group to secede from an established state on a whim.... The breakup of a state is a very radical solution to the political problems arising from cultural diversity, and requires exceptional circumstances to justify it."[5] Walker Connor has noted that while ethnocultural groups have the right to self-determination, states have the right "to preserve themselves, to protect their territorial integrity, to maintain internal order, to legislate against treasonable acts, and so forth."[6] In other words, state survival is paramount, and policies designed to realize this end are a necessity.

States hold that they have a right to survive. Ethnic communities claim that they have a right to self-determination. Herein lies the crux of the dilemma. How can each of these competing claims be satisfied with a minimum of social, economic or political costs? Are the two claims irreconcilable?

The general tendency has been for African states to proceed under the assumption that this dilemma can be ameliorated if not resolved. It is a matter of survival. States that ignore or fail to accommodate ethnic claims to self-determination are almost certainly doomed to political instability and perhaps even balkanization.[7] Albert O. Hirschman has suggested that ideally states would like to command the "loyalty" of their citizens, but discontents do emerge. When this happens the citizen has two types of activist responses open to him, "voice" and "exit."[8] He can either voice his complaint while continuing to see the political community as legitimate, or he can exit, withdrawing from the community. No state could survive for long should there be an overwhelming tendency among its citizens to exit, particularly when they demand to take territory with them. The likelihood of such an extreme occurrence in industrialized societies is relatively minimal, but in some newly independent, nonindustrialized societies like those we find in Africa, exit in the form of secessionism is an ever-present danger. The African state, therefore, must consciously employ certain kinds of public policy mechanisms to reduce the possibility of secessionist claims emerging and to insure its very survival. What is more significant is the fact that policy-makers in most African countries have an extremely narrow range of choice in deciding what policies to pursue. Their political and administrative institutions are fragile, and they lack the redundancy of critical resources such as skilled manpower, technology and capital which characterizes most developed countries.[9]

The intent of this essay is to analyze how African states have attempted to use public policies as a means for mediating the sometimes competing claims of ethnic self-assertion and state survival. My primary thesis is that we must begin to look upon the postcolonial African state as a potentially autonomous political actor whose primary objective is survival.* Efforts aimed at state penetration, national political integration, and social and economic development all reinforce the fundamental objectives of maintaining the continuity and integrity of individual African states. More than any form of political imperialism, the potential for secessionism is the most serious threat to territorial integrity that many modern African states face. This is not to say that all discontent necessarily leads to secessionist movements or even to claims of self-

*For our purpose the term "the state" refers not only to a well-circumscribed geographic "arena" but more importantly to the bureaucratic structures employed in governing this entity, along with the personalities who rule the entity. It is assumed that state "leaders" use state "bureaucratic power" to maintain the integrity of the geographic boundaries of the polity, and to defend the community from alterations in the face of potential internal or external threats. I use the term "state" in various contexts to refer to "leaders," "bureaucracies," and "territory." State leaders, of course, are what animate the state as a self-interested organization. They are the "caretakers," so to speak, of the state's territorial integrity and have the potential to act autonomously and reflexively on its behalf.

determination. What is meant is that if African states are to avoid exposing their vulnerabilities with regard to the self-determination dilemma they must devise and execute policies which minimize ethnic conflicts of any magnitude and intensity and which expand the reservoir of support the state commands from all its citizens.

There is a tendency among students of contemporary African politics to underestimate the role the state plays in politics. It is common among both liberal and Marxist scholars to view the state simply as an arena in which conflicts over scarce resources are fought out.[10] Liberals analyze how competing "interests" register their claims and strategize against one another, and how those claims are processed by the political authorities.[11] The assumption is that the state gains its legitimacy from the normative consensus of the general population and that it maintains that legitimacy by the satisfactory execution of the responsibilities assigned to it. The key variable here is popular consensus. On the other hand, Marxists tend to see the state as an instrument of organized coercion which is controlled and dictated to by the dominant or ruling class.[12] In either case the state is not viewed as a truly autonomous political actor with values and interests of its own.

Recently there has emerged a debate among neo-Marxists over the validity of conceiving of the state as an autonomous political actor.[13] The questions being posed are: Is the state "central" to politics? Is the state autonomous? Is it controlled by the indigenous bourgeoisie, or is it controlled by the metropolitan bourgeoisie? Is the state "overdeveloped" or "underdeveloped?" What consequences does this have for the nature of politics in Africa? A common tendency is not to focus on the state as an "organization for itself," but to concentrate on the state only to better comprehend the nature of real and potential class conflict and class domination.[14]

The predominant currents in contemporary literature on the state in postcolonial societies, then, seem to view the state as nonautonomous and controlled by "domestic" or "foreign" "interests" or "classes." This is true no matter what the ideological persuasion of the writer.[15] It is my contention, however, that if we consider the political functions of the state, as opposed to its economic functions, we can clearly perceive its potentially autonomous nature.

In its quest for survival the state must attend to two basic tasks: maintaining domestic order and competing for power, prestige and/or respect with other states. Rather than being inordinately influenced by certain "interests" or "classes," in this case the state follows its own logic and is guided by its own purposes. To do otherwise would be to abrogate the responsibility of working conscientiously and instrumentally toward the ultimate goals of state continuity and survival. What this implies is that rather than being controlled by certain "classes" and "interests," the state may often find itself having to engage in conflict with them over matters of public policy.

The state employs public policies as instruments for achieving a wide range of purposes. Without the ability to utilize public resources to mobilize or control the general population, the state would never be able to act instrumentally for its own survival. Elite classes or interests might challenge the state, and even the very foundations on which it is built; peasants might refuse to cooperate and thus thwart the objectives of the state. To avoid such problems or to mitigate their effects, states rely upon their reservoirs of material and nonmaterial resources to accomplish their ends. Naturally, strong states are those with a variety and ample quantity of resources which can be converted into public policies which enhance state survival.

The potential for ethnic turmoil which exists in most African states demands that state leaders conscientiously devise policies to avoid or even solve such problems. Brute force alone is not the answer; measured, skillful leadership is often a better tactic. The latter approach invariably includes a contexturally relevant mix of positive and negative sanctions and a blend of material and symbolic resources. In the end the success or failure of the state's policy choices relating to the potential for ethnic conflict determines the degree to which such potential conflict might threaten state coherence.

The efforts of African states to maintain their coherence in the face of the ever-present danger of divisive ethnic claims would seem to provide an excellent opportunity to assess the validity of my assumptions about the potential for the autonomous political behavior of these states. The remainder of this essay is devoted to an examination of how selected African states have attempted to utilize public policies to mediate ethnic conflict and thus to avoid the most serious crises of self-determination. Before systematically analyzing the performance of selected states in this regard, I will turn briefly to suggest a typology for conceptualizing public policy in Africa.

A Typology of Public Policy in African Politics

Theodore Lowi, in a seminal article attempting to clarify early efforts by American political scientists to conceptualize the politics of public policy, devised a scheme which allowed him to organize public policies into functional categories according to their impacts on society. Rather than attempt to consider the whole range of public policies, Lowi wanted to limit the scope of his analysis to a few abstract categories. Another delimiting criterion Lowi used was to ignore policies of nonregulatory, generally universal impact (i.e., national defense policy, social security policy, national communications policy, etc.). Following the rules he set for himself, Lowi identified three functionally specific types of public policy: distributive, regulatory and redistributive policies.[16]

Distributive policies were perceived to be those policies which aim at satisfying individual demands on an individual basis. According to Lowi, distributive policies are operable in a highly "pluralistic" arena characterized by "a large number of small, intensely organized interests...it is a politics of every man for himself (or every community for itself)."[17] Distributive policy-making is dependent upon ample state coffers and it involves individuals, groups or their patrons approaching those in authoritative positions who are amenable to "logrolling" and "striking a deal." It is an arena in which the state operates on the principle "you scratch my back and

I'll scratch yours," or "don't gore my ox and I won't gore yours." Such policies cannot be effective if there is not a substantial "pork barrel," a container filled with unrelated items -- patronage spoils -- on which the state can draw at will. Under these circumstances, Lowi suggests that policies are best understood as co-optation.[18] By dealing with individual patrons and individual communities through bargaining, the state is able to diffuse potential conflict. The beauty of such a policy is that if it works, nobody feels deprived. Everyone has a chance of having his (or his community's) needs satisfied as long as there are enough resources to go around.

In reality there is no such thing as a purely distributive policy. As Lowi correctly suggests, people almost always pay more in taxes than they receive in services from the state.[19] Also, some state policy decisions pre-empt private decisions relating to the use of the same resource. All policies in the long run, then, could be considered either regulatory or redistributive. But for analytical purposes, we can confine our attention to the short run and thereby differentiate among distributive, regulatory and redistributive policies.

Like distributive policies, regulatory policies are relatively specific and individual in their effects. But in applying regulatory policies, policy-makers do not have unlimited possibilities for discretion. Sometimes hard choices must be made and upheld about "who will be indulged and deprived."[20] Certain groups are required to do or not to do certain kinds of things. There cannot be different standards for different groups. For example, regulatory policies might ban drinking alcohol in public places on weekends, or they might decree that Swahili is the official language. As a result of such policies (if they are enforced) certain groups will have to alter their behavior -- namely the weekend drinkers and the non-Swahili speakers. Weekend drinkers and non-Swahili speakers will not be indulged. Bar owners will be deprived of their weekend business, and some cultural-linguistic traditions will also no doubt be negatively affected. The regulatory policy arena is much more prone to conflict than the distributive policy arena under normal circumstances. People are often forced to do things they might prefer not to do, or prevented from doing things they prefer to do.

Whereas the redistributive character of distributive policies is not readily apparent, purely redistributive policies are distinguishable in that they obviously attempt to satisfy the needs or wants of one group at the expense of others. The state under these circumstances makes a conscious decision to dispense valued resources to one set of clients and to deprive others. Once a redistributive policy has been chosen, there can be bargaining over the extent and nature of redistribution. Groups for and against such policies invariably emerge and attempt to affect the impact or extent of the policy. Common forms of redistributive policies are education policies, housing policies, welfare policies and a variety of other social policies aimed at righting past wrongs.

Lowi's typology seems quite useful for categorizing and understanding the types of public policies found in Africa. It centers on public policies designed to have specific impacts, such as those we argue are associated with state efforts to mediate ethnic discontent. But I would suggest that to be truly useful, Lowi's scheme needs to be expanded to include two additional categories: reorganizational policies and symbolic policies. The reason for suggesting these additional categories is that the context of African policy-making differs substantially from that found in the United States or other industrialized societies. African societies are underdeveloped and relatively unindustrialized. They are characterized by fragile political, administrative and economic institutions. Their resources are meager and their ability to tolerate major political instability is limited. Therefore, many African states have resorted to reorganizational or symbolic strategies to diffuse the potential for political instability and to strengthen the hand of the state vis-à-vis groups who would challenge or defy it.

The term "reorganizational policy" refers to any efforts on the part of the state to restructure political or administrative institutions and relationships in order to cope with strains which have the potential for threatening state survival. Political parties, bureaucracies, militaries, and administrative units may all be altered in order to neutralize the potential for ethnic discontent or to strengthen the hand of the state.

All public policies have some symbolic content, but it is argued here that in some cases public policies are predominantly symbolic. Symbolic policies are distinguishable in that they are only minimally dependent on tangible resources. Their major components are intangible and psychological. Such policies are intended to reform human behavior through exhortation and example. Symbolic policies are necessitated by the high levels of popular demand for government action under conditions of resource scarcity.[21] Leaders maintain their credibility through symbolic policy-making -- not by merely "bringing home the bacon" in a material sense, but by creating the impression among their constituents that they are "acting on a problem" or "doing their best," or that they are "active" rather than "passive" leaders.

A key component of symbolic policy-making is ideologizing or sloganeering by agents of the state, especially political parties and chief executives. A leader might declare his primary goal is "land to the tiller" or he might attempt to demonstrate, by his own behavior, that the "correct" brand of Marxism-Leninism must be applied to everyone in society. Commenting on the essence of symbolic leadership, Murray Edelman suggests, "The chief executive may maintain his symbolic leadership through ascriptions of his ability to cope, through publicized action on noncontroversial policies or on trivia, and through dramaturgical performance emphasizing the traits popularly associated with leadership, forcefulness, responsibility, courage, decency, and so on."[22] An African leader might attempt to create the image that he is, let us say, responsible for chasing exploitative Europeans from his country, depriving them of their land, and redistributing the land to needy peasants. The number of European farms confiscated and the amount of land redistributed may be small, but because of the dramaturgy involved in publicizing the leader's actions in the matter, the populace could be positively affected in a symbolic way. However tenuous, this is the real value of symbolic policy. It serves as an additional resource policymakers can use to diffuse discontent.

In reality it is common to have all of these policy types operating virtually at the same time. Life experiences do not conform to neat typologies. However, it is possible to differentiate the five policy types we have identified at least for analytic purposes. This is the tactic I will follow in the remainder of the essay. Where appropriate, I shall make note of obvious overlaps in policy categories as they apply to the selected examples.

Public Policy and the Mediation of Ethnic Discontent

States cannot survive forever simply by resorting to force. They must be able to demonstrate that they are capable of developing their countries economically, insuring the physical security of all citizens, eliminating inequities, and improving the quality of life and life chances of all groups in society. Effective government expands the reservoir of popular support the state can rely on, and it is necessary to maintain that support at a favorable level. In its absence, the state must either bank on coercion or luck to insure its survival. Where the latter has obtained, specific regimes if not whole states have eventually had to contend with challenges to their authority in the form of coups, revolutions and secessionist movements. This is particularly true in the case of states such as we find in Africa which are fragile and lacking in spendable resources.

Not all African states are prone to ethnic conflicts which lead to demands for ethnocultural self-determination. But the extent to which a society is characterized by any form of ethnic inequality or conflict indicates its potential for crises of ethnic self-determination. Some states are highly unlikely to be challenged in this way. For instance, Tanzania, Botswana, Swaziland, Lesotho and Somalia have proven their cohesiveness as nation-states. In contrast to the unlikelihood of intense ethnic instability in those states, one might have predicted for quite some time that countries such as Nigeria, Uganda and Ethiopia would experience ethnic disorders which threatened to dismember them.[23]

Multiple factors seem to contribute to the propensity of states to experience crises of ethnic self-determination. Many of these factors are contextural rather than universal. For example, no African society has exactly the same political history as another, but it is evident that the "weight of history" bears heavily upon the tendency of a state toward ethnic conflict. Other extremely important factors vary from one context to the next. Some of those are: the coercive capacity of the state; the vulnerability of the state at a given moment in history to external or internal political pressures; the scope and intensity of ethnic competition; the leadership qualities and styles exhibited by the chief executive; and the flexibility and sensitivity of policy-makers in dealing with ethnic grievances.

Where the potential for ethnic discontent is high, it is in the interest of the state to devise policies to build loyalty and prevent threats of exit among certain ethnic groups. The nature, quality and applications of these policies, however, must vary from one context to the next, and leaders must be sensitive to the limitations of raw coercion. There are no guarantees that even sincere efforts on the part of leaders to accommodate ethnic demands will be successful.

Therefore, it is extremely important that policies designed to mediate ethnic discontent be as flexibile and as sensitive as possible.

Distributive Policy

Lowi has equated distributive policies with co-optation. These types of policies are by definition selective in impact, affecting only certain individuals or communities. In many cases, it is the squeaky wheel that gets the grease. In deciding how to allocate scarce development resources, a government might be guided by a rational national plan, but in most cases politics is paramount. A chief executive will calculate the political costs and benefits of distributing development spoils in one way or another. Under normal circumstances many of the resources the state has are dispensed on the basis of political considerations.

Among the most common examples of distributive policies used to co-opt certain ethnic elites is the tendency for chief executives to allocate key political and economic positions according to ethnic considerations. This is variously described as "ethnic arithmetic," "ethnic balancing," "regional juggling" and so forth.[24] Examples of this can be found in many African countries regardless of whether they are led by regimes which are military or civilian, socialist or capitalist. Jomo Kenyatta, while president of Kenya, perfected this tactic. In 1963, when Kenya became independent, the cabinet was noticeably skewed toward the two major ethnic groups, the Kikuyu and the Luo. Between them, they possessed ten (59 percent) of the seventeen available cabinet positions. By 1974, even though the Kikuyu continued to dominate Kenyatta's cabinet, having seven seats, there was evidence that other major ethnic groups had been co-opted into the ruling coalition. Ethnic patrons from the Luo, Luhya, Kamba and Kalenjin groups were now equally represented in the cabinet with two seats each.[25]

The mix of ethnic representation in a president's cabinet does not necessarily have to be proportional to the representation of each group in society. But it has to appear to be equitable, particularly when contending groups are relatively powerful. In Zimbabwe, for instance, Robert Mugabe was conscious of the need to apply the rules of "ethnic arithmetic" in selecting his cabinet on assuming office in 1980. He particularly had to be concerned with including certain key patrons from the white community and the Ndebele ethnic group led by Joshua Nkomo. Four cabinet seats were reserved for Nkomo's party and two for whites.[26]

The tendency to attempt to achieve a measure of balance in certain political positions, and in positions in quasi-public firms, does not always seem to hold in filling the upper-level, functional positions in administration. Dennis Dresang found in his 1974 Zambia study, for example, that the Bemba were disproportionately represented in the upper levels of the civil service.[27] He suggested, however, that this was due more to their being better qualified for these key functional roles than anything else. In a later study, focusing on Kenya, Dresang found that rather than always using his appointive powers with ethnic balancing in mind, Kenyatta sometimes used his authority to appoint individuals to East African

Community positions as a form of negative or positive sanction. In 1972, he notes, three Luos were removed from top Community positions and replaced by Kikuyus.[28] This was shortly after some Luos had been implicated in a coup conspiracy and at a time when Luo-Kikuyu relations were at their nadir.

Félix Houphouet-Boigny, the president of Ivory Coast, has also used his ability to dispense patronage both to reward and to punish. He employs the power of his office to appoint individuals to key governmental and other public positions. He also skillfully uses his discretionary authority in doling out or withholding all manner of development resources. Michael Cohen has described how Houphouet-Boigny uses his authority to decide which city, in which region, will host the annual national fete as a form of patronage.[29] When this affair is held, a good deal of money is poured into a community to refurbish it and to supply it with certain showcase projects. Additionally, tourism for the region during the fete guarantees the influx of new money into the community. Rather than allocating the fete according to some rational plan to make sure that all regions get their chance to have it, Houphouet-Boigny has chosen in the past to reward or punish certain elites and their communities through this type of distributive policy.

Important points to bear in mind when considering how distributive policies are applied are how a group's demand is registered and the relative power of a patron or group. Governments sometimes choose to make a public issue of their desire to eradicate inequality without much demand for this by needy communities. But whatever resources are allocated can just as easily be withdrawn if the group or patron is weak or passive. A good example of this can be found in the Kenyan government's decision in 1970 to devote special attention to providing the ethnic minorities of the North-East Province with more educational opportunities. By 1974, the government had decided to withdraw its commitments to the region because the people did not seem to appreciate the value of education.[30] Part of the reason that this could be done without arousing much ethnic discontent has to do with the relative powerlessness of the people (mostly Somali or Borana) of the region.

Under ideal circumstances, patronage can be an extremely effective means of mediating ethnic conflict. The key is the chief executive's ability to co-opt ethnic patrons or to punish them with impunity. Success at using patronage as a positive sanction is critically dependent on the chief executive having something to give. Resources could be in the form of political party positions, government positions, or positions in public corporations. They could also be in the form of special privileges.[31]

Regulatory Policies

Regulatory policies are generally intended to accomplish specific ends. They apply to whole societies, but directly affect only those persons protected or restrained by the regulation. Although the arbitrary use of state power would be a form of regulation, it is not our concern here.

In Africa perhaps the most common form of regulatory policy is "banning." Parties may be proscribed by law or individuals may be detained as seditionists or threats to domestic tranquility. This approach has often been used as a justification for the establishment of de jure one-party states. Leaders imposing a ban on competitive party politics, such as Nigeria's General Ironsi in 1966 or Zaire's General Mobutu in 1965, justified their actions by arguing that competitive parties -- at least in the short run -- interfered with the state's desire to achieve national unity.

Not all one-party states are created by the executive decree of a military autocrat. Some are the creation of civilian leaders and are outgrowths of political parties which became dominant in a competitive party system. Tanzania's party system is the most well-known example of this. Tanzania became a de jure single-party state in July 1965 with the Tanganyika African National Union (TANU) declared the only official party.[32] From this point on, party competition was to be exclusively within the single-party framework. The island of Zanzibar, which joined Tanganyika to form the United Republic of Tanzania in 1964, was allowed to keep its own party, the Afro-Shirazi Party. However, in the spring of 1977 TANU and the Afro-Shirazi Party were unified to form Chama Cha Mapinduzi (The Party of the Revolution).[33] One of the main intentions of establishing the single-party state in Tanzania was to encourage the development of a sense of national unity and to avoid potential class and ethnic difficulties by regulating political competition.

The turn to the single-party system was relatively smooth in Tanzania. In Zambia the same process was fraught with opposition. Zambia had had a multiparty system with one dominant party, the United National Independence Party (UNIP), from just before independence in 1964 until 1972. Traditionally its most formidable competition came from the African National Congress, but in 1971 a new party was formed, the United Progressive Party. The emergence of the UPP raised the specter of interethnic and class-based conflict as UNIP had come to be seen as the party of a certain segment of the Bemba, the ANC attracted mainly non-Bemba speakers, and the new UPP was comprised mostly of disaffected Bemba adherents. President Kenneth Kaunda saw these developments as potentially threatening to internal security, banned the opposition parties in February of 1972, and declared the formation of a de jure single-party state.[34] This move was not a popular one among many Zambians but the opposition was not strong enough or organized enough to effectively resist it.

Banning political parties or establishing a single-party system may not be enough to minimize the potential for ethnic discontent. Ethnic affinities can also be channeled into ethnic associations which are not explicitly political. Realizing this, Mobutu banned all such associations as he tried to build the Mouvement Populaire de la Revolution (MPR) in Zaire.[35] In Kenya, ethnic associations have always existed alongside explicitly political organizations. But in 1980, Kenya's new President Daniel arap Moi, obviously fearing the divisive potential of the large and powerful Gikuyu, Embu and Meru Association (GEMA), at first tried to invite such groups to disband voluntarily, and then gained the support of parliament to bring about this change.[36] Moi was uneasy about the prospects of GEMA undermining his authority in Kikuyuland.

Just as rules can be invoked to ban parties, they can be invoked to establish them, even where they never existed before. Ethiopia is a country which had never had a party system of any kind, but by the late 1970s the revolutionary Provisional Military Advisory Council (PMAC), which had deposed Emperor Haile Selassie in 1974, decreed that a single-party system based on Marxist-Leninst principles would be adopted.[37] A commission, COPWE (Commission to Organize the Party of the Working People of Ethiopia), was then appointed to study how this system might be implemented. As in the cases where parties were banned and refashioned into single-party systems, the PMAC was merely attempting to regulate political participation while at the same time leaving the impression that it was interested in expanding participation.

The banning of individuals who are alleged to pose a threat to the state is as common a regulatory device as banning political parties and ethnic associations.[38] Such African nationalists as Kenyatta, Kaunda, Nkrumah and Banda all spent time in detention during the colonial period. After independence they used this same device to remove certain opponents from the political scene.

The state might also take steps to regulate ethnic discrimination. Mali and Guinea, for example, have laws forbidding ethnic discrimination as well as the propagation of separatist ideas.[39] On assuming control of government in Ethiopia, the PMAC decreed equality for all citizens irrespective of their ethnic affiliation. The Federal Military Government of Nigeria made a similar proclamation, but making public proclamations does not necessarily translate into effective policy. In some cases ethnic inequities have continued in spite of government's public commitment to the contrary.

Other forms of regulatory policies may relate to education, religion or language. In some societies, such as Guinea, Tanzania, Ethiopia and Mozambique, the curriculum in schools is explicitly designed to reduce national parochial sentiments and to instill in students a sense of national identity. It is not certain, however, just how effective such experiments are; it cannot be shown that national identities come to outweigh parochial sentiments.[40] In fact, it has been found that conscious policies of political education are not necessary for an emerging political culture to have an effect on the political orientations of students, but the permanence of this change may vary according to the political circumstances which characterize a polity at a given point in time.[41]

Religious and language policies which demonstrate a respect for linguistic and religious minorities in multiethnic societies are generally more effective than policies which attempt to impose the language or religion of a culturally dominant ethnic group on minorities. Ethiopia's policy of pushing the Amharic language as the national language and Coptic Christianity as the national religion has been deeply resented by ethnic minorities such as the Eritreans, Somalis and Oromos.[42] In contrast, the government of independent Cameroun has followed a language policy strategy of recognizing both French and English as official languages. Indigenous languages and traditional religious affinities have not been legislated against, but the government is consciously attempting to create a bilingual state as well as bilingual individuals.[43] The fact that neither English nor French is indigenous to Cameroun perhaps best explains why language policy has not been a source of serious ethnic conflict there.

Redistributive Policy

In Africa, the demand for a redistribution of society's wealth and opportunities is high, but the capacity of the state to deliver is usually very low. Few governments can completely ignore such demands, even though attempting to meet them can be fraught with difficulties unless there is an enormous reserve of redistributable resources. Nigeria, because of the oil wealth of the Southeast Region, has been able to implement redistributive policies which attempt to compensate for the historical imbalance in educational opportunities in certain regions. With oil money, the government was able to extend universal primary education to all sections of the country[44] and to expand the number of universities from four to thirteen. This would seem to indicate the Nigerian government is consciously trying to spread not only wealth but also opportunities more equitably throughout the country. These policies were facilitated by an ample reservoir of redistributable resources, and the fact that education is universally viewed as an absolute right.

Conflict over redistributive policies, however, is more frequent than the lack of conflict over such measures. For example, in early 1981 the new civilian government of Nigeria submitted a proposal to the legislature to redistribute 75 percent of national oil revenues to the nineteen states.[45] This blanket redirection of national revenues met with intense opposition, particularly from several oil-producing states, and the government had to settle for a compromise of 58.5 percent. Even with this, the controversy continued as the aggrieved states took the matter to court.

Should a government misread the extent of its leverage or miscalculate the adequacy of its redistributable resources, rather than enhance the state's capacity to survive, these policies could well trigger its demise. Haile Selassie, prior to his dethronement in 1974, had been encouraged by numerous foreign aid donors such as the USAID and the World Bank to initiate a land reform program which would reduce the gross inequities which mainly characterize the peripheral areas of southern Ethiopia.[46] Indeed, the emperor did make some feeble attempts to do this, but he was met with stern opposition from the landed aristocracy. As a result, the land reform program stagnated and never became very effective. Historically, Selassie had been able to manage ethnic conflict by relying on his extensive bureaucracy and military. These were _flexible resources_ which he could call upon when needed to buttress the authority of the state. In 1974, however, his most critical resource, the military, failed him, and the emperor was overthrown by a conspiratorial coup d'état. By this time, it was apparent that his reservoir of popular support was extremely meager. This was in part due to his lack of an effective set of redistributive policies.

Redistributive policies are particularly difficult when they cause intense controversy or when they take place in a political environment which is unstable for other reasons. It seems that it is wiser for a political leader to avoid alienating powerful groups unless his own position is secure. At times, chief executives can place a lower priority on redistributive policies when they are not dependent on deprived groups as a base of support. Houphouet-Boigny, for instance, rather than basing his support in the poor regions of Ivory Coast, relies on the dominant, well-to-do regions of the south. They, in fact, would be highly opposed to radical policies for redistribution because this would threaten their own privilege. Wisely, at least in the short run, Houphouet-Boigny has chosen not to alienate the well-to-do regions. In the long term, however, it could well be that the continued exacerbation of regional inequalities could lead to ethnically based turmoil.

Leaders are more likely to use their redistributive policies as mediating agents for ethnic conflict when they can target these policies for underdeveloped regions where they have little support. The North-Western and Western provinces of Zambia are among its most underdeveloped regions, and they are also areas where support for UNIP, the ruling party, has traditionally been weak. Kenneth Kaunda, in the late 1960s and early 1970s, conscientiously followed a policy of parceling out development funds and projects to the most needy areas, regardless of their political attitudes toward him or his government. In this period, the North-Western and Western provinces received the highest levels of national development expenditure per capita, and the Copperbelt and Northern provinces, where Kaunda had solid support, received the lowest.[47] There is also evidence that by utilizing development resorces among the Tonga in Southern Province, Kaunda was able to erode the support of the opposition ANC significantly. Between the 1964 and 1968 national elections in the region, the level of UNIP support increased from a mere 21 percent to 45 percent. This has largely been attributed to Kaunda's strategic allocations of development resources.[48] While Zambia is still prone to ethnic conflict, redistributive policies have no doubt significantly lessened the potential for intensely divisive conflict.

In former settler colonies such as Zimbabwe and Kenya, the most salient redistributional policy is that relating to land. In both cases, the colonial era was characterized by the alienation of African land to Europeans and the subsequent establishment of a commercial agricultural sector based on large-scale, European-owned farms. A key issue in the nationalist movements in both countries was the land question.

In Kenya the independence settlement was centered around the land issue. According to the agreement, European farms would be purchased with money the Kenyan government borrowed from Great Britain, and individuals would be allowed to resettle on land previously used as European mixed farms.[49] By 1970 more than 48,000 families had settled on 1.37 million acres of land, and the government had plans for purchasing at least another million acres over the next decade. About four million acres of ranch land and coffee, tea and sisal plantations still in European hands were seen as virtually untouchable because of their high capital value.[50]

The main beneficiaries of the resettlement program were Kikuyu farmers of all classes. They were generally seen as deserving since it was their land which had, for the most part, been confiscated by the settlers and since it was the Kikuyu who were the most heavily represented in the Mau Mau rebellion. Any plan which did not specifically address Kikuyu land-hunger would have been doomed to failure. Although some objections were raised by certain groups, these protests were not over the resettlement program per se and were not generally based on ethnic criteria.[51] This is not to say that the land redistribution policy was uncontroversial. In fact, controversy did grow out of the fact that the land was being purchased instead of nationalized. Oginga Odinga and Bildad Kaggia became the most vocal critics of the land transfer program.[52] They favored distributing land to the landless in a sweeping program which was based exclusively on need. As it was, those resettled had to have the means to purchase the land. Kenyatta was not prepared to accept the chaos which would certainly ensue if the government, after having promised to redistribute all European lands, found itself in the position of having severely disrupted its economy and still not having enough land to go around. The conflict over the land policy led to the ouster of Odinga and his followers from KANU, and to the formation of the opposition KPU. But by this time Kenyatta had already built up a considerable amount of leverage because of his redistributive and symbolic policies, and even an issue as potentially divisive as land could not significantly threaten his regime.

In Zimbabwe land-hunger was given as one of the most significant reasons for the nationalist revolution which challenged the white minority regime between 1973 and 1980. Prior to independence, 3 to 5 percent of the country's population occupied about 50 percent of all land, which included most arable land and all known mineral deposits. The independence settlement had been similar to that arrived at in Kenya. The new government was to borrow money and use it to buy former European land for redistribution to landless African farmers. As in Kenya, the government of Prime Minister Robert Mugabe eschewed the idea of wholesale confiscation of European farms.[53]

Rather than redistributing land on the basis of private ownership, the new government considered collectivization, but this plan met with stiff resistance from many of those who would be needed to support such a program. As a result, the government had to resign itself to practicing collectivization only on an experimental basis.[54]

The Tribal Trust Lands (native reserves) were now held on a freehold basis, but the problem of resettling refugees remained enormous. Many had hoped that the one to two million refugees of the liberation struggle would be quickly settled on confiscated or bought land, but by late 1980 only about 200,000 had been resettled.[55] The Mugabe regime had purchased some 320,000 acres of land, but this was not enough to accommodate the pressing demand.[56] Consequently, resettlement was much delayed. Part of the reason for this delay was that the government initially experienced difficulty in raising the funds needed to buy European farms. There were those among the African elites who viewed Mugabe's hesitance to simply "grab" European property as "ridiculous."[57] Additional problems related to the fact that squatters had moved onto vacated land without waiting for the government's reallocation program.

Reorganizational Policy

The main functions of reorganizational policy are either: 1) to strengthen the hand of centralized state power; or 2) to redistribute political power so as to accommodate the demands of certain groups. Reorganization may entail restructuring political parties, party systems, constitutions and administrative arrangements. Needless to say, some reorganizational policies might include regulative and/or symbolic elements as well. In any case, the potential success or failure of a given reorganizational strategy depends a great deal on the context in which it is attempted and on the relative strength of the state vis-à-vis contending groups.

Party reorganizations in Africa have generally taken the form of alterations in the ruling parties of single-party states, or in situations where a party has become dominant in a competitive system. Julius Nyerere, shortly after he was elected prime minister of Tanganyika, stepped down in order to refashion the dominant Tanganyika African National Union. In 1967, the party was again reorganized, this time to become the only party in a de jure single-party system. Ten years later the party constitution was amended again so as to unite TANU with the Afro-Shirazi Party as Chama Cha Mapinduzi, and to introduce new electoral arrangements.[58] In each case, the intention was to strengthen the hand of the central authorities while enhancing the ability of the party to provide a forum for popular participation.

After banning political parties after 1965, Mobutu proceeded to reorganize the Zairian party system and to introduce a single national party, the MPR. The new party was designed to penetrate to the lowest levels of administration. Local administration became wedded to the party, with local administrative officials -- who were usually foreign to the area -- becoming the ex officio local heads of the party. The national political bureau had to approve all nominees for elective office before they could run. Membership on the political bureau itself was decided by election, and those serving on the bureau were required to run in constituencies other than their own.[59] Simultaneously, all ethnic associations were being banned. In the process, Mobutu was able to significantly weaken centers of power in various regions and to enhance his own personal power. He could now deal with local and regional elites in a very personalistic way.

Milton Obote also attempted to reorganize Uganda's party system and the political system as a whole in hopes of weakening regionalist tendencies and enhancing his own personal power as well as that of the central government. In 1966 Obote presented the national parliament with an interim constitution which called for transforming Uganda into a unitary state.[60] This move resulted in immediate opposition from several traditional kingdoms which had maintained some regional autonomy after independence. The most powerful of these was Buganda. On May 20, 1966, the Buganda parliament committed the ultimate act of defiance of central authority by approving a de facto motion of secession. Obote responded by declaring a state of emergency and ordering the military to use force to crush the rebellion.[61]

Amid heightened ethnic tensions, on September 8, 1967, Obote was able to get the national parliament to approve a new republican constitution. The new constitution strengthened the unitary political system and abolished all traditional kingdoms. The central government now reserved the right to appoint district councils, and complex formulas for electing members of parliament and the president himself were introduced.[62] The net effect of these moves was to provide the central government with the capacity for much greater penetration and control throughout the country.

In late 1969, Obote's Uganda People's Convention, which by now had become virtually the only viable political party, resolved that the republican constitution be amended so that Uganda would officially become a one-party state.[63] All of these measures were allegedly intended to weaken ethnic affinities and to strengthen national unity and thereby to prevent self-determination claims from threatening the survival of the state. As part of this party reorganization scheme, Obote also endorsed a proposal to restructure electoral competition so as to promote participation in elections based not on regions or ethnic affinities but on the nation as a whole. However, this ingenious scheme was never actually implemented as Obote was deposed by General Idi Amin before it was finally put into effect.[64]

The lesson of Obote's Uganda seems to be that reorganizational efforts which take place when ethnic or regional tensions are severe are doomed to failure. Also, it could be said that amid intense ethnic or regional controversy involving relatively powerful dissident groups, movement from a decentralized arrangement to a centralized administration cannot be accomplished without the severest of political costs.

In Kenya reorganization from a relatively decentralized to a centralized administrative arrangement proceeded rather smoothly. At independence, Kenya was not a completely unitary state; the _majimbo_ Constitution allowed for regional assemblies which theoretically could have provided a base for ethnic or regional power with the potential to undermine the initiatives of the central government. Within a year, however, Kenyatta was able to declare pre-emptively that regionalism was dead and that Kenya was officially a republic.[65] What seems to explain Kenyatta's success is the character of the context in which he operated and his particular approach to the task of state-building. Kenyatta was an extremely popular figure at the time; he had the support of powerful ethnic patrons such as Oginga Odinga and Paul Ngei, and by 1964 he was able to co-opt just about all of his opposition.

Transition from a relatively centralized to a more decentralized administrative arrangement has often been more successful than changes from a decentralized to a more centralized organizational framework. Sudan, for example, chose a decentralization strategy when it reorganized its state administration in 1972 in the wake of a secessionist civil war which had raged for sixteen years.[66] The war had pitted the forces of the predominantly Arab and Muslim central government, with its base of support in the north of the country, against rebel forces from the south, Christian-led and comprised mostly of blacks adhering to traditional religions. The settlement called for the establishment of three southern provinces which could be granted regional autonomy over most of their affairs. Arabic was retained as the state's official language, but English was accepted as the working language in the south. Indigenous languages were also respected and promoted. The agreement allowed the state to remain intact, and a national legislature was given the responsibility of supervising legislation relating to national issues.[67]

The strategy of granting cultural autonomy to ethnic communities in those most deeply divided African societies would seem the best solution for averting the emergence of claims to self-determination which threaten the survival of the state. Sudanese federalism, for instance, might offer a key to resolving Ethiopia's self-determination crisis. This strategy allows for a redistribution of power, and minimizes the sense of cultural insecurity which often accompanies a movement toward more and more centralization. Where cultural insecurity or ethnic hostilities are not so intense, other policy strategies might be attempted.

Symbolic Policy

Both concrete and symbolic elements are common features in everyday politics, and each of the policy types above possesses symbolic elements. But every invocation of symbolism by a politician is not a manifestation of symbolic policy in the making. The term "symbolic policy" becomes applicable only after symbolic acts are used by political leadership consistently enough and systematically enough to be considered institutionalized and when it is clear that these gestures elicit voluntary popular reaction of one form or another. The mere appearance of a head of state at a rally or parade accompanied by the chanting of slogans and the singing of patriotic songs is not necessarily an indication that symbolic policy is at work. The populace must be moved to voluntarily behave in ways which are suggested by their leadership. They may be asked to participate in communal living or they may be asked to forsake "tribalism" and to learn the virtues of "good citizenship." To the extent that 1) the response of the population is voluntary -- that is, behavior is not coerced to any significant degree, 2) the benefits are more psychological than material, and 3) this leads to a regularized pattern of popular behavior in the desired direction, we could say that symbolic policy is operable. These preconditions eliminate most African states from consideration. All authoritarian regimes which rely mainly on coercion and regimentation in eliciting certain forms of popular behavior can be excluded. Other regimes which would seem likely to fall into this category do not, simply because the use of symbolism has not become institutionalized or because certain other preconditions are not met.

John Nellis has suggested that it is possible for regimes to "purchase" the behavior of a population. "By manipulation of wealth production and distribution mechanisms," he argues, "a government can, to a certain though undetermined extent, induce either behavior change or maintenance of existing behavior patterns."[68] This seems to be the rationale behind distributive and redistributive public policies. Material rewards are used to manipulate the political behavior of a given population. But as was already suggested, most African states do not possess a redundance of material resources which could be distributed or redistributed. Faced with this dilemma

-- the need to be able to dispense material resources in order to gain support for state policies or to minimize the potential for social conflict while at the same time possessing limited resources which could be used for such purposes -- some African governments where possible have sought alternative distributable benefits. Classical examples of the state resorting to the dispensing of a large measure of nonmaterial, mostly psychological resources are Kenya under Kenyatta and Tanzania under the leadership of Julius Nyerere. Political symbolism, in an institutionalized way, has come to be used as nonmaterial "currency" because the use of material "currency" is often infeasible or inappropriate. It is important to note that only when political symbolism used in this way elicits the desired response can we say that it qualifies as symbolic policy. The symbolic currency is valid only when it is accepted as a stimulus to certain forms of behavior. We will return to this discussion below.

Political symbolism and ideology do not necessarily go together, but they might. In either case, however, they cannot be effective without proper leadership. Leaders might be patrimonial, bureaucratic, or charismatic in terms of their political style. What is important is that their method of addressing the masses be appropriate for the context in which they operate. In order to employ nonmaterial currency effectively, leaders must master the art of political symbolism, or they must devise an ideology and supporting institutions to make this approach effective.[69]

As with how we conceptualize leadership, the way we evaluate political ideology also needs to be reconsidered if we are to determine its utility as an instrument of public policy in Africa. African ideologies should not necessarily be judged exclusively on their doctrinal "correctness." There are few African ideologies to be found, for example, which could be considered "scientific" ideologies. Yet, there are several which could be called "functional" ideologies or "working" ideologies, in that they elicit a specific popular reaction.[70] Functional ideologies might violate many, if not all, of the rules of political philosophy relating to a definition of ideology. Leaders looking for nonmaterial currency however, do not always have time to measure their utterances by the standards of political philosophy.

For populist leaders like Nyerere and Kenyatta, what is important is what ideology does: it is meant to justify or criticize one value or norm preference possibility as against some other possibility. As such, concrete ideologies may contain a mixture of positive and negative statements. For example, the value of "hard work" may be juxtaposed against "idleness" or "laziness." "Communalism" may be given a much higher value than "individualism." Whether implicit or explicit, the mixture of positive and negative statements is required if the functional ideology is to work. Where one value or norm preference is being justified, its opposites or alternatives have to be criticized to strengthen the positive justification.

If we consider an ideology to be functional rather than scientific, ideological sophistication is not an absolute requirement. This allows us to consider Tanzania's Ujamaa Socialism and Kenya's Harambee movement as being based on functional ideologies, rich in political symbolism. The former is much more sophisticated and well-articulated than the latter, but both rely heavily on slogans and cliche phrases in communicating with the masses.

Tanzania's Ujamaa Socialism is Nyerere's creation.[71] Ujamaa in Swahili literally means "familyhood." This principle was introduced by Nyerere as a mechanism for creating and reinforcing a sense of national unity among the more than 100 ethnic groups in Tanzania. There are few people in Tanzanian society who fully comprehend the broader implications of this ideology as it is articulated in Nyerere's writings, but many people have consistently been inspired to follow his lead.

The ideology is not scientific in the strict sense of the term. It means different things to different people, and holding it does not preclude other ideological positions. In fact, within Nyerere's government itself there is a broad range of ideological opinion based on the great "isms."[72] But Nyerere's most loyal lieutenants are agreed on the appeal they should use in attempting to mobilize or restrain the passions of the masses.

According to the tenents of Ujamaa Socialism, Tanzanian society is founded on the principles of familyhood, mutual respect, freedom and work. These principles are promoted by the president, members of Chama's leadership and the bureaucracy extolling the virtues of hard work and communal living. Also, to insure that the right kinds of citizens are produced, there is a conscious effort to design Tanzania's educational system so that it instills the right African socialist values in young people and prepares them for productive rural lives.[73] Certain representatives of the state and party also attempt to reinforce these principles by demonstrating what Ujamaa means in practice. For the most part, the value of this approach is thought to be in the sense of psychological gratification it engenders in some of the public.

Nyerere attempted to operationalize the Ujamaa concept in 1967, submitting a proposal to TANU for the nationalization of the major means of production and exchange, for the acceptance of a socialist code of conduct for both local and national leaders, and for redirecting the country's development from the urban sector to the rural sector.[74] The cornerstone of this approach was a plan to move all rural residents into communally based Ujamaa villages on a voluntary basis.

By 1972 it seemed that as many people who would be attracted to Ujamaa villages voluntarily were already living in them (15.6 percent of the total rural population).[75] However, shortly thereafter, the government began to use compulsion as a means for peopling Ujamaa communities. It is here that symbolic policy began to overlap with regulatory policy. Some people still benefited from symbolic policy, but others did not. This highlights the tenuous and ephemeral nature of symbolic policy. Its effectiveness is highly dependent on skillful leadership and favorable political conditions.

By 1976 residence in these villages had swollen to 91 percent of the total rural population. Some farmers in the agriculturally rich northern regions resisted this policy, and so did some of the bureaucrats who were supposed to help implement it.[76] There is no evidence, however, that this protest was based on ethnic grievances. The important point here is that the ideology of Ujamaa was effectively used in a functional manner -- at least initially -- to advance the value preferences of communalism and hard work over preferences for individualism and laziness. It was also used to reinforce a sense of Tanzanian national identity.

In comparison with Ujamaa, Kenya's functional ideology under Jomo Kenyatta was soft and diffuse. Its most obvious manifestation was in the national motto "Harambee," or "Let's all pull together." Kenyatta often invoked this appeal in an effort to get rural populations to engage in self-help activities which provided them with needed development infrastructure and services when the government was unable to meet the high level of demand for these things. Schools, clinics, cattle dips, wells and community centers were among the sorts of facilities peasants were encouraged to provide for themselves. Kenyatta and other ethnic patrons would journey to distant villages and lead fund-raising rallies amid a great deal of dramaturgy and hortative language.[77] All of this political symbolism was functional in that it created the sense among rural communities that they were responsible for doing for themselves because the government could not satisfy all demands. Self-help was also a form of indirect tax as it resulted in providing communities with social services normally provided by the government. This policy was strengthened by the fact that local communities, at the encouragement of ethnic patrons, came to compete with each other and came to place the blame for the lack of certain social services upon themselves rather than on government. Kikuyus, Luos, Kambas, Luhya, all the major ethnic groups were persuaded to choose "harambee" as a logical vehicle for providing social services for themselves when government could not. It could be said that this strategy served to diffuse the potential for ethnic discontent, at least in the short run. In the long run, however, it could well exacerbate the potential for ethnic conflict, since, contrary to Ujamaa, Harambee feeds parochialism instead of encouraging people to become actively engaged in collective, interethnic projects. A critical difference between Ujamaa and Harambee as forms of symbolic policy is that the objectives of the latter are relatively narrow, while the former attempts to govern all aspects of human activity.

Not all African ideologies are functional ideologies, although they may be rich in symbolic content. Ideologies like Humanism as articulated by Kenneth Kaunda and Léopold Senghor's Negritude have few if any functional qualities.[78] A fundamental aim of Humanism is to transform Zambia into an egalitarian society, and Kaunda tries to use it as an agent of national unity. However, Humanism has not been actively and consistenly utilized as an effective instrument for mobilizing the population for specific development activities as have Ujamaa and Harambee. Negritude is also devoid of any programmatic qualities which could be applied to elicit mass support for development activites. It only serves to heighten the sense of African pride among its adherents. Not only are Humanism and Negritude not functional ideologies, based on their impacts we could also say they have little if any merit as symbolic policy. Kaunda and Senghor might wish to use them as such but there is scant evidence that their nonmaterial, symbolic currency is spendable in the manner we have suggested is necessary.

Political symbolism may be used as symbolic policy even without a functional ideology. Under these circumstances, leadership style becomes paramount. A good example of this could be found in the way in which Félix Houphouet-Boigny has used formal "Dialogues" as a policy instrument. In late 1969, he held a series of thirty-three public meetings to allow numerous cross-ethnic, functional groups (e.g., traders, parents, students, civil servants, soldiers, etc.) to communicate directly with him and to appeal to him for policy changes.[79] Several of the sessions revealed intense ethnic friction. At the conclusion of the 1969 Dialogue the president pledged to institutionalize the process of holding a Dialogue at least once a year for four consecutive weekends and to decentralize this process so that more government officials would pay attention to the needs of the people.[80]

The Dialogue was hailed as marking a new era in Ivory Coast political life, resulting in the heightened participation of more people in the discussion of national problems. What was perhaps more significant was the calculated use of these meetings by Houphouet-Boigny as an element of symbolic policy. Although material benefits did accrue to some groups after these sessions, the main kind of resource dispensed was the symbolic and psychological assurance that the president was concerned with the problems of certain groups and committed to resolving them. Like Harambee, however, the Dialogue has more short-term than long-term benefits and does little to resolve deep-rooted socioeconomic and ethnic grievances.

Conclusions

The potential for divisive ethnic conflict in African states has not been muted by the passage of time. Most African states are characterized by cultural pluralism. This, coupled with the sometimes intense competition among various groups over power and scarce resources, has the potential in many cases to lead to explosive political conflict. Not all incidents of ethnic conflict involve the issue of self-determination, but left unattended by leaders of the state, there are many such incidents which could lead to demands for cultural and/or territorial autonomy by disgruntled groups. Therefore, African states must consciously devise strategies and policies aimed at minimizing the potential for divisive ethnic conflict. This need is particularly severe in cases where the demand for autonomy by certain groups threatens to dismember the state.

The intent of this essay has been to critically analyze various attempts by African states to manage ethnic conflict and preserve the continuity and integrity of state boundaries. I argued that rather than viewing the state as simply an arena in which conflicts between various classes or interests take place, or as an organization which is manipulated by dominant interests or classes, we must begin to regard the state as a potentially autonomous actor, as "an organization for itself," with its own values and interests. The most basic of these values and interests is state continuity -- that is, the tendency to value the very survival of the state above all else. Well conceived and executed public policies designed to foster state-building, penetration, national political integration and economic development could enhance the state's ability to preserve itself. On occasion, state leaders ignore the need to work consciously to minimize the potential for ethnic conflict, or they attempt to implement deficient programs; in either case the result can be an exacerbation of the potential for ethnic conflict.

States have a variety of public policy strategies available to them in their efforts to manage ethnic conflict. This essay identified five types of policies which have been most frequently employed by African policy-makers for this purpose: distributive, regulatory, redistributive, reorganizational and symbolic policies. Ideally the state can manage ethnic conflict better when it has a large amount of distributive and redistributive resources. Distributive policies are supported by a ready supply of valued things (money, technology, expertise, patronage, etc.) the state's leaders can draw upon in order to co-opt elites who might provide the leadership necessary to mobilize ethnic discontent. These resources do not have to be locally derived; they could come from external sources (i.e., foreign aid, loans, technical assistance), but they do have to be readily available for use by state leaders when needed.

Periods of worldwide economic crisis tend to reduce the distributive and redistributive capacities of developing states such as those we find in Africa. Worldwide inflation in the late 1970s and early 1980s seemed to force African governments which had relied heavily on distributive, redistributive and symbolic policies to turn increasingly toward the severest of regulatory alternatives, coercion and oppression. Which policy strategy seems to dominate in one state as opposed to another is often determined by such factors as the quality and style of leadership, prevailing political and economic conditions, and the relative power of the state.

Lowi has described distributive policy as co-optation based on patronage. In order to be effective, skillful leadership is almost as important as an ample coffer. Co-optation may not always buy ethnic harmony, but as several examples in this essay seem to indicate, under usual circumstances and when skillfully applied, distributive policies of this kind do minimize the potential for ethnic conflict considerably. This approach, however, seems to work best when ethnic tensions are not already high.

Redistributive policies are inherently conflictual. They are particularly problematic in Africa in that African governments generally have so little they can redistribute. Few countries are like Nigeria, with its huge oil revenues which allow the central government considerable leverage in policy-making. Needless to say, redistributive policies, like distributive policies, seem to work best where there is an abundance (even a redundance) of state resources, and where the political environment is relatively quiescent.

Symbolic policies may be employed when material resources are few. The currency of symbolic policy is intangible and psychological. It is characterized by rhetoric, symbolism, slogans, hortative language and regularized acts of political dramaturgy on the part of state leaders. Skillful leadership is a requisite in this approach, and unless a formula is discovered which fits the given context in which it is applied, political symbolism might never attain the status of symbolic policy.

This form of policy as a pure type is extremely rare in Africa, but it does exist. Its utility varies from one context to the next, and its effectiveness varies from one leader (or leadership group) to the next. Where it has worked, symbolic policy has enabled the state to engender a generalized feeling of support among disparate ethnic communities toward certain state policies, and toward the political community and its leaders. In the process the state has been able to compensate for a dearth of material distributable or redistributable resources.

Regulatory policies may be used as a means of maintaining integrative behavior or as controls against disintegrative behavior. As most African states are relatively recent, artificial creations characterized by varying degrees of ethnic heterogeneity, a new, expanded national political culture must be developed and accepted by constituent groups in each state. Some regulatory policies establish new rules of the political game, so to speak, and communicate to the population what forms of behavior will be indulged or prohibited. The hope is that over time people will develop integrative habits and comply with the new rules of the game as a matter of course. Ideally, the authority of the state to use force in support of these rules is readily accepted. Again, the intensity of ethnic competition and conflict is critical. Under circumstances involving low ethnic tension levels, the legitimacy of the state might be readily accepted and compliance with rules might be forthcoming automatically. However where ethnic tensions are high the chances are great that the legitimacy of the state as it is presently constituted will be in doubt. Ibos, for example, hardly considered secession until the ethnic tensions in Nigeria reached a point where they felt insecure as a group. It was only then that they considered ethnic autonomy seriously. Severely strained ethnic relations, and a state whose regulatory capacity had been seriously diminished, provided a climate which led to secessionist moves in Uganda and post-Selassie Ethiopia. When ethnic tensions are moderate and when the state's regulatory capacity is intact and effective, some conflict might be managed within the framework of regulatory policies which control popular behavior rather than coerce it.

Those regimes which rely less on raw coercion and more on subtle control are more likely to experience long-term success in managing ethnic conflict. Force is a weapon which the state always has at its disposal.[81] However, it is best used sparingly and selectively. The excessive use of force by the state can be seen as a sign of weakness and in the long run it can do more to threaten state coherence than to aid it. Excessive use of force in an effort to prevent secessionism always presents the image of a state which is low in legitimacy and desperately struggling to survive.

A better alternative to using regulatory policies which rely heavily on brute coercion in deeply divided societies would seem to be some form of reorganizational policy which devolves power from the center to given ethnic communities. This is the intention of movements toward federalism and regional autonomy in societies attempting to avoid or overcome internal crises of self-determination. When ethnic hostilities are deep-seated and broad in scope, forced containment of claims for ethnic autonomy seem merely to diffuse potential conflicts in the short run. In the long run, ethnic conflict and even demands for self-determination always loom just over the horizon, waiting for an opportunity to assert themselves.

In societies with low to moderate ethnic tensions, the state is often able to reorganize itself in a pre-emptive way, centralizing and strengthening its absolute authority. When this approach is taken, however, state leaders must busy themselves with devising policies which enhance their popular legitimacy. Should they fail to do this, ethnic and/or class problems could lay the foundations of their undoing. In Ethiopia immediately after World War II, the climate was right for Selassie to centralize his empire and to strengthen his own authority. Traditional elites were either supportive or acquiescent. A new elite class was at the earliest stages of development and dependent on the emperor for its prosperity. But the emperor emphasized centralization without attending to the need for building a reservoir of support among ethnic minorities. By 1974 the state was embroiled in multiple conflicts with various elite groups who proceeded to withdraw their support one by one.[82] In this climate the state was extremely vulnerable to divisive ethnic and class conflict. The opportunity for such a scenario was provided by the coup which toppled the Selassie regime. Before the new regime could consolidate itself, ideological as well as ethnic conflicts came to threaten the dismemberment of the state. Given the scope and intensity of the claims represented in the Eritrean, Oromo, and Somali movements in Ethiopia, it seems that the only rational alternative open to the state is to come up with a scheme which allows for some regional autonomy for those groups.

State leaders must be sensitive to the contexts in which they operate. They must have a sense for the nature and extent of potential ethnic conflict, and they must devise policies which enable them to manage these conflicts with a minimum of cost. This requires skillful leadership given the meagerness of expendable resources available to the state. Under the most extreme circumstances, the very survival of the state is dependent on the successful management of conflicts which at least potentially involve claims of self-determination.

Such crises are rare. Milder forms of ethnic conflict, however, are extremely frequent in Africa, and they could well grow into more severe conflicts. Therefore, African states must conscientiously address ethnic conflicts of lesser magnitude as forthrightly as they do more serious ones to present their becoming threats to state coherence.

Footnotes

Thanks are due to Cynthia Barnes, Ed Brown, Jackie Vieceli, Cy Reed, and to my colleagues at Bellagio for their contributions to my thinking out and writing this essay. My gratitude is also extended to the African Studies Program at Indiana University for its clerical assistance.

1. See Vernon Van Dyke, "The Individual, the State and Ethnic Communities in Political Theory," *World Politics*, 29, 3 (April 1977): 343-369; and Rupert Emerson, "The Prospects for Democracy in Africa," in Michael Lofchie, ed., *The State of the Nations*

(Berkeley and Los Angeles: University of California Press, 1973), p. 248.

2. See Zdenek Cervenka, *The Unfinishd Quest: African Unity and the OAU* (London: Friedmann, 1977); and *OAU Charter*, Article III, paragraph 3.

3. Ali Mazrui asserts that the only meaning self-determination has ever had for African nationalists is what he calls "pigmentational self-determination." The implications of this view are that self-determination claims are seen as legitimate only when there is interracial domination. This might explain why Amhara colonialism in Ethiopia and Americo-Liberian colonialism in Liberia escaped the mandate of the OAU. See Ali A. Mazrui, *Towards a Pax Africana: A Study of Ideology and Ambition* (Chicago: University of Chicago Press, 1967) pp. 14, 62.

4. See, for example, Crawford Young, *The Politics of Cultural Pluralism* (Madison: University of Wisconsin Press, 1976), p. 523; and Samir Amin, *Class and Nation: Historically and in the Current Crises* (New York: Monthly Review Press, 1979), p. 177.

5. Onyeonoro S. Kamanu, "Secession and the Right of Self-Determination: An OAU Dilemma," *Journal of Modern African Studies*, 12, 3 (1974): 360-361.

6. Walker Connor, "Self-Determination: The New Phase," *World Politics*, 20, 1 (October 1967): 47.

7. See Young, *Politics of Cultural Pluralism*, p. 524.

8. Albert O. Hirschman, "Exit, Voice and the State," *World Politics*, 31, 1 (October 1978): 105-106.

9. See Lofchie, ed., *State of the Nations*; Naomi Caiden and Aaron Wildavsky, *Planning and Budgeting in Poor Countries* (New York: John Wiley and Sons, 1974); and Y. Dror, *Public Policymaking Reexamined* (San Francisco: Chandler Publishing Co., 1968), pp. 1-125.

10. For an insightful treatment of this issue, see Theda Skocpol, *States and Social Revolutions: A Comparative Analysis of France, Russia, and China* (London: Cambridge University Press, 1979), pp. 24-32.

11. Representative of the liberal school would be Donald Rothchild and Robert L. Curry, Jr., *Scarcity, Choice, and Public Policy in Middle Africa* (Berkeley and Los Angeles: University of California Press, 1978).

12. See, for example, Karl Marx, *Manifesto of the Communist Party* (London: Lawrence and Wishart, 1943), p. 12.

13. See Hamza Alavi, "The State in Post-Colonial Societies," *New Left Review*, 74 (July-August 1972): 59-82; Roger Murray, "Second Thoughts on Ghana," *New Left Review*, 42 (March-April 1967): 25-39; John Saul, "The State in Post-Colonial Societies: Tanzania," in Ralph Miliband and John Saville, eds., *The Socialist Register 1974* (London: The Merlin Press, 1974), pp. 349-372; W. Ziemann and M. Lanzandorfer, "The State in Peripheral Societies," *The Socialist Register 1977* (London: The Merlin Press, 1977), pp. 143-177; and Colin Leys, "The 'Overdeveloped' Post-Colonial State: A Reevaluation," *Review of African Political Economy*, 5 (January-April 1976): 39-48.

14. See Richard Sklar, "The Nature of Class Domination in Africa," Journal of Modern African Studies, 17, 4 (December 1969): 531-552.
15. A noteworthy trend away from this direction is Thomas M. Callaghy's "State-Subject Communication in Zaire: Domination and the Concept of Domain Consensus," Journal of Modern African Studies, 18, 3 (1980): 469-492. Callaghy's excellent article clearly articulates the extent to which Mobutu has gone to preserve this state, and how he has done so in a manipulative, rationally calculating way.
16. Theodore Lowi, "American Business, Public Policy, Case Studies and Political Theory," World Politics, 16, 4 (July 1964): 677-715.
17. Ibid., pp. 692-693.
18. Ibid., p. 693.
19. In the case of rurally based, largely non-market-oriented African societies, taxes could be indirect (i.e., government induced self-help activities). See Philip Mbithi and R. Rasmusson, Self-Reliance in Kenya: The Case of Harambee (Uppsala: Scandanavian Institute of African Studies, 1977).
20. Lowi, "American Business, Public Policy, Case Studies and Political Theory," p. 690.
21. Murray Edelman, The Symbolic Uses of Politics (Urbana: University of Illinois Press, 1967).
22. Ibid., p. 81.
23. See Robert Melson and Howard Wolpe, "Modernization and the Politics of Communalism: A Theoretical Perspective," American Political Science Review, 64, 2 (1970): 1112-1130; Crawford Young, "The Obote Revolution," Africa Report 11, 6 (June 1966): 8-14; and Edmond J. Keller, "Ethiopia: Revolution, Class and the National Question in Ethiopia," African Affairs, (October 1981): 519-549.
24. Donald Rothchild, "Ethnicity and Conflict Resolution," World Politics, 22, 4 (July 1970): 597-616.
25. Vincent B. Khapoya, "Kenya Under Moi: Continuity and Change," Africa Today, 27, 1 (1980): 23.
26. New York Times, July 25, 1980, p. 2.
27. Dennis L. Dresang and Ira Sharkansky, "Ethnic Politics, Representative Bureaucracy and Development Administration: The Zambian Case," American Political Science Review, 68, 4 (December 1974): 1605.
28. Dennis L. Dresang and Ira Sharkansky, "Public Corporations in Single-Country and Regional Settings: Kenya and the East African Community," International Organization, 27 (1973): 316.
29. Michael Cohen, "The Myth of the Expanding Center," Journal of Modern African Studies, 11, 2 (1973): 227-246.
30. Republic of Kenya, Development Plan 1974-78 (Nairobi, 1974), p. 412.
31. See Michael Cohen, Urban Policy and Political Conflict in Africa: A Study of the Ivory Coast (Chicago: University of Chicago Press, 1974), pp. 21-114; Joel D. Barkan, Bringing Home the Pork: Legislative Behavior, Rural Development and Political Change in East Africa, Comparative Legislative Research Center Occasional Paper no. 9 (Iowa City: University of Iowa, 1975);

Edmond Keller, "*Harambee*! Educational Policy, Inequality, and the Political Economy of Rural Community Self-Help Organization in Kenya," *Journal of African Studies*, 4, 1 (1977): 86-106; and Henry Bienen, *Kenya: The Politics of Participation and Control* (Princeton: Princeton University Press, 1974), especially pp. 66-108.

32. Henry Bienen, *Tanzania: Party Transformation and Economic Development* expanded edition, (Princeton: Princeton University Press, 1970).

33. See Goran Hyden, *Beyond Ujamaa in Tanzania: Underdevelopment and an Uncaptured Peasantry* (Berkeley and Los Angeles: University of California Press, 1980), pp. 37-38.

34. Jan Pettman, "Zambia's Second Republic -- The Establishment of a One-Party State," *Journal of Modern African Studies*, 12, 2 (1974): 231-244. In 1968 the government had banned the Union Party as a "threat to peace."

35. See Jean-Claude Willame, *Patrimonialism and Political Change in the Congo* (Stanford: Stanford University Press, 1972).

36. "Courageous Move," *Weekly Review*, August 1, 1980, p. 15; "Gema Comes Under Pressure to Wind Up," *Weekly Review*, October 3, 1980, pp. 9-10; "End of the Road for Tribal Unions," *Weekly Review*, October 10, 1980, p. 18; and "The Final Stages," *Weekly Review*, October 17, 1980, p. 10.

37. See Edmond Keller, "The Revolutionary Transformation of Ethiopia's Twentieth-Century Bureaucratic Empire," *Journal of Modern African Studies*, 19, 2 (1981): 307-335.

38. For an excellent discussion of how this device has been used in Africa, see Aristide Zolberg, *Creating Political Order* (Chicago: Rand McNally, 1966), pp. 77-79.

39. See R.N. Ismagilova, *Ethnic Problems of the Tropical Africa: Can They Be Solved*? (Moscow: Progress Publishers, 1978), p. 137.

40. See Kenneth Prewitt, "Political Socialization and Political Education in New Nations," in Kenneth Prewitt, ed., *Education and Political Values: An East African Case Study* (Nairobi: East African Publishing House, 1971), p. 3.

41. See Edmond J. Keller, "Education, Ethnicity and Political Socialization in Kenya," *Comparative Political Studies*, 12, 4 (January 1980): 442-469; and Keller, "The Political Socialization of Adolescents in Contemporary Africa: The Role of the School in Kenya," *Comparative Politics*, 10, 2 (January 1978): 227-250. Employing regression analysis, these essays suggest that the salience of ethnicity as a political factor is fluid, intermittent, and experiential, and that education does seem to enhance the development of integrative attitudes, but given the proper circumstances, ethnic affinities might override the effect of education on such attitudes.

42. See, for example, P.T. Baxter, "Ethiopia's Unacknowledged Problem: The Oromo," *African Affairs*, 77, 38 (July 1978): 283-299; and Richard Sherman, *Eritrea: The Unfinished Revolution* (New York: Praeger Publishers, 1980).

43. B. Fonlon, "The Language Problem in Cameroon: Historical Perspective," in David Smock and K. Bentsi-Enchill, eds., *The Search for National Integration in Africa* (New York: Free

Press, 1975), pp. 189-205.

44. See John Ostheimer, Nigerian Politics (New York: Harper and Row, 1973), pp. 155-156.
45. "What Happened at the Assembly," West Africa, 3315, February 9, 1981, pp. 261-262.
46. See John Cohen and Dov Weintraub, Land and Peasants in Imperial Ethiopia (The Hague: Van Gorcum, 1975); and Peter Schwab, Decision-Making in Ethiopia (Rutherford, New Jersey: Fairleigh Dickenson, 1972).
47. Dennis L. Dresang, "Ethnic Politics, Representative Bureaucracy and Development Administration," pp. 1612-1613.
48. Robert Molteno, "Cleavage and Conflict in Zambian Politics," in William Tordoff and Robert Molteno, eds., Politics in Zambia (Berkeley and Los Angeles: University of California Press, 1974), p. 97. A similar practice was followed between 1969 and 1970 in Solwezi District of North-Western Province.
49. For a good discussion of the land transfer program see Gerald Holtham and Arthur Hazlewood, Aid and Inequality in Kenya (London: Croom Helm, 1976), pp. 104-144.
50. See Colin Leys, Underdevelopment in Kenya: The Political Economy of Neo-Colonialism (Berkeley and Los Angeles: University of California Press, 1974), pp. 28-117; and Henry Bienen, Kenya: The Politics of Participation and Control (Princeton: Princeton University Press, 1974), pp. 161-182.
51. See John Harbeson, Nation-Building in Kenya: The Role of Land Reforms (Evanston, Illinois: Northwestern University Press, 1973), p. 67.
52. See Cherry Gertzel, The Politics of Independent Kenya (Evanston, Illinois: Northwestern University Press, 1970), pp. 32-72.
53. See "Zimbabwe Conference on Reconstruction and Development," (Salisbury, March 23-27), pp. 37-39.
54. New York Times, November 2, 1980, p. 1. The Mugabe government purchased two large European farms and planned to develop them into sample collectives.
55. "Peter Yates," "The Prospects for Socialist Transition in Zimbabwe," Review of African Political Economy, 18 (May-August 1980): 79-83.
56. New York Times, November 2, 1980, p. 1.
57. Letter from R. Shugudzomwoyo to the Herald, November 14, 1980, quoted in "Yates," "Prospects for Socialist Transition in Zimbabwe," p. 79.
58. Hyden, Beyond Ujamaa in Tanzania, p. 137-138.
59. Young, The Politics of Cultural Pluralism, pp. 211-212.
60. Young, "The Obote Revolution," pp. 9-10.
61. N.W. Provisier, "The National Electoral Process and State-Building: Proposals for New Methods of Elections in Uganda," Comparative Politics, 9, 3 (April 1977): 314.
62. Ibid., p. 315.
63. Ibid., p. 316.
64. Ibid., pp. 314-317.
65. Bienen, Kenya, p. 35.
66. B. Ammar, "Regional Autonomy Brings Peace to Southern Sudan," New Middle East (March-April 1972), pp. 12-13; and T. Betts, The Southern Sudan: The Ceasefire and After (London: Africa

Publications Trust, 1974).

67. Ammar, "Regional Autonomy Brings Peace," p. 12.
68. John Nellis, A Theory of Ideology (London: Oxford University Press, 1972), p. 14.
69. See Edelman, Symbolic Uses of Politics, p. 73.
70. Bernard Barber, "Function, Viability, and Change in Ideological Systems," in Bernard Barber and Alex Inkeles, eds., Stability and Social Change (Boston: Little, Brown and Co., 1971), pp. 244-260; and Douglas Ashford, Ideology and Participation (Beverly Hills: Sage, 1972).
71. Hyden, Beyond Ujamaa in Tanzania, pp. 98-100.
72. Bienen, Tanzania, pp. 206-208.
73. See G. Von der Muhll, "Education and Social Revolution in Tanzania," in Kenneth Prewitt, ed., Education and Political Values, pp. 23-52.
74. Hyden, Beyond Ujamaa in Tanzania, pp. 100-128.
75. Dean McHenry, Jr., "The Struggle for Rural Socialism in Tanzania," in Carl G. Rosberg and Thomas M. Callaghy, eds., Socialism in Sub-Saharan Africa: A New Assessment (Berkeley and Los Angeles: Institute of International Studies, 1979), p. 43.
76. See P.L. Raikes, "Ujamaa and Rural Socialism," Review of African Political Economy, 5 (May 1975): 33-52; and McHenry, "Struggle for Rural Socialism in Tanzania."
77. See Keller, "Harambee!"
78. See Kenneth Kaunda, "Humanism: A Guide to the Nation" (Lusaka: Government Printer, 1967); Stephen Quick, "Socialism in One Sector: Rural Development in Zambia," in Rosberg and Callaghy, eds., Socialism in Sub-Saharan Africa, pp. 83-111; and Irving Markovitz, Leopold Sedar Senghor and the Politics of Negritude (New York: Atheneum, 1969).
79. Cohen, Urban Policy and Political Conflict, pp. 115-144; and Callaghy, "State-Subject Communication in Zaire," pp. 472-476. Callaghy notes that "dialogue" in Zaire tends to be only "one way" from Mobutu to the people in command fashion. He also notes that MPR prefets use proganda to try to shape popular opinion, but the involuntary nature of most popular participation in Zaire leads us to exclude symbolic politics there from the symbolic policy category.
80. Cohen, Urban Policy and Political Conflict, p. 159.
81. See Aristide R. Zolberg, "The Structure of Political Conflict in the New States of Tropical Africa," American Political Science Review, 62 (March 1968): 70-87.
82. See Keller, "The Revolutionary Transformation of Ethiopia's Twentieth-Century Bureaucratic Empire."

14
Federalism and Politics of Compromise

J. Isawa Elaigwu
Victor A. Olorunsola

Introduction

In the introduction to this book, it was asserted that the lack of fit which is sometimes evident between the claims for state sovereignty and ethnic autonomy derives from the fact that each of these two principles can, in certain circumstances, involve legitimate, albeit contradictory, objectives. It was maintained that "each principle displays an intrinsic consistency and each is championed by committed proponents." The editors suggested that such characteristics lend an inner dynamism to claims made on their behalf, providing the basis for a possible protracted and destructive conflict. "Either an exclusive nonnegotiable demand for ethnic autonomy outside the state or an overweening centralist thrust within the state may cause conflict to surface between these principles. Clearly, such discontinuities must be narrowed to ensure creative relations between peoples and levels of government."

Dismemberment may or may not be a solution to the dilemma of ethnic self-determination and state coherence. If Nigeria and Zaire are any example, bringing about such a dismemberment, or, if you like, a divorce, is a most difficult and perhaps impossible task. Although the reasons for this may not be crucial at this time, explanations advanced by Crawford Young in his chapter "Comparative Claims to Political Sovereignty" are most enlightening. The point is that Nigeria either has learned from its own experience or has concluded, however reluctantly and expensively, that coexistence and accommodation between the tendencies toward state coherence and the predisposition toward ethnic self-determination are necessities.

After approximately six years of so-called Westminster-style government and thirteen years of military rule, Nigeria has opted for "presidential federalism." Nigeria watchers might have noticed that the change of October 1, 1980 smacked of a trans-Atlantic trip from London to Washington in a quest for a model government.[1] While Nigeria inherited parliamentary democracy from Britain, the federal experience she yearned for could not be learned from a unitary Britain. In its second attempt, Nigeria still clung to the federal principle, but rejected the parliamentary model. She turned to another reference group -- the United States (also an ex-British colony). The United States has a presidential and federal system. The extent to which Nigeria's experience will be enriched by lessons from the U.S. experience is yet to be seen. Or will Nigeria become a blueprint of America? This will be politically tragic given ecological, political, and sociocultural differences between the two countries.

The late Martin Diamond correctly observed: "Whether federalism survives, how it can be adapted and modified so as to deal with contemporary problems, etc., all require knowing what it is we want from federalism and what federalism by its nature can supply."[2] Federalism has survived so far in Nigeria. Will it survive in the future, especially in the light of the country's experiences in the first year of her new experiment? What do Nigerians want from federalism and to what extent has federalism been able to satisfy these wants?

Basically federalism as a system of government emanates from the desire of people to form a union without necessarily losing their various identities.[3] It is an attempt to reflect the diverse social, political, cultural and economic interests within the framework of a broader national unity.[4] Nigeria's heterogeneity, as well as the political problems which partially emanate from this, have become politically proverbial. The late 1950s witnessed ethnic and geoethnic parochialism as groups competed to inherit political power from the colonial authorities. The security provided by parochial political platforms for competition led to pressures for the adoption of federalism and the gradual dismantling of unitary institutions -- such as the Nigerian Marketing Board, etc.

For many Nigerians, federalism is an important mechanism for allaying their fears -- fears of political and economic domination. Federalism in Nigeria emerged as a result of social forces at work within the country. These social forces have highlighted the issue of imbalance in the system. If federalism was expected to assuage fears among Nigerian groups, if it was regarded as a partial protective political device, it was also embedded with its own seeds of discord. One of its most politically cogent problems is that of attaining federal balance. The strategy of federal balance is, in our opinion, Nigeria's formalistic attempt and hope for the eradication of fears and insecurities which heretofore have dominated her intercommunity relations.

What do we mean by federal balance? What is Nigeria attempting to balance in her constitution? In adopting "presidential federalism," are Nigerians balancing imbalance or are they generating new forms of imbalance? Are the issues involved in Nigeria's political system balanceable?

We agree with Adebayo Adedeji that a "federal system of government is the result of a compromise. It is a compromise between centrifugal and centripetal forces."[5] This compromise is a form of balance between two opposing forces. There is no doubt that all federal systems experience adjustments, at different points in time, between these two extreme pulls. But the extent to which a federal system survives very much depends on the ability of the political elites in the country to maintain a delicate balance between centrifugalism and centripetalism. Excessive pulls in favor of centrifugal forces may herald disintegration, as Nigeria has painfully learned. Yet excessive pulls to the center may challenge the very existence of federalism and the cocoon of security it provides for members of society. These stresses can also lead to disintegration, as the bloody massacres in Northern Nigeria of 1966 confirmed.[6] It is the delicate balance or compromise between these two extremes that we define as federal balance. In short, it is the maintenance of a delicate balance between ethnoregional autonomy and state control in Nigeria's intercommunity relations.

This chapter attempts to highlight the problems of federal balance in Nigeria. We shall look at the federal structure and the dynamics of presidential federalism so far, and make tentative observations on Nigeria's performance under the new constitution. In this regard, we suggest that:

1) imbalance in Nigeria's federalism is historical and multidimensional;
2) gross imbalance is fraught with as much danger as reckless attempts at balancing the unbalanced;
3) the 1979 constitution (like West Germany's Basic Law) was very much a reaction to Nigeria's past, even though it has attempted to create a new platform for political actors;
4) echoes of the past are still very much around, even if in the context of different structures and at lower levels of intensity; and
5) despite the wishes of Nigerians, certain issues in their federal system defy balancing and pose great dilemmas for their leaders.

We shall expatiate on these issues as we move on. Let us look at the first suggestion. To what extent has imbalance in Nigeria's federalism been historical and multidimensional?

Imbalance and Nigerian Federalism: A Historical Perspective

Nigeria's federal system had problems of balance which had, at one point or another, threatened the existence of the political system. Imbalance in Nigeria emanated from the federal structure and the nature of the impact of Western civilization on the country. These created fears and suspicions among Nigerians which, in turn, helped to accentuate rather than correct imbalance.

While it may not be useful to beat the colonial dead horse any more for the ailments of Nigeria, it may be argued that ambivalent integration under colonial rule was partially responsible for generating fears and suspicions among Nigerians. After the amalgamation of Nigeria in 1914, the colonial authorities made no effort to encourage horizontal interaction among the various groups, for understandable reasons. Such encouragement would have heralded the good riddance of the colonial masters from the scene as groups developed confidence and anticolonial psychology.

Colonial rule encouraged vertical relationship between the local administrative units and the colonial centers of power. The result was that Nigerians of Northern and Southern provinces never had an opportunity to interact politically until 1947 (under the Richards Constitution). This led to suspicions and fears as people of the various political units found themselves interacting with one another as strangers. They had had no opportunity to build up mutual confidence among themselves through horizontal forms of interaction. This parochialism among Nigerian groups we shall call the parochialism of ignorance -- for each group was ignorant of the other.

But between 1946-54, it had become evident that the colonial administration would soon fold its political umbrella. Naturally, Nigerians "battled" for the inheritance of the colonial legacy -- to fill the political vacuum. In this competitive setting Nigerians withdrew, based on suspicions among Nigerian groups, to organize with their ethnic or ethnoregional bedfellows in order to inherit political power from the colonial masters. This parochialism in the terminal colonial period we may call the parochialism of awareness -- for it was based on awareness of others in the competitive setting. It was this form of parochialism that goaded Nigerians to the choice of a federal system of government.

Related to the above are two important factors which escalated fears and suspicions among groups: 1) the structural imbalance in Nigeria's federal system; and 2) the differential spread in the pattern of Western education.

It was John Stuart Mill who once said that in a federal system, "there should not be any one state so much more powerful than the rest as to be capable of vying in strength with many of them combined."[7] Similarly, the doyen of federalism, K.C. Wheare, warned that "it is undesirable that one or two units should be so powerful that they can overrule the others and bend the will of the federal government to themselves."[8]

We would like to suggest that in the First Republic, the lopsided federal structure generated fears and suspicions among groups. The Northern Region was in a position to hold the whole country to ransom, as shown by the figures below.

Region	Area (%)	Population (%) (1963 census)
North	79.0	53.5
East	8.3	22.3
West	8.5	18.4
Midwest	4.2	4.6
Lagos	-	1.2

Given the above figures, it was not surprising that in the southern part of the country, there was always the fear of domination by virtue of the Northern Region's large population -- the tyranny of population. To the Southerners the federal structure as it existed made it virtually impossible for the South to control political power at the center given the ethnoregional politics in the country.

Similarly, given the Southern headstart in Western education (which had become a passport into occupational roles in the modern sector of the Nigerian political system), the North feared Southern domination in the economic and public service sectors of the society. The possibility of a tyranny of skills from the South was fresh in the minds of Northern leaders. The North thus sought to protect its civil service from being swamped by the South.

It may be suggested that there was a relative division of functions between the North and the South which maintained some delicate balance in the political system. The Northern control of political power was counterbalanced by the South's monopoly of economic power in the country.

We may even go further to argue that, contrary to Sklar's contention,[9] the military coup in January 1966 tilted what had been a delicate balance on which Nigeria had been able to survive since independence. The concentration of both political and economic power in the hands of Southern leaders altered the delicate Nigerian balance. Political power had been the North's safeguard against the South's economic and educational advantages. The South's advantage in the bureaucracy, which if anything was strengthened in authority by the coup, was greatly augmented. The North reacted violently as it saw its last card -- the political card -- suddenly taken away or rendered ineffective.

These imbalances created problems for the federal system. Centrifugal forces continued to haunt Nigeria's federal balance. Threats of secessions by various regions in 1950, 1953 and 1964 climaxed in the abortive secession of Eastern Nigeria in 1967. This was a manifestation of extreme centrifugalism, and a challenge to the process of state building. It took a civil war to return Nigeria to a position of relative balance between centrifugal and centripetal pulls.[10]

In order to dilute the autonomy of the regions which had taken on pretensions of virtual sovereignty, General Gowon created twelve states in 1967. With greater demands for additional states, his successor, Murtala Mohammed, created nineteen states in 1976. Nigeria's present federalism operates in the context of nineteen states. In addition to the creation of states, the various military administrations took a number of measures to drastically cut down the powers of subnational states.[11] From a federal system with a weak center, Nigeria emerged as a federal system with a strong center by October 1979. Many politicians complain now about concentration of powers at the center. What seems very clear at this point, however, is that while the process of nation building is still going on, the Nigerian state has come to stay. No single state is in a position to hold the whole country to ransom or threaten secession.

In what ways has the October 1979 constitution tried to cope with problems of autonomy and control?

The 1979 Constitution and Political Balance

Like the Bonn constitution of the Federal Republic of Germany, the Nigerian Constitution is very much a reaction to its historical past. Nigerians seem to be quite conscious of the very elements which nearly disintegrated that country. A few examples come to mind. The constitution provides for a federal structure. Section 2(1) gives the formal name of the country as the "Federal Republic of Nigeria."[12] This means that a change from its federal nature would entail a constitutional amendment. Article 20, Section 1 of Germany's Basic Law similarly states clearly that Germany is to remain a federal state.

Section (3)1 defines the structure of Nigeria -- as comprising nineteen states -- while Section 8 enumerates the procedures for creating additional states.[13] This at least provides an opportunity for correcting whatever structural imbalance there is in the present federal structure.

The relationship between federal center and the states is taken care of in the exclusive and concurrent legislative lists (second schedule, parts I and II respectively), as in most federal systems. Residual powers go to the states. A new element is the presidential system, introduced in the new constitution. Again, this was a reaction to the past. There are some Nigerians who strongly believe that the failure of the First Republic can be traced to the Westminster model of government.[14]

Other examples of such past problems which the new constitution set out to correct abound. For example, in order to avoid the emergence of parochially based political parties, sections 201-209 provide, among other things, for the formation of national political parties which must reflect the "federal character" of Nigeria and which must be open to "every citizen of Nigeria irrespective of his place of origin, sex, religion, or ethnic grouping." Each party's headquarters must be located in the capital of the Federation (Section 202).

In a similar vein, the issue of suspicions and fears of domination of public services expressed by various groups in the past surfaces in Section 14(3), which provides that:

> The composition of the government of the federation or any of its agencies and the conduct of its affairs shall be carried out in such manner as to reflect federal character of Nigeria and the need to promote national unity, and also to command national loyalty thereby ensuring that there shall be no predominance of persons from a few states or from a few ethnic or other sectional groups in that government or in any of its agencies.

Section 14(4) has the same clause for the governments of the various states -- as part of the fundamental obligations of all governments of the federation.

In essence, the new constitution is very much a reaction to the past which still haunts Nigeria. It will be a long while before the bogey of the past fades, like a cloud, from Nigeria's political horizon. But a new platform has been created on which political actors can perform. How far have they succeeded since October 1979 in maintaining the delicate balance we discussed earlier?

Autonomy and Control Performance Record Under the 1979 Constitution

One must hasten to say that the 1979 constitution came into effect on October 1, 1979, and a proper evaluation of performance seems a bit precocious at this time. But tentative observations can be made in order to highlight some of the possible lessons already learned in the process of implementing the 1979 constitution. We shall look at: 1) the structure of the federation; and 2) issues emanating from the dynamics of the political process.

The Federal Structure. The Nigerian state, as described under Section 3(1) of the constitution, is composed of nineteen states.[15] But there have been numerous demands for the creation of additional states, such as Taraba, New Benue, Kogi, New Oyo, Delta, New Cross River, Anioma, Oshun, and others. It is really interesting to watch on television the number of delegations which have undertaken political pilgrimages to the centers of power in Lagos to put forward their cases. Each delegation is led by some of the most respectable citizens of the country. Each delegation creates the impression that its case is the most important case. Lack of space limits our treatment of reasons put forward by each delegation for creation of states. The important point is that these reasons are no different from those advanced in earlier exercises in state creation.[16]

As discussed earlier, the North-South conflict had been accentuated by the imbalance in the federal structure. Gowon attempted to rectify this imbalance by creating twelve states -- six states in the North and six in the South. As the former secretary to the federal military government, Allison Ayida once pointed out:

> The most sensitive potential threat to the stability of the Nigerian Federation was...a North-South confrontation; and it was of strategic importance that the number of states in the "Northern" parts of the country should be seen to be equal to the number of "Southern" states (this was an important consideration which could not be made explicit in the days of the "gathering storm" in early 1967).[17]

Although some Nigerians reacted negatively to Ayida's statement -- that it was important to ensure that Northern and Southern parts of Nigeria had equal numbers of states -- the issue is emerging again. Many publications in newspapers, especially from the Southern states, reacted against the imbalance created by the nineteen-state structure. There are now ten states in the Northern part of the country and nine states in the South. Why does this bother anyone?

It bothers some Nigerians because of the way in which the states are perceived by politicians and electorates alike. States serve a useful political function. The logic of federalism harps on equality of states. Section 126(2-6) as well as the amendment to Section 126 (by Decree No. 104 -- Constitution of the Federal Republic of Nigeria [Amendment] Decree 1979) makes it conditional that a candidate is elected only if in addition to the number of votes acquired, he or she "has not less than one-quarter of the votes cast at the election in each of at least two-thirds of all states in the federation." The number of states in each geopolitical area thus acquires political salience. Hence, many agitators for the creation of states have argued in favor of establishing states in a way that gives the North and South an equal number. This is as much an echo of the past as it is a present political reality.

Another reason for the demand for additional states is to be seen in the economic and democratic interpretation of the role of states in the federal system. The economic logic behind agitations is that if Anambra state, for example, is split into two states, more resources would be allocated to the area. By the revenue distribution formula bequeathed to the politicians by the military, resources are distributed among states on the basis of 50 percent in equal shares to each state and 50 percent on the basis of each state's population.

In the conflict around the economic issues of state creation, old rivalries rise to the surface occasionally, even if in different contexts. For example, the Ibo-Yoruba rivalry seems to have taken a new turn in the demands for resources. The former Iboland has two states, while Yorubaland (excluding Kwara) has four states. In the competitive process of resource acquisitions, the Yorubas are seen to have had a better deal by the Ibos. Hence K.O. Mbadiwe recently called on Ibos in all political parties to rally together on the issue of state creation in order to sort out their differences. There are many cases similar to the above.

Yet there is the democratic logic that additional states would offer opportunities for greater participation -- through representation (as in the senate) or through appointment (as in public services) for more people and groups. Translated into practical terms, this means that a Katsina state would give people from present Kaduna state five more representatives in the senate. Each state is expected to have five representatives in the senate (Section 44).

The content of Section 14(3) also makes it imperative for any presidential incumbent to appoint people from various states in order to meet the requirements of reflecting "federal character" in appointments. This includes ministers, board chairmen, ambassadors, etc.

The issue of creation of additional states has continued to pester the presidential system. Like all medicinal prescriptions, the creation of states which helped to resolve the problem of structural imbalance has had its own side-effects -- some political rashes. These political rashes are the emergence of "new majorities" and "new minorities." The fear of domination becomes more salient as groups resort to avoidance mechanisms to protect their interests. The creation of more states would lead to greater dependence of states on the center. Federal-state power watchdogs should take note of this. Given the numerous demands for states and the procedures for creating additional states, it is most unlikely that any state will be created before 1983. This is likely to become a real campaign issue in the 1983 election. How these demands are handled will manifest the political maturity of Nigerian leaders.

Federal-State Relations. The second issue is federal-state relations. The problem of federal balance is not only to be seen in the structural composition of the federation. The nature of federal-state relations is very important. How have these operated since October 1979?

The nature of federal-state relations has changed. In the first place, Nigeria now has nineteen states instead of the four regions she had up to 1967. Secondly, Nigeria now operates a presidential system of government. Unlike the parliamentary system, the state chief executive as well as his federal counterparts are all directly elected to office by the populace.

In addition, the military structure imposed on Nigeria's federalism since 1966 tied state military governors to their commander-in-chief, who was also the head of the federal military government. This meant that the central government could take actions to which the states could not object.

With the return to civilian government under the 1979 constitution, many state governments came under different political parties.[18] Because the government party in the state is, in some cases, different from the party in power at the center, Nigeria has witnessed aggressive state nationalism. The demands for state autonomy which have escalated in recent times could be seen as attempts to avoid state identities being "swamped" by the political party in control of the federal government. This is very clear in the relations between the president and the governors of some states. The appointment of presidential liaison officers[19] raised much concern among many Nigerians. In the point of view of many state government officials, while the president's right to appoint officers under Section 5(a-b) of the constitution is accepted, there is no provision for the appointment of presidential liaison officers (PLOs) who are not members of the public service.

Many non-NPN government states rejected the PLOs. Part of the objection is tied to the belief that these PLO jobs were patronages for loyal party men. In addition, some states genuinely believed that the PLOs would erode the autonomy of states. Many governors did not hide their antipathy for the implication that their governments could only deal with Lagos through these PLOs. The issue of the PLOs is directly traceable to the problem of communication gap between the president and the governors in the states.

Another example of actions taken by the federal government which take on political coloring is the issue of Alhaji Shugaba's deportation.[20] It does seem here that the president was poorly advised. The Borno state governor, as the chief security officer of the state, was not informed about Shugaba's deportation. The hurried, clandestine operation in the deportation of Shugaba gave political tint to an action the president is entitled to take, but which has become politically stigmatized because of the crudity of implementation. The GNPP (and even Shugaba himself) accused the NPN of having masterminded the deportation order. Furthermore, it raised the issue of citizenship in Nigeria, as allegations that Akinjide, Paul Unongo, and Alexander Fom were not Nigerians seem to confirm. Yet in politics, stumbles of this nature are not easily ameliorated without loss of credibility.

In their relations with the center, there is no doubt that many state governors have admirably reasserted the autonomy of states which is seemingly desirable in a federal system. However, it does seem that many of them are still finding it hard to shed some political traits bequeathed to them by the military. The militarization of political culture in thirteen years of military rule has entailed the militarization of political style and language of most Nigerians. The reference often made to "old brigade" and the "new brigade" in Nigerian politics is a sign of the militarization of political language.

The civilian governors and politicians in the states, having been schooled for thirteen years under the military, are having problems in making a smooth transition from a military regime to a democratic civilian regime. The ultimatums and threats they issue are the psychological legacies of military governors which have been hard to shed.

The governments of Plateau, Oyo, and Kwara states have threatened takeover of federal television stations in their states. Most of these television stations were owned by the states before the federal takeover in 1975-76. The states had not been paid any compensation. The states correctly complain that they have no control over the programs their people watch and that state government activities are often blacked out.

There is no doubt that state governments have a case. The manner in which they have gone about presenting it, however, has been one of confrontation rather than discussion and compromise. Some governors should have realized that given Nigeria's peculiar conditions, state takeover of Nigerian Television (NTV) does not help them. There is the need for greater decentralization of federal institutions. For example, NTV can be decentralized such that while the stations are still federally owned, the states could control recruitment and programs to help reflect the interests of the local people who watch the Nigerian television programs. The present zonal administrations should be scrapped and state governments given greater authority as mentioned above.[21]

The federal government should coordinate activities of the various stations in Lagos. After all, suppose the UPN-controlled states decide to "black out" a presidential address, there is little that can be constitutionally done since the stations would have been state-controlled. A governor of a state from any party today can become president tomorrow. It was Shridath Raphal who observed that:

> for a federation to be able to resist failure, the leaders and their followers must "feel federal" as they must be moved to think of themselves as one people, with one common self-interest, capable, where necessary, of overriding most other constitutions of small group interest.

While it is necessary to maintain state autonomy, Nigeria's past experiences show that a very weak center can be disastrous. True enough, there is a need for separate identity of states, but there is also greater need for the unity of Nigeria, which is greater than any of its parts.

As the issue of educational institutions becomes more salient, threats to federal institutions have also mounted. Bauchi state government was so angry with officials of Federal Polytechnic, Bauchi, for not being sensitive to the needs of the state that it gave them a quit order from the facilities the state had made available to them as a temporary site. In Plateau state, Governor Lar threatened to take over the University of Jos if it did not show greater response to the urgent needs of its catchment area.[22] No state can take over a federal institution unless laws which established them are repealed.

In other cases, some states have made ridiculous threats to the federal government. Benue state, frustrated after an expected opening of the airstrip in Makurdi, threatened to launch an airline if the federal government did not react quickly to the opening of the airstrip. Under the constitution, "aviation, including airports, safety of aircraft and carriage of passengers and goods by air" are in the exclusive legislative list of the federal government.[23] Yet Benue state has, like the federal center, an NPN government. The states are so conscious of their autonomy that at times it does seem as if their officials have not properly read the constitution.

Or perhaps, like other politicians, state officials are merely testing the new constitution to see how far they can go. This is very clear in the case of Ondo state, which had indicated its intention to establish a ministry of petroleum and oil resources.[24] Again, item 36 of the Second Schedule of the Nigerian Constitution states that "mines and minerals, including oil fields, oil mining, geological surveys and natural gas," are in the exclusive legislative list.[25] This is an interesting case of abrasive tickling of the federal government even when the constitution is very clear on the issues involved.

There are other cases in federal-state relations where differences in party programs have created policy problems. In Lagos state, Governor Jakande's UPN program includes free education and government takeover of all primary and secondary educational institutions. As far as Shehu Shagari's NPN government is concerned, the "qualitative" education envisaged in the party's manifesto permits the existence of private schools. While Bishop Okogie has taken the Lagos government to court, it does seem that such conflicts are inevitable in federal systems. The National Assembly can make a law with respect to private ownership of schools. Once made, any state laws which violate federal laws would become null and void. The UPN would also have had the opportunity, just like the NPN, of convincing National Assembly members of the worth of its stands.

Another interesting aspect of federal-state relations is the attitude of the incumbents themselves. At the federal center, there is no doubt that the present incumbent, President Shagari, is a humble and moderate man in politics. He is taking up his duties slowly but surely. It is a gigantic task. One can see the executive president like a big jet plane having difficulties in taking off because of technical problems with its wings or some other parts. Passengers may get frustrated while anxiously waiting at the lounge. But when it finally takes off, it easily attains high speed. People seem frustrated by Shagari's care and slow speed in this takeoff stage of the presidential system. This caution is necessary. One can envisage in the nearest future when Nigerians will be complaining that the president is acting too fast and getting too powerful. The presidency is a big and powerful bird and must fly with caution at the takeoff stage.

However, Shagari has shown his political astuteness at least twice since he came to office. While state governors basked in the prestige of prefixes such as "His Excellency," the president's office announced that the incumbent was now to be known as Mr. President. He drew the carpet from beneath the feet of the governors, many of whom have not realized this yet. In the symbolic political dimension, Shagari is engaging in political capital accumulation for the office he holds now and for himself personally. Again, this was vividly demonstrated in his public rejection of the ₦50,000 salary proposed for the president by a committee of the National Assembly. While the members of the assembly were furious with him and even accused him of turning the country against the assembly, Shagari came out of the rumpus with huge political good will to be stocked in his reservoir for the future.

Many Nigerians are, however, worried that there is too much mixture of Tafawa Balewa's meekness and Gowon's overconsideration of the feelings of others in Shagari's character for decisive presidential action in policy making and implementation. But perhaps the greater fear is of some members of the political entourage in his presidential bandwagon who are believed to be more interested in acquisition for the self than in service for the nation.

On the other hand, apart from the residual militarism in their styles, governors seem to be having difficulties in adapting to their new political roles. A good politician is one who is capable of adapting to the role of the politician and the statesman as situations demand. Many governors have not been able to change from their role as a soap-box politician (during the election campaign) to that of a statesman in Government House. As a successful politician in election, a governor automatically becomes the "father" of all in his state, whether citizens are his party bedfellows or not. The political blunder by Governor Ambrose Alli of Bendel made during President Shagari's visit to that region illustrates the above point. In a way Governor Lar of Plateau state did make amends for Alli's behavior (in state-federal relations) by his display of statesmanship during Shagari's visit to Jos in March 1980.

Part of the problem of federal-state relations is related to the "hawkish" stance of some federal government political executives and officials. In many cases they always seem to try to interpret the constitution in such a way as to favor the federal government. Two cases illustrate this point.

In reaction to the notorious power cuts by National Electric Power Authority,[26] Lar and the House of Assembly of Plateau state considered the possibility of switching to the Nigerian Electricity Supply Corporation (NESCO), which had been supplying Jos and its environs with power until recently. In response, the federal minister for mines and power contended that no state had any authority to generate electricity. Lar came out to challenge the constitutional legality of the minister's position. While one recognizes that constitutional provisions are subject to interpretations, it does seem that the constitution is clear on this.

The Second Schedule, Part II, the Concurrent Legislative List, item F-14(a-e) recognizes the powers of the federal government to generate and transmit electricity throughout the federation, and between one state and the other. Thus it is clear that the minister is correct to assert federal authority in this area. But item F-14(a-e) of the same schedule stipulates that the House of Assembly of a state "may" make laws for: a) electricity and the establishment in that state of electric power stations; and b) the generation, transmission, and distribution of electricity to areas not covered by a national grid system within that state (Amendment Decree No. 104, 1979). Unless it can be proved that state laws on generation and distribution of electricity such as those suggested by Plateau state contradict federal laws (which by themselves are assumed to be unrepealable or unamendable, and violate the constitution by operating in areas covered by a national grid system, the states do seem to have a case here.

The second illustration is the case of creation of more local governments in Nigeria. Some federal sources in government and in the National Assembly have at various times publicly contended that a state government has no right to create more local governments in its area. The main argument of these officials is that the First Schedule, Part I, which describes the states of the federation, mentioned specific local areas. Unless these aspects in the constitution are amended, they argue, states cannot create more local government councils.

Section 7(1) of the constitution states:

> The system of local government by democratically elected local government councils is under this Constitution guaranteed; and accordingly, the _government of every state shall insure their existence under a law which provides for the establishment, structure, composition, finance, and functions of such councils._[27]

It may be argued that creation of more local government councils comes under the function of the state government with regard to the "establishment, structure, composition, finance, and functions of such councils." The argument that the schedule which describes the states of the federation referred to existing local government areas which cannot be altered without constitutional amendment seems hollow. The schedule does not refer to local government areas. It merely enumerates areas which comprise a state.

In the case of Kaduna state, for example, Saminaka is one of the areas mentioned. The creation of two local governments out of that area does not alter Saminaka as a component area of the state. Similarly, while the areas mentioned might have coincided with local government council areas, there is nowhere in the constitution where reference to these areas is interpreted as "local government council areas." In fact, Section 3(a) of the constitution reads:

> Each state of Nigeria named in the first column of Part I of the First Schedule to this Constitution _shall consist of the area shown opposite thereto in the second column of that Schedule._[28]

It is, therefore, clear that the power to create local governments belongs to states. If for one reason or the other the second column of that schedule creates a bottleneck in policy-making, the National Assembly should amend the constitution and remove that part. It is pretentious to say that any other level of authority beyond the state can deal with issues of local governments in Nigeria.

The federal civil service should not be overbearing in relating to states. The recent circular by the head of federal civil service that all public officers who have taken up political positions should resign their posts in state services is not in order. The state service is autonomous of federal center. The tradition of centralization under the military seems to be very much on the scene for some federal officials. If the center wants this directive adopted by the states, it should send it as a bill to the National Assembly, and a "law" can be passed to that effect by the Assembly.

Finally, a ticklish and ongoing problem in federal-state relations is finance. Since the creation of twelve states in 1967, states have become more heavily dependent on the federal center for their budgetary needs.[29] This dependence differs from one state to the other -- from 40 percent in some states to 90 percent in others. The Okigbo Revenue Allocation Commission has had a welter of demands by state governments for more funds to be allocated to the states. Only recently a committee of the House of Representatives recommended that as an interim measure 30 percent of all federally derived revenues be allocated to states.[30] The Senate has now passed a bill allocating funds as follows: 58.5 percent to federal government, 31.5 percent to states, and 10 percent to local government.[31]

The problem of state autonomy is compounded by the inability of states to raise revenues from internal sources. In some states, there is little or no effort made to do this beyond community tax and personal income tax. In some others, the creation of additional states has reduced their resource bases and rendered them hopelessly dependent on federal sources. In addition, while oil revenue has been very helpful, it has increased enormously the powers of the purse of the federal center at the expense of states -- again a legacy of the Lagos-centric actions of the military regime. This has obviously its merits and demerits. The oil boom of the early 1970s also created a dangerous sense of complacency and a false sense of security Nigerians have not been able to overcome.

It thus seems very clear that as more states are created in Nigeria, the tendency toward greater authority at the center would become more glaring. It is most likely that the more states, the weaker the new states in their resource bases and, therefore, the greater the power of the federal center. Finance is an important aspect of federal autonomy. In addition, the nature of modern government works against traditional concepts of subnational unit autonomy. In all federal systems the power of the center grows because of the demands for uniformity of action and the complexity of modern government. This has been the case in the United States (we doubt that President Reagan will succeed in his attempts to reverse the trend), Germany, and (despite its peculiar problems) Canada. Nigeria is no exception.

Thus, in a paradoxical fashion, while most Nigerians tend to be very much attached to federal principles of state autonomy, they are most vocal about the creation of additional states with its implication of erosion of state autonomy.

The structure of the federal system and the distribution of powers in this structure have posed problems for the new politicians. They are learning gradually. The system is new, the political platform is quite new, and the political actors are a mixture of the old and the new.

Let us turn to those issues which create problems for the strategy of federal balance as a result of the dynamics of the political process.

Echoes of the Past: Continuing Sources of Fears and Suspicions

The very dynamics of the political process in Nigeria indicate that there are many echoes of the past which are still heard today, even if within new and at times different structures.

The North-South bogey which hung like a cloud over the First Republic has not vanished, regardless of the creation of states. Occasionally the North-South dichotomy simmers to the surface in creation of states, revenue allocation, etc. Similarly, the ethnoregional aspects of the old political parties are becoming more apparent than in 1979 when the contemporary political parties emerged.

The Nigerian Peoples Party (NPP) is very often referred to as the reincarnation of the old National Council for Nigerian Citizens (NCNC) -- with its base in Iboland. The National Party of Nigeria (NPN) is very much a replica of an enlarged Northern People's Congress (NPC), with its strong Hausa-Fulani base. In the same vein the Unity Party of Nigeria (UPN) is seen by many Nigerians as the rejuvenation of the old Action Group (AG), grafted onto its Yoruba base. Of course, the People's Redemption Party (PRP), despite its internal fissures, is often seen as the Northern Elements Progressive Union (NEPU), updated and enlarged, and having more luck this time than it had in the First Republic; the PRP presently has control over two state governments.[32] Perhaps the only really new political party is the Great Nigerian Peoples Party (GNPP), formed by a splinter group from the NPP. It is very much a protest party in the Northern state -- against the Hausa-Fulani establishment. Each of these parties has members from other ethnic groups to satisfy the provisions of the constitution, but each creates certain impressions which are reminiscent of the past.

Given this situation of the parties and the open declarations of decamping by members of one party to the other (usually more propitiously placed in Nigeria's calculus of "starvation politics"),[33] many Nigerians have begun to wonder if we are not to have a repeat performance of the politics of intolerance, impatience, and opportunism of the First Republic. The old cliché of "cross-carpeting" in the legislatures is now barred, but nonlegislators certainly do cross-carpet, with a Machiavellian calculus of convenience.

In addition, the minority-majority ethnic group cleavages of old are still very much around and are rife in many political parties. At the NPN National Convention in June 1980 in Kaduna, the issue of election of candidates to national offices brought minority-majority group conflicts to the surface. Governor Okilo championed the cause of the minority states.[34] He accused members of his party of playing "selfish and sectional politics" and declared that some Nigerians were not yet rising above their "selfish and tribal outlook." He challenged his partymen: "We talk of federal character, why are you afraid of equality of states? Come forward now and accept federal character and equality of states."[35]

On another occasion, the fear that the minority group might gang up against the majority group led Mallam Adamu Ciroma, the minister of industries, to issue a warning. Ciroma warned that any gang-up by minorities would elicit drastic reactions from majority groups. While there was no evidence of such a gang-up by minority groups, even if it were possible it is not clear that unity among majority groups could be easily attained to counterreact, given their intrinsic and diverse interests. More often each majority group has found it easier to ally with some minority groups against other majority groups. Cases abound in Nigeria's political history, especially in the ethnoregional politics of 1960-65. Yet perhaps more important is the fact that these fears are expressed. It shows very clearly that a potent source of conflict still lingers on: even if pushed under the carpet within new structures, it occasionally comes to the surface.

As Ivo Duchacek once put the dilemma of majority-minority in federal balance:

> The problem for most ethnic minorities is that they are permanent minorities and the ruling group a permanent majority. In interethnic relations, therefore, the convenient democratic game of numbers -- majoritarian decision-making in the framework of a broad consensus -- does not work since the unalterable power symmetry between permanent majority and permanent minorities impedes the formation of a consensual community. The quota system satisfactorily guarantees jobs, educational opportunities, and presence in decision-making bodies; but no quota system can be really so generous as to transform a minority into a majority. No quota system can, therefore, fully protect a minority against the tyranny by the majority in a legislative body unless it is coupled with a veto power over central issues. Then, however, as the history of Cyprus shows, government may become impossible.[36]

The issue defies balancing in a federal system, yet a federal system provides for more compromise than a unitary system in this situation. The problem is not only at the national level. At the level of subnational states, minority-majority issues go on. Perhaps only a sense of fairness learned and practiced over time can assuage the fears and suspicions which this political issue creates. Creation of additional states only creates "new majorities" and "new minorities," even though it lessens the intensity of conflicts.

The differential pattern in acquisition of Western education (now a passport to social mobility) was always a source of fear in the old republic. It is very much present in the Nigerian political arena. The crises over admission of students into universities led the military government in 1977 to establish a central body -- the Joint Admission and Matriculation Board. Yet by 1979 there were widespread protests in parts of the country against the pattern of admissions into universities under this agency. The widespread demands for the abolition of this body led to the president's bill for the establishment of a University Entrance Examinations Agency.

The bill returns admission of students to the universities again. The demands for a more equitable admission formula still go on and will still go on. It relates to our earlier analysis of the differential pattern of the spread of Western education in Nigeria as a source of fear and suspicion among groups. Yet this issue, by its nature, defies balancing.

Revenue allocation was a real bone of contention in the First Republic. It is still very much a problem. As mentioned earlier (see footnote 31), the president has signed a revenue bill which gives the federal government 58.5 percent, states 31.5 percent for state governments, and 10 percent for local governments. As in 1964, states have reacted with anger to this formula, arguing that it gives the federal government too much money at their expense. In fact, twelve governors have taken the president to court. These governors are, understandably, not NPN governors.[37] The problem of revenue allocation is an echo of the past that still bugs the Nigerian political system. But it is a problem of all federal systems. It only reminds one of Ayida's warning in 1973 that

> the next crisis is most likely to have its origins in basic economic issues and social conflict -- the equitable allocation and proper management of the increased disposable resources of the federation and the familiar conflict between the have and have-nots.[38]

There is no doubt that states are heavily dependent on the federal government for their finances. While a case can be made for more funds accruing to states, most states seem very frivolous in their expenditure and hardly attempt to raise revenue from internal sources. Moreover, most Nigerians would want to see a strong federal center in the light of Nigeria's historical past, which is very green in people's memories.

Finally, generational conflicts between "older" and "younger" politicians come up to the surface very often. The PRP crisis between the Aminu Kano/S.G. Ikoku group on the one hand, and the governors Abubakar Rimi and Balarabe Musa group on the other, is partly related to this issue. While the PRP is the most ideological of all Nigerian parties, the extent of its radicalism as represented by Rimi and Musa goes counter to the grain of the elders of the party such as Aminu Kano and Ikoku (both seasoned politicians). While Aminu and Ikoku see relations with UPN and GNPP as antiparty and as posing a threat to their leadership, Rimi and Musa cannot imagine themselves functionally interacting with the NPN. While there are more reasons for PRP crises, the generational issue is very cogent.

Similarly, the NPP caucus in the Senate found itself embroiled in a crisis of leadership, and a member of the "old brigade," Senator Jaja Wachukwu, lost his leadership position to a member of the "new brigade," Senator Obi Wali. The NPN has similar problems with its Kaduna-based caucus, usually referred to as the "Kaduna Mafia" -- generally younger and very active members of that party. If Nigeria succeeds in conducting the 1983 elections peacefully, there is no doubt that younger men are likely to dominate the political arena. Already older politicians are beginning to find that there are advantages to retiring to the background, with a thumb on the political remote control command.

Generally, these issues are echoes of the past which still pester the Nigerian body politic. Solutions to them are not easy. Some of these issues defy any attempts at balancing within Nigeria's federal network and may be regarded as unbalanceables in Nigeria's political equation. This is our next focus.

Challenges: Balancing Unbalanceables?

Federalism is "about cooperation, that is, the terms and conditions under which conflict is limited."[39] It is, therefore, a mechanism for effecting compromise. The art of compromise involves delicate balancing of conflicting claims. While some of the issues mentioned earlier are part and parcel of Nigeria's attempt to balance imbalance, some issues by their nature defy balancing and pose real challenges to the decision-makers. These issues create political dilemmas that are difficult to resolve. Let us take a few examples.

The differential pattern of the spread of education has been identified as a very potent source of fears and suspicion among Nigerian groups. Yet one cannot balance the situation by holding static the educationally advanced areas until the less advanced areas catch up. Quota systems will keep out some of the most brilliant Nigerians from educationally advanced areas while offering no absolute guarantee for closing the gap. Yet a widening gap between these two areas in one country only creates a reservoir of problems for greater crises later. The general argument is that more money should be spent in the less educationally advanced areas to enable them to correct the imbalance. But as theorists of change know, a head start in education can hardly be slowed down. This issue defies balancing, and yet it is closest to the hearts of Nigerians for understandable reasons. Perhaps Nigerians may set a minimum level of educational standard for everyone and hope that, with encouragement in disadvantaged areas, the gap will be appreciably narrowed -- narrowed to a point where disparities in educational acquisitions no longer pose grave threats to the nation.

Related to this is the attempt to insure that the composition of public institutions reflects "federal character" as outlined in section 14, subsection 3 of the constitution. There is no doubt that representation creates a sense of belonging to the nation. Yet, laudable as it is, Nigerians have not yet been able to define appropriately the mechanism for effecting federal character. Does it mean that every state must be represented? Even then some minority groups may never get represented in federal or state institutions. If all states or local governments are to be represented, what happens to efficiency? What happens when there are fewer vacancies than the number of states? If "federal character" does not mean equal representation of states but an attempt to insure that one or two groups do not dominate a particular public institution, to what extent can one balance the situation without alienating others and inserting seeds of frustration into the institution. Yet "federal character" is problematic because it is linked to Western education.

Another issue which defies balancing is the federal structure. As mentioned earlier, the minority-majority issue is contextual and situational. A minority group yesterday becomes a majority group today in the context of a new and smaller state. This problem is likely to go on tickling the federal system. But perhaps even more dangerous is the calculus of state creation in the background of the old North-South line. Can Nigeria really have as many states in the North as in the South? Will the creation of more states not tilt the balance in federal-state relations much more in favor of the center, and possibly lead to the unitary system?

Conclusion

To begin our conclusion, let us examine our findings in light of the editors' comments on decisional models in Chapter 12 of this volume. In that chapter two decisional models were identified, the hegemonial and the bargaining. The chapter asserts: "Broadly speaking, hegemonial models seek to control conflict from the top downward; bargaining systems are based upon a mutual adjustment of conflicting interests. The former strengthens authoritative institutions at the center, with some risk of system overload and low responsiveness; the latter decentralizes control, increasing autonomous participation at some cost in coordinated decision-making."

Nigerians opted for federalism out of interethnic insecurity in the terminal colonial period. They failed to generate any commitment to federalism as goal or ideal and lacked comprehension of the very values of federalism. For the politicians, federalism provided an opportunity "to have ours, and do it our way!" It is not clear, despite the resurgence of federal sentiments, whether many of its operators understand and feel committed to it, even in the Second Republic.

This observation is important because it reflects a deliberate choice of federalism as a "bargaining" mechanism as opposed to a "hegemonic" mechanism for resolving political conflicts. The constitution makes elaborate arrangements for sharing power, adjusting conflicting interests and permitting autonomous participation.

The most prevalent bargaining strategies mapped out by the 1979 constitution have been "protectionist," "redistributive," and "share oriented" within the context of the Rothchild-Olorunsola models. (Consider sections 14(3) and (4) and sections 201-209 of the 1979 Nigerian Constitution, which we have discussed in this paper.) These bargaining strategies have been rated variously as having medium-high to high potential for moderating conflict. Thus, if the constitution were given full sway, we should be immensely optimistic about the future of Nigeria. Although it is premature to arrive at a final judgment about the performance of a constitution less than two years since the initiation of its operation, tentative observations can be made. Unfortunately, we cannot be as optimistic as we would have liked to be. We have demonstrated many cases of preferences for confrontational styles or tactics rather than for discussions and compromise. Let us recapitulate a few: 1) some state governments threatening to take over federal television stations in their states; 2) governors threatening to issue vacate orders to federal institutions domiciled in their states or threatening to take over federal institutions (universities) unless they show greater responsiveness to parochial constituents; 3) federal officials and political executives with equally hawkish stances, as evidenced by the NESCO and federal civil service/state civil service authority dispute; 4) the federal government's handling of the Shugaba case.

These may only be problems inherent in any "try-out situation." In the life of every federal system there is a continuing process of adjustment and balancing. There is hardly any federation that has attained a perfect balance. Adjustments in one area may affect another area in the federal scale. If there is any contribution this paper has made, it is in the attempt to remind scholars that the issue of federal balance is not always structural -- that is, among component governments. There are certain forms of imbalance in the federal system which can be corrected in relative terms to permit the system to operate above certain levels of efficiency. On the other hand, certain issues, by their very nature, defy "balancing" in the short term. Too much pressure for balancing may break down the system. The demand for the creation of more new states, with its implication for the erosion of state autonomy, may be a case of too much pressure for balancing. The creation of states within the federal structure was a device to reduce the perceived overcentralizing predisposition of the federal government. Unfortunately, Nigerian states may be in the throes of a dependence paradox. There is inadequate understanding that there is a point of diminishing return in the utility curve of the creation of new states. Indeed, as we pointed out, the creation of additional states has reduced the resource bases of many states and rendered them exceedingly dependent on the federal government. Thus, Lagos-centric actions of the past are unwittingly being encouraged by tactics and strategies calculated to bring about an acceptable equilibrium between state coherence and ethnic self-determination. Finally, we must observe, regrettably, that the minority-majority ethnic group cleavages of old are still around and salient in political parties. Nevertheless, it is consoling to note that on at least one occasion the legislative branch of the federal government has united to challenge the executive branch and has done so without regard to ethnicity.

Footnotes

1. Some of us have argued that the military also operated a federal system even in the absence of popular channels of participation. See J. Isawa Elaigwu, "The Military and State Building: Federal-State Relations in Nigeria's Military Federalism, 1966-76," in A.B. Akinyemi, P.D. Cole and W. Ofonagoro, eds., Readings on Federalism (Lagos: Nigerian Institute of International Affairs, 1979), pp. 155-181.
2. Martin Diamond, The Federal Polity, quoted by Stephen L. Schechter, "The Present State of American Federalism," Publius, 8, 1 (Winter 1978): 6.
3. Literature on federalism includes (among others) K.C. Wheare, Federal Government (London: Oxford University Press, 1964); Carl Friedrich, Trends of Federalism in Theory and Practice (London: Pall Mall, 1968); Daniel J. Elazar, The Politics of American Federalism (Lexington, Mass.: D.C. Heath, 1969); Ronald Watts, Administration in Federal Systems (London: Hutchinson Educational, 1970); and James Sundquist, Making Federalism Work (Washington, D.C.: The Brookings Institution, 1969).

4. It is interesting that while Nigeria was politically unitary under the British, it was very much administratively decentralized. While Britain always prefers to bequeath a form of federalism to its ex-colonies, there was no doubt that Nigerians opted for federalism between 1951-59 as a result of mutual suspicions and fears of domination among Nigerian groups.
5. Adebayo Adedeji, "Federalism and Development Planning in Nigeria," in A.A. Ayida and H.M.A. Onitiri, eds., Reconstruction and Development in Nigeria: Proceedings of a National Conference (Ibadan: Oxford University Press, 1971), p. 103.
6. The May 1966 massacres of citizens of Eastern Nigerian origin (especially Ibos) was Northern reaction to Decree No. 34, which made Nigeria a unitary state. For the Northerners, federalism gave a relative control over their own affairs. For additional treatment on the imbalance, see Victor A. Olorunsola, The Politics of Cultural Sub-Nationalism in Africa (New York: Doubleday Press, 1972).
7. J.S. Mill, Representative Government, Everyman edition, pp. 367-368, quoted in Wheare, Federal Government, p. 51.
8. Wheare, Federal Government, fourth edition, p. 50.
9. According to Sklar, with the January 1966 coup, "political power had shifted away from the Northern rulers and their allies to a more progressive section of the population. The dangerous imbalance between legal and technological power had been corrected." In other words, the January coup corrected existing imbalance. Richard L. Sklar, "Nigerian Politics in Perspective," in R. Melson and H. Wolpe, eds., Nigeria: Modernization and the Politics of Communalism (East Lansing: Michigan State University Press, 1970), p. 50.
10. Even though Nigeria operated a military system, it was a military federal system of government. The military governments of the regions/states were very powerful and autonomous, especially under the Gowon administration.
11. See discussion of these measures in Elaigwu, "Military and State Building."
12. Note that even under the thirteen years of military rule (except for the brief period between May 24 and August 1966), the term "federal" has always been part of the name given to the central government, e.g., "The Federal Military Government of Nigeria."
13. It is also important to note that although the 1963 constitution had provisions for the creation of additional regions, only one region (the Midwest) was created within the period of the First Republic.
14. These men (in the Constitution Drafting Committee), and many military rulers at the time, felt Nigeria should experiment with a new system of government for a change.
15. Anambra, Bauchi, Bendel, Benue, Borno, Cross River, Gongola, Imo, Kaduna, Kano, Kwara, Lagos, Niger, Ogun, Ondo, Oyo, Plateau, Rivers, and Sokoto states.
16. These reasons include: 1) discrimination in the present geopolitical setting; 2) possible instability if such a state were not created -- almost a threat; 3) lack of opportunities for participation and access to scarce but allocatable resources; 4) personal ambitions of individual elites, which are cloaked behind

group interest, etc.

17. A.A. Ayida, The Nigerian Revolution, 1966-67 (Ibadan: Ibadan University Press, 1973), p. 6.
18. The Great Nigeria People's Party (GNPP) controls Borno and Gongola states; Nigerian People's Party (NPP) forms the governments of Plateau, Imo and Anambra states; National Party of Nigeria (NPN) is in full charge of Cross River, Rivers, Benue, Sokoto, Niger, Bauchi and Kwara states; the People's Redemption Party (PRP) is in power in Kaduna and Kano states; while the Unity Party of Nigeria (UPN) effectively controls Ogun, Lagos, Oyo, Bendel and Ondo states.
19. These were supposed to be presidential to the states to coordinate agencies and projects. Some states (Imo, Plateau, Anambra) initially refused to have anything to do with the Federal Housing Project if these were handled through PLOs. New Nigeria, May 20, 1980, p. 1.
20. Shugaba is the majority (GNPP) leader of the House of Assembly of Borno state. He was deported by the federal government as an alien and was later brought back into the country by court order. The matter is still in court. Shugaba won the first round, but the federal government has appealed against the Maiduguri court decision.
21. New Nigeria, May 29, 1980, carried a report of the federal government's plan to review the issue on NTV. They would still be federally controlled but increasingly decentralized.
22. The Nigeria Standard (Jos), December 11, 1980, p. 1.
23. Federal Republic of Nigeria, The Constitution of the Federal Republic of Nigeria, 1979 (Lagos: Government Printer, 1978), Second Schedule, item 2, p. 96.
24. National Concord, May 17, 1980, p. 3.
25. Constitution of the Federal Republic of Nigeria, item 36, p. 98.
26. NPEA has given many reasons for its constant power cuts. These have included, at one time, drought and its effects on River Niger, and at other times, flood in River Niger.
27. Underlined to emphasize relevant part of the quoted section of the constitution.
28. Underlined for emphasis.
29. Elaigwu, "The Military and State Building," pp. 175-177; see also Rafiu Akindele, "Federal Grants-in-Aid to State Governments: A Note on an Aspect of the Federal Spending Power in Nigeria," in Akinyemi et al., Readings on Federalism, pp. 182-189.
30. See Elaigwu, "The Military and State Building," p. 176.
31. Daily Times, May 21, 1980. Under this Senate approval, 21.5 percent was to be transferred from federal funds of 58.5 percent for the purposes of underwriting the cost of initial developments at the Federal Capital Territory at Abuja, 31.5 percent for state governments, and 10 percent for local governments. The House of Representatives had earlier voted 50 percent for federal government, 40 percent for state governments, and 10 percent for local governments. The Senate voting pattern was

very interesting. It went thus:

	FOR	AGAINST
NPN	--	29
UPN	27	--
NPP	9	8
GNPP	--	8
PRP	--	5

The National Assembly established a joint committee of the two houses to reconcile their differences. See _The Punch_ (Lagos), January 16, 1981, p. 1; _The Nigeria Standard_ (Jos), January 16, 1981, p. 1; and the _New Nigerian_, January 16, 1981, p. 16. The Senate amendments to the bill carried by 13 to 11 votes. _New Nigerian_, January 30, 1981, p. 1. The president has since signed the revenue bill amidst protests from some quarters. Two legislators and twelve governors have taken the case to the courts. See _The Nigeria Standard_, February 4, 1981, p. 1; and _The Punch_, February 8, 1981, p. 16.

32. PRP has expelled its only two governors from the party. There are court cases pending on this expulsion.
33. "Starvation politics" is used to refer to a situation in which a party member changes his party with the hope of harvesting some perquisites (contracts, status, etc.) from the new political party he joins.
34. Minority states are those composed of minority groups in the old four regional structures such as Niger, Benue, Plateau, Cross River, Rivers and Bendel states.
35. _The Nigerian Tide_, July 1, 1980, p. 1.
36. Ivo Duchacek, "Antagonistic Cooperation: Territorial and Ethnic Communities," _Publius_, 4 (Fall 1977): 23.
37. The president, Alhaji Shehu Shagari, is an NPN man.
38. Ayida, _Nigerian Revolution_, p. 7.
39. Aaron Wildavsky, "A Bias Towards Federalism: Confronting the Conventional Wisdom on the Delivery of Governmental Services," _Publius_, 6, 2 (Spring 1976): 95.

15
Geoethnicity and the Margin of Autonomy in the Sudan

Dunstan M. Wai

Introduction

The basic principle of the right to self-determination is "the simple one that alien rule should give way to rule by the people of the country concerned."[1] Self-determination is essentially a political right; and the decolonization process in Africa and Asia was launched after the recognition of this right by imperial rulers in both regions. The issue at stake now is whether it is "a right which once exercised is therefore extinguished, or is it still available to those who may again seek to lay claim to it?"[2] In other words, is the right to self-determination still operative within independent African countries?

There are different answers to and views on the question: some argue that the right to self-determination is extinguished after independence from colonial rule has been achieved; others take the view that self-determination is not an absolute right and that it can be applied only with due regard to circumstances. Alfred Cobban, who for the most part supports the latter view, has contended that there are, in the first place, such things as rights, and, secondly, that these rights can be attributed to a collectivity such as a nation.[3] "The rights which we derive from the general conception of right are, however, not absolute, but are susceptible of, and indeed demand compromise, whenever they come into practical conflict with one another as experience proves that they constantly do."[4]

The basic issue here is when does the right to self-determination become a source of conflict when claimed by an ethnic group or region which is part of an independent African country? It seems that before a group of people makes the demand for the right to self-determination and to political secession, they must believe in the existence of at least six conditions. First, they must believe that the security of their lives and property cannot be maintained if they are subject to control of a government as it is constituted. They must think that the state no longer offers them psychic orientation and security. Second, they must believe that peaceful and orderly processes of negotiation which aim at the re-establishment of a workable pattern of political relationships between them and the rest of the country have been effectively frustrated and suspended by the government and cannot fruitfully be resumed. Third, the state, in their view, must no longer respond to their specific popular and material needs and desires, and they must perceive a condition of social and economic deprivation relative to those who control political and economic power. Fourth, they must believe or assume that secession is widely recognized throughout the country as a politically legitimate step, and would be accepted in, if not actually supported and/or initiated by the rest of the country. Fifth, they have to believe that the move to independence has overwhelming support in their area of domain. Sixth, if they were not involved in the negotiations prior to political independence, they must also believe that they did not exercise their right to self-determination at the time of their country's independence.

The last point was embraced by Southern Sudan. The third condition would apply to Biafra, where it was believed that declaration of secession would be supported and followed by other regions, particularly the Western Region. The first, second, and fourth conditions would apply to the Biafrans, the Southern Sudanese, the Eritreans, and the Moslem Chadians.

It must be remembered, however, that strife and communal conflict within nation-states are not necessarily the consequences of a desire for self-determination; more often, they result from a desire to resist it. For indeed, if the states involved are prepared to accept a result based on self-determination, then there is no reason to presuppose that violence will ensue, no more than it did over British-administered Togoland in 1956 or Cameroun in 1961. Hence, both the demand for and the denial of the right to self-determination can provoke civil violence.

This chapter attempts to examine the North-South conflict in the Sudan, and the margin of regional autonomy which the South achieved following the successful conclusion of the Addis Ababa peace talks between the representatives of the Khartoum government and those of the Anya-Nya armed forces.[5] It is our contention that the Sudanese conflict is in many respects different from the other African secessionist conflicts dealt with in this volume, and therefore provides a different perspective on issues of self-determination in postcolonial Africa. There are at least eight factors supportive of this view.

First, it is a conflict between Africanism and Arabism reinforced by Muslim zest for Islamic proselytization of Southern Sudanese. Second, a deeply rooted historical antagonism has long characterized societal relations between the South and North of the Sudan. Third, separate colonial administrative units were set up for the two regions in full awareness of racial, cultural and religious differences, and wittingly or unwittingly, colonial policies resulted in unequal economic and political power configurations. Fourth, the Southern Sudanese did not join the Northern Sudanese in demands for a united independent Sudan of North and South. Fifth, the South was denied participation by imperial Britain, Egypt and Northern Sudan in negotiations on instruments of independence for the Sudan. Denying Southern Sudanese the opportunity to express their views on their future destiny in an independent Sudan meant denying them their right to self-determination. Sixth, the Northern Sudanese-controlled state imposed an internal colonial regime in the South. The preferential economic policies pursued by Khartoum toward the North and against the South immediately aroused among the Southern Sudanese feelings of relative social and economic deprivation. Furthermore, the behavior and attitudes of the Arab administrators who had replaced the British in the South confirmed fears and suspicions of Southerners about a North-South union. Seventh, the pattern and intensity of the war received less publicity in the international press than comparable or other internal African struggles for self-determination (for instance, Biafra and Katanga). Eight, the manner in which the Sudanese conflict was temporarily eased and politically transformed remains conspicuously unique in Africa.

Hence, most analogies which have been drawn between the African-Arab conflict in the Sudan and secessionist efforts in the rest of Africa are inappropriate. A comparison of the Sudanese conflict with the Nigerian-Biafran war will clarify some of the issues involved. For this purpose, we shall use the conflict variable of geoethnicity -- the sense of separate identity of an aggregate of culturally and historically related ethnic groups occupying a particular geographic region of a country, differentiated from other similar geographical aggregates, which when activated or escalated can be the source of interregional tension and conflict, including outright secession.[6] Isolating this variable may help us to distinguish between conflicts which arise from the dynamics of hostile interaction between regionally situated peoples of different cultures, as experienced through the perceptual prism of their respective heritages and value systems, and those hostilities which are generated through the evocation and manipulation of communal sentiments by elites competing for power at the national level. Although this distinction is easily blurred by the tragedy of violence committed along the structural lines of pluralism and by the cultural biases of propaganda, it neverthless asserts that the sources of motivation for the emergence of cultural loyalty as a medium for secessionist political mobilization can be differentiated along a broad continuum. The following comparative discussion will therefore strive toward a clear understanding of the influence of geoethnicity and cultural hegemony on nation-bulding by examining the configuration and experience of pluralism in each case, and particularly the types of roles played by elites in expressing, enhancing or even creating cultural grievances.[7]

Geoethnicity as a Basis for Claims to Self-Determination

The Sudanese war was a struggle against internal colonialism. The specifically cultural orientation of the Arab, Northern-dominated government's policies toward the South constituted a form of oppression which enjoyed considerable legitimacy among Northern leaders and in the cultural context from which they evolved, and which clearly hurt the entire Southern population. In this case, cultural pluralism in itself defined and precipitated secessionist conflict and the element of elite manipulation never proved decisive in the expression of grievances by the Southern peoples. Their cause was essentially one of self-preservation in the face of Arab efforts at cultural-assimilationist conquest.

The particular configuration and history of cultural relations in the Sudan opposed two geographical regions in which two generically different sets of world-views prevailed. Although numerous ethnic groups lived both in the North and in the South, centuries of migrations and conquests by Arab Muslims had obscured many of the cultural disparities among the Northern section and had produced instead an extensively Arabized and Islamized collectivity which, regardless of particularistic political views and varying degrees of assimilation into the Muslim-Arabic context, tended to consider itself more a part of the Arab world than of the African political and cultural spectrum.[8] The bonds of language and religion, and the social changes brought about by the penetration of Islamic law and institutions, carried with them disdain for the Southern Sudanese and helped justify the systematic cultural oppression by government. The unification of North and South during decolonization revived a pattern of Northern expansionism which had merely lain dormant during the colonial interregnum. Furthermore, the slave-raiding of the North upon the South remains a historical fact of significance in terms of interregional perceptions and attitudes, at least in the psychic realm.

Historically, the Northern cultural influences on the South have been inhibited by ecological barriers, effective armed resistance or flight by the Southern peoples and, during the colonial epoch, by the intervention of the British colonial force, which drove out the Northern aggressors and excluded them (as well as their language, religion, clothing and other cultural forms), quite rightly so, from the South for thirty years.[9] Furthermore, while the Southern region was administered through a system of indirect rule which sought to preserve traditional structures of authority in the various cultural sections though offered little in the way of a development program, the North was engaged much more actively in the educational and economic opportunities accompanying modernity, in large part on the strength of its kinship with the Muslim-Arabic world, particularly Egypt. Thus, at the time of decolonization, the trend toward cultural amalgamation had been reinforced in the North and had produced a large educated elite capable of articulating Arab nationalist sentiments. All but one of the political parties which emerged during this period shared a fundamental orientation toward the Arab world. The single exception was the Southern party which assumed a strictly defensive posture and attempted to dissuade the British authorities from uniting the two regions under a single unitary government.

The collective awareness of the Southern peoples of their handicap, their anticipation of cultural oppression, and patterns of rebellion and flight which accompanied the realization of these fears, bear witness to the wide diffusion of a sense of political entanglement in the South. As the scheme of independent government unfolded, and as Northerners streamed into the region, a common sentiment of the futility of forced union with the North and deeply rooted fears of brutality at the hands of an alien and hostile force prevailed among the Southern peoples and their leaders, who attempted to voice their cause against the overwhelming Arab nationalist and expansionist fervor of Northern leaders.

Southern representatives were ignored during the various conferences which set the terms of independence, as were British administrators in the South, who, anticipating the pattern of internal colonialism which could evolve after the transfer of power, protested vehemently. The Southern demand for federalism as a compromise was rejected after cursory examination and the minimal safeguards which they requested were not incorporated into the Sudan's constitution. This frustration was aggravated by the composition of government: the Sudanization process gave only six of 800 key administrative posts to Southerners, on the basis of their lack of education and preparation for political responsibilities.[10] The same imbalance was extended to parliamentary and judicial institutions. Consequently, Arab Northern Sudanese were given free rein in ruling the region, and social tensions created by their presence in the South began to mount.

The period between 1956 and 1963 saw the tenuous establishment of Northern control over the South and the confirmation of the worst Southern fears regarding their prospects for contributing to the Sudan government's policy-making for their region and for the country as a whole. The few social development projects which the South had acquired during the colonial era were systematically phased out or subjected to cultural assimilation designed to force Southern elites to act within the Muslim-Arab cultural context. The most prominent discriminatory policies involved the required use of Arabic in schools and the imposition of various Islamic customs; even the weekly holiday was changed from Sunday to Friday. While the Cabinet Department of Religious Affairs organized a massive Islamic proselytization program, an open campaign was waged against Christian missionaries who had developed and managed the Southern school system. After years of progressive paralyzing of their pedagogical and religious activities, they were expelled altogether in March 1964. By then, a handful of Islamic institutes had been set up and Islamic studies merged with other academic fields. Thus, knowledge of Arabic and adherence to Islam had become indispensable for scholastic success and individual career opportunity and mobility in the Northern-dominated system. Southern bids for greater political power were hampered by culturally oppressive structural reforms. These policies were all enforced with a great deal of coercion: student strikes against Islamic proselytization and civilian protests against Northern administrators brought on harsh reprisals by the Northern army and police. Such a pattern of rule fully corresponds to the idea of internal colonialism.

The Southern reaction demonstrated the salience of geoethnicity serving as the direct source of rebellious and secessionist activities. The Southern corps of the Sudanese army had mutinied five months before independence, setting off a wave of violence in the South. The spontaneity of this initial outburst, provoked by the fear on the part of the Southern military personnel that they were being sent to the North on false pretexts to be killed, was further reflected in the immediate flight of many Southerners to neighboring black African countries and the beginning of calls for secession by politicians in exile. However, although refugees developed a number of organizations designed to advance this cause, including several provisional "governments," the actual struggle was waged inside the country, often despite divisive tendencies inspired by ethnicity and conflicting claims of exiled spokesmen. The effort to undermine

Northern control was sporadic and unfocused, although the flow of arms into the country increased in the mid-1960s and a more determined and organized guerrilla movement, the Anya-Nya, was founded in 1963. Despite these fluctuating resources and crises of leadership, the struggle continued. The support of the rural population was crucial in sustaining the struggle for so many years. Somehow, the brutality of Northern reprisals seemed to inspire rather than discourage resistance.

The fact that politicians in exile raised racial and cultural issues, and at times used them to justify objectives such as immediate international recognition of the right of Southern Sudanese to govern themselves, emphasized the reality of the cultural oppression experienced by all Southerners. Both the underlying philosophy of the Arab Sudanese government, displaying a stubborn unwillingness to recognize the South's grievances and to seek a peaceful solution to the crisis by moderating institutionalized discrimination in any or all spheres of modernity, and the tenacity of such practices which overcame cultural distinctions and past rivalries, bear witness to the vitality of cultural self-righteousness and fear in prolonging resistance. Although the Southern case was never effectively presented in a spirit of unity and common purpose, despite short-lived attempts at establishing secessionist political structures within the region, the fact that their common aggressor was motivated by cultural expansionist intentions and that the Sudan government continued to identify the entire country as an Arab-Muslim entity was held by all Southerners as the primary grievance. Until Southerners were allowed to develop without having to compete within the framework of Muslim-Arabic institutions, the struggle continued as one of self-defense and liberation.

The literature of the war emphasized the African-Arab schism; and the most important concessions made to Southerners, founded on an acknowledgement of the military stalemate and of the need for good will in solving the Southern problem, involved their cultural grievances.[11] Aside from the material benefits and structural provisions for increased political power, at least at the Southern regional level, the Khartoum government sought to draw refugees back into the country, to appease the Anya-Nya through conciliatory and apologetic statements, and to discontinue policies of cultural discrimination. The eventual legal curtailment of Arab hegemony in the South under the Addis Ababa Agreement of 1972 and the new Southern regional government which was set up in recognition of the Southern distinctiveness provided an infrastructure for self-rule in the South. The Addis accord also affirmed that the North recognized the South's dual cultural nature and would no longer project itself as an exclusively Arabic and Islamic state.

Sub-Saharan African states in general are divided horizontally by language, culture, and religion. The extreme cultural fragmentation may, however, be in itself a form of insurance against ethnic conflict, if, and only if, there is no dominant ethnic group capable of exploiting smaller tribes. But when differences in culture coincide with sharp economic inequalities and are coupled with historical hostilities and crises of political leadership -- as they are in the Sudan -- they can produce conflict of the most intense sort. It would be in order here to examine the Nigerian-Biafran war and determine the extent to which cultural pluralism was a critical factor in fostering communal conflict culminating in the war. Were there any parallels with the African-Arab conflict in the Sudan?

The Nigerian-Biafran war was widely publicized as a genocidal conflict brought on by ethnic differences between competing elites.[12] The cultural grievances which strained attempts to create a democratic polity were in fact similar in many ways to the positions of the Sudanese factions. As in the Sudan, a North-South tension between a primarily Islamic society and diverse African cultures was reflected in conflicting views regarding the proper form of independent government. Unequal development of the regions during the colonial era and age-old cultural stereotypes both nourished the fears of the Hausa-Fulani people in the North. They complained that they had been deprived of the basic elements necessary for fair competition with the others in the new Nigerian state as a result of British separate and indirect rule and passivity in developing their region. The difficulty in achieving consensus on the proper structure of government and fundamental orientation toward nationhood culminated in the breakdown of civil order in 1966, marked by the tragic expression of ethnic hatred. Thus initial tensions in political debates between the leaders of opposing cultural groups had, as in the Sudan, escalated into bitter interethnic violence challenging the validity of a unified state. During the anarchic months of 1966, self-preservation became the foremost concern among the Ibo. And, as their secession attempt wore on, "the initial passion for vengeance gave way to a gnawing fear that the grisly slaughters in the North would be repeated in the Ibo heartland, if it was overrun by federal troops. This chilling specter of genocide kept the struggle going long after there was any prospect of military victory. Only the flickering hope that world opinion might eventually bring about international intervention kept the war alive against increasingly long odds."[13]

Despite such similarities to the Sudanese war, the origin and thrust of geoethnic conflicts in Nigeria fall far short of the phenomenon of internal colonialism which prevailed in the Sudan, where cultural characteristics and objectives and state coercion and policy-making combined to create a unique form of political and cultural oppression. In Nigeria, the dissolution of national institutions and eruption of armed conflict along the lines of ethnicity came as a result of prolonged competition between elites who attempted to shift the balance of power in the country in favor of their ethnic clientele. These politicians were able to draw on numerous spontaneous manifestations of cultural incompatibilities, especially those occurring in ethnically segregated cities and marketplaces, to excite their popular bases of support and to establish the legitimacy of their strategies. Thus, the crucial catalytic forces and events behind the Biafra crisis were produced by elite groups for competitive purposes which were largely unrelated to the value differences inherent in the cultural pluralism of the state.

The development of ethnic conflict at the national level can be partly attributed to the British policy of separate administration for the North and South, which limited interaction between the dominant regional-cultural sections to the various contexts of modernization -- whether the urban environment itself, which attracted members of all cultures and tended to be segregated accordingly, or the more specifically organizational forums, such as the universities, the lower echelons of the civil service and army, and the growing market system. The mass of the Nigerian population remained within its local traditional context until the decolonization era. As in the Southern Sudan, political organization was generally limited to villages and feuding tribal or national factions. The Yoruba Oyo kingdom and the Hausa-Fulani Sokoto Empire had broken up by the mid-nineteenth century as a result of internal feuds aggravated by the steady expansion of the slave trade, while the Ibo linguistic group had never developed a significant structure of authority beyond the village. Thus ethnicity was not a particularly organized or manifest presence in the relationship among the Hausa/Fulani-Yoruba-Ibo triumvirate, and the British assured continued separateness and localism with their system of indirect rule.

The revival of a wider communal consciousness among these groups followed the emergence of modern incentives and the criteria of progress and security. The Ibo were the most active in developing a sense of solidarity beyond their traditional base. Their migration to other parts of the country during the colonial era embellished their political and cultural antecedents so as to create a dynamic conception of Ibohood embodied in numerous ethnic associations in the modern context, ranging from trade unions to scholarship programs. As the diasporan community became more solidly entrenched in cities in the North and West, particularly Lagos, it came to be increasingly identified as a homogeneous force to be reckoned with by competitors from other ethnic backgrounds. The creation of rival associations established the grounds for conflict among the Ibo, Yoruba, and Hausa. With the beginning of political mobilization in the 1950s, ethnically homogeneous political parties became the chief media for the introduction of geoethnicity into the arena of elite competition and for the extension of the basis for differentiation and conflict to the masses. The Hausa-Fulani leaders were particularly intent on exploiting anxieties caused by the Ibo presence in their cities to rally their followers and neighboring cultural groups around the "One North, One People" slogan. As the pace of competition accelerated with the establishment of elective offices and the prospect of national government, the rhetoric and maneuvering of these elites rested increasingly on alleged exigencies of cultural self-preservation. Complaints about the impingement of modern politics on traditional values and institutions, separatist threats, and accusations of expansionist intentions, aimed above all against the Ibo by Northern leaders, became an integral part of the political style of the conflicting parties.

This dialectic of party development along ethnic lines was nourished by the British management of decolonization. The first major step involved the "Nigerianization" of the senior civil service, which consisted of the promotion of the top African members of the administration. Since the Ibos and the Yorubas emerged as the chief beneficiaries of this policy, fears of unequal opportunity and permanent handicap were voiced by Northern elites, who claimed that they represented by far the majority of the Nigerian population and deserved a much greater share of power. Although this stage in the transfer of power was largely experimental, as was the installation of complete regional administrations in 1951, the anticipation of frustration took form in response to specific gains and losses in the modern sector. The consequent intensification of party rivalries made the question of cultural pluralism crucial to the determination of the proper form of government for Nigeria, and party leaders used alleged concerns for the fate of small minority ethnicities to justify their respective visions of the structure of independent government. The result of this interaction was a failure to design an effective central government and an intensification of regional politics and identities. Thus sentiments evoked by the continued development of cultural stereotypes outside of a progressive scheme of social integration between cultural sections, and by the cultivation of fears of eventual subordination or visions of ethnic success, were generated in the context of regional power plays by the three major communal actors. Although the politicians were able to excite the population at times, such as on the occasion of an Action Group (AG) delegation tour of the North, where mob riots resulted in indiscriminate killing of the visitors, the contest was less a direct confrontation between political forces than a serious effort at mobilization of ethnic sentiments in preparation for the ultimate establishment of effective central governmental institutions.

The early independence era leading to the Biafra crisis saw a progressive breakdown of civil order which led to the two military coups of 1966. Interregional disputes at the federal level were compounded by internal dissent within the regions. The West was particularly chaotic, as Akintola attempted to erode Awolowo's following in the AG by creating his own party and forming an alliance with the Northern People's Congress (NPC), which dominated the federal executive and legislative organs. The tactics of Akintola's group involved bribery and coercion of the masses to achieve electoral goals. The turmoil in the West spread to other regions and gradually to some national institutions.

The army emerged as the dominant force on the strength of its own internal divisions. The January coup resulted in the killing of senior officers and key Northern political figures. This left a significant number of Ibos in control of the army and of the country. However, the aftermath of their coup was a period of extreme civil disorder, intensified further by a countercoup in which Northern and Western members of the army purged the organization of its Ibo personnel. Killings continued in the fall, and a massive repatriation of Ibos to the East occurred when it became apparent that there was no protection or immediate appeasement forthcoming. By 1967, the Supreme Military Council under General Yakubu Gowon had restored a semblance of order to the country. But the Ibos had abandoned their large share of power, and their leaders prepared for secession, scorning Gowon's attempts at negotiating a resolution to the tense situation. The rush toward secession was a difficult political choice made by General Odumegwu Ojukwu. He later sought to justify the secession of the Eastern Region by claiming that the Ibo were unanimous in their belief that no political solution could be found to the environment of hatred which had precipitated the massacres, and that other non-Ibo ethnic groups in the East favored such a strategy.

The spontaneity of the Southern Sudanese struggle finds no parallel in the systematic build-up of arms and propaganda channels for the Biafra effort. Thus, although both conflicts can be classified as genocidal and born of a history of antagonism between geoethnic groups, the role of elites in managing these struggles must be distinguished: the Nigerian crisis reflected conflicting ambitions among a small segment of the population, the various elite groups; the Sudanese conflict was, however, inherent in the coexistence of expansionist Muslim-Arab forces and African minorities. The claim of cultural differences made by the Biafran leadership became a self-fulfilling manipulation of communal sentiment as the war escalated and all prospects for accommodation disappeared.

The internal colonial system in the Southern Sudan and the tripolar ethnic contest in Nigeria can be given distinct places along the analytical spectrum of cultural pluralism as an impingement on nation building. The cultural oppression experienced by the Southern Sudanese was a natural outgrowth of independent Muslim-Arab government and revived a historical pattern of assimilationist conquest. The Biafra case, on the other hand, merely drew momentum from Nigeria's pluralism: competing elites mobilized their respective communal followings for electoral campaigns and used incidents of interethnic antagonism to break apart the state in accordance with their strategies for personal and factional aggrandizement. Indeed, the Nigerian state had been produced by a political contract between leaders of different ethnic groups who were all aware of the state which each had in unity or disunity and who were all able to participate in national politics. From open and fair competition the Northern Nigerians, whose status during the colonial era was closest to that of the Southern Sudanese in terms of unequal development, emerged as the dominant force in the independent federal government, capable not only of expressing their viewpoints through the political process but of weakening their adversaries through a complex series of alliances and political engineering. Thus, there is no parallel in the Nigerian case to the handicap of the Southern Sudanese, who were forced to submit to naked power and processes of cultural assimilation solely on the basis of their geoethnic and cultural distinctiveness from the Northern Sudanese Arabs.

Cultural pluralism is not in itself sufficient to cause eruption of violence, and indeed, it was not the only source of war in the Sudan and Nigeria. But for our purposes here, we will not explore other conflict-laden issues. Nor shall we discuss the patterns of the wars, the involvements of external actors, the various abortive attempts to find solutions, and the eventual forces that facilitated the changes from confrontation to accommodation and submission in the case of the Southern Sudan and Biafra respectively. Suffice to note that historical hostilities appear to have preconditioned the outcome of independence politics in the Sudan and Nigeria to an extent corresponding to the length and intensity of these conflicts.

The nineteenth-century African-Arab struggle in the Sudan determined the orientations of North and South toward political unification much more conclusively than did the sporadic precolonial modernizing competition between the dominant forces in Nigeria. Ethnic-regional identities in the Sudan were most strongly expressed in the context of interregional relations, since a significant body of historical experience had defined these relations as inherently antagonistic, a source of pride to the Northerners, and of fear to the Southern peoples. Thus Arab expansionism and anti-Arab clustering, carried over from the nineteenth century, characterized the political identities of the respective regions. In Nigeria, the modernizing environment, the urban and political world of the elites, was the historical source of conflict, and it was only with the management of aggressive tendencies by the elites in the 1960s that the masses became actively concerned with the ethnic-regional power struggle and could come to identify themselves as adherents to a particular movement. The immense swelling of the Nigerian army during the civil war indicates the effectiveness of violence in the formation of political identity. While historical hostilities had not even come close to generating this sort of participation, the military contest worked to unite previously isolated sections of the Nigerian rank and file and seemed to purge many of the elite-generated pretenses from the political arena. Most Nigerians except the Ibo and a few non-Ibos from the former Eastern region were drawn into identification with the state. Thus, the Nigerian situation, in view of its historical background, can be described as being more fluid than that in the Sudan, its geoethnic identities more plastic than those of the Arabs and Africans in the Sudan, to mobilization and identity engineering as wrought by elites and by circumstances.

The crisis of leadership in the Sudan was primarily a Northern phenomenon, since no channels for leadership other than those related to liberation, subversion or exile politics were available to the Southern elites. The bitterness of the sectarian rivalries, the inability of the army and civil service to provide disciplined leadership or to discourage resistance in the South, a general disregard for Southern needs and for the stresses which the Southern problem as a whole were placing on the country -- all this characterized elite instability. Politicians were too absorbed in their exclusively Northern power struggles to consider the responsibility of finding a solution to the violence in the South, and their personal concerns impeded the development of a stable and imaginative administration.

The Nigerian elites were equally insensitive to the effects of their competition both on the structure and process of government, and on the potential for nation building embodied in the establishment of central governmental institutions and in other spheres of regional and ethnic interaction. As a result, the regional and ethnic polarization of political concerns which culminated in the secession of Biafra were never contained by responsible attempts to coordinate politics at the center.

The secessionist Ibos of Biafra constituted an educated, skilled, aggressive people who were feared economically and politically in the rest of Nigeria. In the Sudan, the Arabs and the Arabized people of Northern Sudan have the monopoly on education, trade, and government. English is widely spoken in Eastern Nigeria, and, indeed, in the whole of Southern Nigeria, but it is employed under conditions of extreme linguistic fragmentation in the Southern Sudan. In Northern Nigeria, Hausa has functioned both as a *lingua franca* and as an acculturative device in a fashion similar to that of Arabic in the Northern Sudan.

The Northern Nigerians' fear of domination by Southerners in the political, cultural, and economic domains was analogous to Southern Sudanese fears. Parallels can be drawn as well between the confidence, arrogance, and disdain of many Southern Nigerians and that of most Northern Sudanese toward their respective geographical opposites.

Secession was recognized by both the Southern Sudan and Biafra as a risk-laden adventure. But the communal groups in the two regions concluded that their safety could only be secured through complete withdrawal. In both cases, the secessionist goal enjoyed overwhelming support from the people involved. On the other hand, power holders in Khartoum and Lagos genuinely believed that they had a sacred duty to crush the secessionist efforts with all the military might at their disposal.

But whereas at the time of its secession Biafra was under full control of a well-developed set of regional institutions, with its own staff manning the hierarchy and with a well-equipped army, the Southern Sudan had no regional institutions, no cohesive army, and no effective leadership. It was, however, in the words of Crawford Young, "paradoxically the amorphous character of the Southern Sudanese uprising, structurally, that made it inextinguishable: No decisive battle could be fought, no critical leader could be captured; the hydraheaded Anya-Nya were everywhere and nowhere. For Biafra, on the other hand, its very structural resources were in the end a weakness; once its institutional core had been defeated, nothing remained."[14]

It must also be remembered that whereas the Biafrans were at the vanguard of Nigeria's decolonization politics and participated fully in the negotiations which led to the independence of Nigeria, and were, indeed, partners with Northern Nigerians in the first Federal Republic, the Southern Sudanese were not involved in the independence movement, did not want a union of the North and South, and were denied participation in the discussions which gave final shape to the constitutional framework prior to independence. The Biafrans entered into a political contract with the rest of Nigerians on the eve of independence, but the Southern and Northern Sudanese did not. That explains why the Southern Sudanese launched their war of independence before complete British imperial withdrawal from the Sudan.

Legitimacy, Consensus and the Sudanese State

The growth of nation-states has shown that the process of national integration is difficult to achieve without political legitimacy and national consensus. Legitimacy is the quality acquired by a political system through the justification of power and obedience in the power hierarchy which it represents. It constitutes "the foundation of such government power as is exercised with a consciousness on the government's part that it has a right to govern and with some recognition by the governed of that right."

The legitimacy of the state is dependent on the tacit consent of the governed. As Herbert Kelman has pointed out, legitimacy is primarily a perceptual concept.[15] A state is only legitimate to the extent that its subjects perceive it so, and there are two basic perceptions that make people loyal to the nation-state. One is the perception that the state represents an ethnic and/or cultural identity. This perception creates sentimental attachment to the nation-state. The other is the perception that the state promotes their needs and interests, which Kelman calls an instrumental attachment.[16] A stable relationship between the state and the governed emerges when the latter feel some attachment and loyalty to the former and have, therefore, accepted its authority as legitimate.

Legitimacy is related to the concept of national consensus -- collective agreement on some general principles. One could view this national consensus as the extension of the individual perception of political legitimacy, for consensus involves the collective recognition of a government's legitimacy. The subjects of the state must collectively, through some democratic process and mechanism, consent to be governed. There must be a general consensus that a government is legitimate, or it cannot be effective. Consensus can, however, develop independently of political legitimacy, among a group of people without any government to channel it.

However, once legitimacy is established in a state, crises may still erupt if there is a change in the way governmental authority is conceived or a change in the way the government is supposed to act. If leadership alters the ideals it professes to represent and/or takes an unacceptable policy course, some people may become disenchanted. Generational change may also produce sharp changes in attitudes concerning the limits and potentialities of political action.

In the Sudan, consensus developed on two regional levels, North and South. Where in the South there is a widely shared distrust of Arab Northern Sudanese and in the Arab North a sense of superiority to the Southern Sudanese exists, no appreciable consensus existed between the Northern and Southern peoples (at least before the Addis agreement) on such issues as state coherence and legitimacy, national identity and relationships with the international community. The Arab Northern Sudanese-controlled state is not perceived as legitimate by the Southern Sudanese, who feel neither sentimental nor instrumental attachments to the Khartoum-controlled "state." Now, <u>if the identity of the state reflects the multinational or national identity, then it is certainly true that the "Sudanese state" has continuously since independence reflected the identity of the Arab Northern Sudanese</u>. All successive Khartoum governments failed to realize that cultural imperialism, as manifested in the policies of Arabization and Islamization pursued in the South and backed by brute military force, would unite the victims and solidify their resolve to fight for their right to self-determination. For, indeed, as Edward Shils has noted, "a society which inflicts the distress of a sense of exclusion and inferiority cannot wholly succeed in assimilating into its affirmative consensus those whom it wounds."[17] <u>The "Sudanese state" with its loci of identity and power in the North was systematically wounding the Southern Sudanese: it didn't recognize and respect their values, and worked against such values as well as against the material and security needs of the South</u>.

Various attempts at North-South reconciliation failed and the war continued for almost two decades. The civilian governments of the 1960s used the coercive instruments of the state to the maximum but with strictly negative results. In 1969, a group of young army officers, most of whom had served in the South, led by then Colonel Jaafar el Nimeiry, staged a successful coup and assumed the reins of power. The war in the South was one of the main postcoup rationalizations for the overthrow of the old, tired, inefficient, corrupt and conservative politicians.[18]

The Nimeiry regime acknowledged in its first month in office the basic differences between the North and the South, the Arab and Arabized Northern Sudanese and the black African Southern Sudanese, and announced a long-term commitment to a policy of regional autonomy for the Southern provinces within a united Sudan.[19] After three years of vacillation, Nimeiry's regime agreed to negotiate with the Anya-Nya armed forces under the auspices of an OAU member country. The peace talks were held in Addis Ababa, Ethiopia, in the early months of 1972. Both the North and the South abandoned their hardline positions and agreed to a regional autonomy scheme for the Southern Region.

Regional Autonomy and Territorial Unity

The Addis Ababa peace accord preserves the territorial unity of the Sudan but grants specific powers and functions to the Southern Region.[20] Legislation in the South is exercised by a People's Regional Assembly, members of which are elected by Sudanese citizens residing in the South through direct secret ballot. The members of the regional assembly in turn elect one of their colleagues speaker. Legislative authority in the South rests with the regional assembly and areas under its power for legislation include "preservation of public order, internal security, efficient administration and the development of the Southern Region in cultural, economic and social fields." Regional legislation must, however, conform with national plans. Functions which are exclusively a preserve of the central government and are, therefore, outside the jurisdiction of both the People's Regional Assembly and the High Executive Council (HEC) include national defense, external affairs, currency and coinage, air and interregional river transport, communications and telecommunications, customs and foreign trade except for border trade and certain commodities which the regional government may specify with the approval of the central government, nationality and immigration, planning for economic and social development, educational planning, and public audit.

Business is conducted in the regional assembly in accordance with rules of procedure which are laid down by the assembly during its first meeting. The regional assembly has the legal authority to elect the president of the regional government. Specifically, in Article 19 of the Addis Ababa agreement, it is stipulated that the president of the High Executive Council shall be "appointed and relieved of office by the President on the recommendation of the People's Regional Assembly." The regional president and members of the HEC are answerable to two authorities, the national president and the People's Regional Assembly. The HEC is composed of members appointed and relieved of office by the national president on the recommendation of the president of the HEC (Article 20). The regional president and members of the HEC "may attend meetings of the

People's Regional Assembly and participate in its deliberations without the right to vote, unless they are also members of the People's Regional Assembly." The central power of the state rests with the national president, who regulates the relationship between the HEC and the central ministries. However, the existence of the regional government in the South is guaranteed in the constitution and cannot legally be altered unilaterally by the central government. The organic law which guarantees and protects the regional government "cannot be amended except by a three-quarters majority of the People's National Assembly and confirmed by a two-thirds majority in referendum held in the three Southern Provinces of the Sudan."[21] This legal provision distinguishes the Southern regional government from a merely decentralized government whose local organs can be altered at will by either the national president or the national assembly.

The Addis Ababa Agreement stipulates that there should be a Southern army command to be composed of 12,000 officers and men, half of whom would originate in the South, the other half to come from elsewhere in the Sudan. Most of the troops from the South were drawn from the former Anya-Nya army. There was, however, a provision that there should be integration between the two halves; each unit of Southern soldiers would receive elements from the old Sudan government forces in the form of an officer, to act as second in command, and various technical and administrative personnel.

As I have written elsewhere,[22] the Addis Ababa Agreement appears to be a hybrid of federal and nonfederal features which are narrowly adapted to the specific circumstances of the Sudanese situation. On balance, it is more than a simple decentralized arrangement subject to the unilateral whim of the central government (the substantive guarantees of regional autonomy assume that) and yet less than a full-fledged federal structure (along the lines of the United States model) given the degree of control of internal affairs of the Southern Region by the national president. There is, therefore, the danger that if the national president interprets the document at will, the written guarantees may be rendered meaningless.

The successful conclusion of the peace negotiations between the North and the South and ending of the military confrontation brought a considerable relief in the Sudan, particularly in the Southern Region following the ratification of the agreement. In the short term, unity between the two disparate regions of the Sudan is assured; in the long term, relations between the Arab and Arabized Northern Sudanese and the African Southern Sudanese seem uncertain. A brief look at the life of the regional government and its relations with the center in the last ten years may give us a clue as to whether a process in North-South relations is taking place which will be difficult to reverse in any basic way in the years ahead.

Reconciliation and Continuing Tensions

During the seventeen years of war, the Northern Sudanese were divided into two camps on the Southern question. There were those who genuinely believed that the Sudan could not exist without the Southern Region, and there were those who suspected that the "nation" could not function properly and effectively with the Southern Region.[23]

The anti-Southern Region elements are mostly the pan-Arabists (the Muslim Brothers, Arab Socialists, and traditional sectarians -- the Ansar and the Khatimiyyah leaders). This group seeks closer links with the Arab world and closer identification with the problems and fantasies of the Arab Middle East; of course, they resent the African orientation they think the South Sudan wants to give the country as a whole. The other Northern group which looks unsympathetically on Southern concerns is the constitutionalist element. It argues that the granting of regional autonomy to the South will lead to "balkanization" of the Sudan. Related to this group is one claiming that the Northern Sudan is subsidizing eventual Southern secession by helping the South become a viable economic unit. And at the lowest level are those Northerners who believe that the Southerners are "just Africans" and would be better left to Africa.

Outside this anti-Southern camp, there is the other clique, which believes that the South and the North are both integral parts of the Sudan, and that attempts should be made to keep the South at peace with the North but without making the South strong enough to challenge the perpetual hegemony of the North at the central decision-making level. This last camp won the argument and is credited with bringing about the regional autonomy agreement. The various elements within the anti-Southern camp conceded, but remain skeptical of Southern intentions, and therefore have resorted to various manipulative strategies to render the regional autonomy arrangement ineffectual. They have created and contributed to some of the stress and strain in North-South relations since 1972 by sabotaging the effective institutionalization of autonomy in the South.

First, the overlapping of powers between the central and the regional governments through the operation of the People's Local Government system has created some difficulties. For example, the provincial commissioners (PCs) report directly to the national president, even though their work necessarily concerns matters that the Addis Ababa Agreement assigns to the region. Although the PCs are nominated by the president of the HEC and appointed by the national president, their reports sent directly to Khartoum are seen as potential threats to regional policy-making discretion. So far their selections have been politicized to ensure that they are not stooges of the central government. Another related issue of overlapping of powers is the public service in the South, which is supposed to be administered by the regional government. Yet the assistant commissioner of police for the Southern Region (the highest-ranking police officer) reports to the commissioner in Khartoum. Even though such reports are only routine and the security of the region is technically under the control of the president of the HEC, the position of the police chief in the South needs to be clarified within the Addis Ababa regional scheme.

Second, although Article 5 of the Addis Ababa peace agreement stipulates that English be the principal language for the Southern Region, and Arabic the official language of the Sudan, more emphasis has been put on the teaching of Arabic in schools -- at the expense, to be sure, of English and local languages. Related to the language issue is the resurgence of Islamic fanaticism and moves in the North to legislate Islam as the state religion in spite of explicit acceptance, in both the Addis Ababa Agreement and in the permanent constitution, that the Sudan will not project itself as an Islamic and Arabic state. Pursuit of Arabization and Islamization policies in the South have unfortunately surfaced, and the timidity of the Southern leadership is being interpreted by Muslim fanatics as acquiescence.

Third, the Southern Region is neither proportionally nor adequately represented at the national level. As of May 1982, the only Southerner who is a member of the central cabinet is Vice President Abel Alier. While Alier was also president of the HEC, he spent most of his time in the South managing the affairs of the region and occasionally attended cabinet meetings in Khartoum. Even if he had wanted to attend all of them, he could not have done so because, more often than not, there is no transport from Juba to Khartoum. Thus, at present, the North has exclusive monopoly and control of the critical decision-making process in the country. The South plays no role in the formulation and implementation of domestic and foreign policies.

Fourth, the central bureaucracy is entirely Northern Sudanese and so national planning and allocation of domestically generated revenue as well as foreign aid remains in the hands of the North, both at political and administrative levels. State security organs and diplomatic service also remain a preserve of the North. Southerners are recruited at an average rate of three each per year into the military and police academies, each of which takes about a hundred candidates annually.

Fifth, the Southern region has not been receiving a fair share of revenue and development project allocations. Moreover, money allocated for the South is seldom released on time by the Central Ministry of Finance. Sugar projects in Mongalla and Melut in the South were abandoned in preference for Aalaya, West Sennar and Kenana Sugar schemes in the North, and Tonj Kenaf and the Wau Fruit Canning factories were replaced by the Abu Naama Kenaf project and the Karima Fruit processing project in the Northern Region. The Kapoeta Cement Factory in Eastern Equatoria was abandoned when the central government built a series of cement factories in the North. The only development projects in the South are still at a pilot stage and are those sponsored and funded by the World Bank and the UN Development Program, whose respective respresentatives in the Sudan are constantly under pressure from leadership in the North to reduce their activities in the South.

The South remains at the periphery of central decision-making. A feeling of relative economic deprivation is growing, and the intransigent refusal of the Arab Sudanese to share real political power and revenue from within and aid from without with the Southern Sudanese will gradually erode any desires in the South to identify with the "Sudanese state."

Yet, despite these shortcomings in the implementation of the regional autonomy in the South, and the lack of sensitivity of the Northern leadership to Southern desire for participation in national issues, some Northerners claim that the Southern region has gained a privileged position in the Nimeiry regime and that decentralization should be extended to the North. Proponents for creation of regional governments in the North rested their arguments on the claims that the size of the country and its poor infrastructure militated against responsive and efficient administration from Khartoum -- uneven social and economic development and ethnic diversity required a decentralized system to satisfy local demands.[24] Above all, it was argued that a decentralized political system would ultimately achieve political aims and meet democratic expectations by involving the public at large, thus weakening elements of opposition. It is worth recalling here that such arguments for devolution of power -- regarded as dangerous to the national interest during the time of the war between the North and the South -- gained credibility among most Northern Sudanese elites.

A long constitutional procedure was followed by the central government before the idea of decentralization was implemented.[25] Five regions -- Northern, Eastern, Central, Kordofan and Darfur -- were created. The major difference between the regional autonomy in the South and the autonomy of the regions in the North lies in the fact that while the president of the High Executive Council is responsible to the Southern regional assembly and can be removed from office by a two-thirds no-confidence vote of the same assembly, the governors in the Northern region are presidential appointees.[26] Opponents of the Addis Ababa Agreement in the North were not happy with this important difference between the Southern Region and the other five regions. They immediately embarked on a series of attempts to provoke a confrontation between the central government and the Southern Region.

In late 1980, the Northern members of the national assembly connived with the central government ministers and a few of Nimeiry's advisers to redraw the boundaries between the South and the North. Major parts of the Gogrial district in Bahr el Ghazal Province and all the oil areas in the Bentiu district of Lakes Province were incorporated into Kordofan Province in the North, while parts of Renk district in Upper Nile Province were added to the adjacent Northern provinces. The national assembly, after a walk-out by the Southern members, passed legislation to validate the so-called new map.

The Southern reaction was instantaneous: there were mass demonstrations in the region against the new boundaries; the People's Regional Assembly met in an emergency session and passed a resolution that condemned the new boundaries and affirmed the Southern commitment to respect the provincial boundaries acknowledged at the time of independence and recognized by the Addis Ababa Agreement. The regional government in the South supported the Southern stand and argued that it was unconstitutional for the North, through its members in the national assembly, to alter the boundaries of the Northern and Southern regions. Furthermore, it was up to President Nimeiry either to uphold the constitution respecting the old boundaries and reverse the decision of the national assembly or to support the new boundaries and explain his decision to the peoples of the South.

Nimeiry was allegedly taken unawares by the actions of his ministers and of the Northern members of the national assembly. He acted swiftly to contain the growing tension between the South and the North: he appointed a committee composed of both Northerners and Southerners to examine the boundary issue and make recommendations to him. The committee recommended the reversal of the decision of the

Northern members of the national assembly and the president promptly accepted it.

Second, the Central Ministry of Energy and Mining decided without consultation with the South to direct Chevron Oil Company to commit itself to building an oil refinery in Kosti in the North instead of Bentiu in the South, where the company's prospecting had been most successful. The decision of the central government was unpopular in the South, and the regional government tried in vain to persuade Nimeiry to reverse the unilateral decision of the Central Ministry of Energy and Mining. Instead of responding to the appeals of the South to have the refinery in Bentiu for sound economic reasons as well as to make the South feel pride in contributing to the national economy, Nimeiry threatened military force to coerce the South to accept his decision. The Southern regional government avoided confrontation and indicated that it would respect the decision of the president.

Third, failing to provoke a confrontation between Nimeiry and the South, opponents of the autonomous status of the South began campaigning for creation of more regions in the South. The idea of dividing the South into regions (two or three) was first suggested by Northern political parties in the 1960s, but was rejected by the Southern parties. The North has always felt that treating the South as a single entity would eventually lead to development of a strong, cohesive and progressive region that would easily secede to form an independent African state. President Nimeiry had ignored that line of reasoning and accepted the Addis Ababa peace accord as negotiated by both his own selected team and representatives of the Southern Sudan Liberation Movement. However, the critics of the peace accord are now Nimeiry's political allies and hold key positions both in the central government and in the Sudan Socialist Union (SSU).[27] They have successfully persuaded Nimeiry that redivision of the South and creation of regions out of the present region is necessary for perpetual control of the South by the North. They cited the South's resistance to the boundary question and location of the oil refinery in Kosti as indicators of growing Southern intransigence to decisions emanating from the central government. To add political ballast to their argument, they persuaded a Southerner, General Joseph Lagu, who had been living in Khartoum since Nimeiry removed him from the presidency of the South, to spearhead the call for redivision of the South.[28] Nimeiry assured General Lagu that he fully supported the idea of creating several regional governments in the South which would have the same powers as those in the North. With financial support and administrative facilities at his disposal, Lagu called for redivision of the Southern region on grounds that it would save the South from institutionalized tribal domination by the Dinka, protect the Addis Ababa Agreement from politicians with tribal inclinations, effect decentralization of power in the South, and guarantee the concept of a united Southern Sudan by accommodating the aspirations of the various communal groups in the South.[29]

Lagu's call for creation of several regions in the South was received with shock and bewilderment. An overwhelming majority of the Southern members of the central committee of the SSU cabled President Nimeiry requesting that he exclude the issue from discussion in a forthcoming meeting on the grounds that it was unconstitutional.[30] Nimeiry, ignoring the public outcry, included the issue on the agenda of the central committee meeting, where it was sweepingly condemned. Realizing that the South was opposed to the idea, Nimeiry referred the issue to the Southern units of the SSU for further discussion.

Meanwhile, attempts were made to amend the permanent constitution to enable the president to create more regions in the South. The regional government maintained that to amend the permanent constitution, Nimeiry would have to comply with the provisions of Article 34 of the Self-Government Act 1972, which specifically protects the South as a single region and stipulates that amendment can be made only by a three-quarters majority of the People's National Assembly and the approval by a two-thirds majority of the citizens of the Southern region in a referendum to be carried out in that region. Nimeiry's legal advisers counseled adherence to the constitutional procedure.[31] Proponents of redivision, however, began to urge President Nimeiry to overlook the legal restrictions on his powers and to order the creation of more Southern regions on grounds of public interest. Once again, the Southern regional government appealed to Nimeiry to protect and respect the permanent constitution of the country. After failing to manuever the Southern leadership on the sensitive issue of redivision of the South, Nimeiry summoned an emergency meeting of the SSU Politbureau to discuss the subject. For three consecutive days, the Southern members won the overwhelming support of their Northern colleagues against redivision of the South.[32] It was bad economics, the Southerners persuasively argued, because the central government was not in a financial position to maintain additional regional bureaucracies, and new regional governments in the South could not sustain themselves. It was bad politics, they maintained, because proponents of redivision were appealing to local sentiments among rival ethnic groups, and such politics would be detrimental to the health of the Southern region and of the Sudan as a whole. The critics of redivision pointed out that implementation of decentralization programs (which had already been announced) within the Southern region would accommodate the aspirations of the local communities and allay their fears of domination of larger ethnic groups.[33] The overwhelming majority of the members of the SSU Politbureau appealed to President Nimeiry to abandon the idea of redivision completely. Nimeiry was reported to have assured his colleagues that he would appoint a committee to examine the issue in detail and report back to the SSU Politbureau.

However, ten days later, Nimeiry dissolved both the national and regional assemblies as well as the regional government in the South. His rationalization was specious: the establishment of five regional governments in the North required a change of regional representation in the national assembly to reflect the new balance of power relationship between the regions and the center. Regarding the dissolution of the regional assembly, he argued that the issue of redivision of the South required public debate and that the Southern masses should express their opinion through a referendum if the newly constituted national assembly approved the proposition for redivision. Of course, there were no grounds at all for Nimeiry to dissolve the regional government. By acting unconstitutionally, he lost Southern trust and confidence overnight. The South, however, decided to avoid a confrontation with Nimeiry, and pledged to follow the constitutional process to express its position on the diversionary issue of redivision.

After dissolving the assemblies and the Southern regional government, Nimeiry appointed an interim military government in the South to supervise elections for a new national assembly and a referendum on the proposition of redivision of the South. All the nominees for the interim High Executive Council headed by Major General Gasmallah Rassas were proposed by Lagu and were mostly from among his supporters for creation of several regions in the South. Nimeiry also reduced the South representation in the national assembly to less than one-fifth of the total membership (twenty-eight out of 151) to ensure passage of the redivision issue. These actions reinforced the impression that Nimeiry was set to create more regions in the South regardless of the overwhelming Southern opposition. In order to mobilize effective defeat of the idea of redivision in the forthcoming referendum, its opponents formed the Council for the Unity of South Sudan in December 1981 under the chairmanship of the Sudanese elder statesman, Clement Mboro, and composed of people of different political persuasions. Its executive committee sent a letter to Nimeiry informing him of the formation of the council and assured him that the council would work within the organs of the SSU -- after all, all of its members were prominent SSU members. Nimeiry reacted swiftly by ordering the arrest of the entire executive of the council (twenty-one persons). Subsequent pressure and protest led to the release of all those arrested but one: Samuel Aru Bol still remained behind, languishing in detention. Surprisingly, the South remained restrained in spite of all these provocations.

However, Vice President Alier and a number of other influential Southerners who were opposed to the idea of redivision launched an effective quiet diplomacy with President Nimeiry to avert a seemingly explosive situation in the South. Their appeals to Nimeiry bore positive results when Nimeiry's foreign friends and a group of individual Northern Sudanese who feared the breakout of another civil war mounted pressure on him to resolve his self-inflicted crisis with the South by abandoning the issue of redivision. In mid-February of 1982, Nimeiry announced to the newly elected members of the People's National Assembly that the issue of referendum in the South on redivision was withdrawn from discussion and that Southerners should reconcile their differences. The announcements stunned the proponents of redivision both in the North and South. But by the stroke of a pen, Nimeiry reversed himself and moved swiftly to placate the majority concerns of most Southerners. Once again, he demonstrated his adeptness as a political chameleon. For the present, he has shifted his concentration to his opponents in the North.

Conclusions

Postcolonial commentaries on Africa have shrouded the political dimensions of self-determination in theories of political integration that reify existing state boundaries. Hitherto, a unique situation such as that of the Sudan -- where two collectivities, one rooted in Africanism, the other in Arabism, find themselves lumped together under one territorial state as a result of both colonial miscalculations and lack of foresight -- has been denied a special treatment that it so deserves in the literature on state and nation building. The contribution in this chapter is to help close that gap by focusing on issues of geoethnicity and state coherence with particular reference to the Sudan.

Applying the concept of geoethnicity to the Sudan, we find that the differences existing within the North and the South are not as fundamental as the schism between the two regions. The Arab and Arabized Sudanese in the North identify with Arab culture and the larger Arab nation, whereas the Nilotic and Sudanic peoples of the South have a sense of identity rooted in African authenticity -- culture, religion, cosmology, etc. Deeply rooted historical antagonisms of the two groups led the Anglo-Egyptian colonial regime to establish two separate administrations reflecting their cultural concentrations and locations. However, once the imperial referee withdrew, and the Northern Arab group attempted to impose its hegemony over the Southern Sudanese, war erupted. Refusing to be integrated into the cultural, political and administrative fabric of the Arab North, the South retained its distinctiveness. It did so, though, at the expense of political participation on the national level. As far as the North is concerned, all conceivable means ranging from brute military force to subtle political manipulation and control must be used to bring the South and its "infidels" into the Arab and Islamic fold. For the Southerners, the postcolonial state, controlled by the Arab North for a quarter of a century, has become a symbol of tyranny and oppression. It has failed to offer psychic orientation and security or to bring material improvement to Southern people. The conflict produced a military stalemate between the two contestants, and subsequently a peace agreement was reached in Addis Ababa under the auspices of the late Emperor Haile Selassie.

The Addis Ababa Agreement granted the South autonomous status within the Sudanese territorial state. It is essentially a political settlement and power arrangement between the North and the South, and it may or may not succeed in facilitating the development of a new set of institutions. It provides a framework for conflict regulation between the Southern and Northern Sudanese. Unfortunately, since the establishment of an autonomous regional government in the South, various attempts have been made by Northern bureaucrats and politicians to subvert the institutionalization of the peace accord. Recently, Nimeiry has been persuaded by its opponents to abandon the accord to create several regions in the South and to give these regions the structure, power and authority of regions in the North. Such an arrangement would eliminate further consolidation of the Addis Ababa peace agreement, render the South powerless, and place it under the indefinite control and at the mercy of the Northern-dominated central government. It is inconceivable that the South would submit to such a drastic curtailment of its present margin of autonomy.

What then is the future of geoethnic tensions between the North and the South? And what will come of the Sudanese state (controlled by the North) in pursuit of absolute authority? The present tension between the two regions prompted by President Nimeiry's unconstitutional action in dissolving both the regional assembly and the government does not provide an optimistic scenario. Suffice it to indicate here the kleptocratic aspects of Nimeiry's regime and its vulnerability as a result of its failure to institutionalize state power and its relations with society.

Justification of power in the Sudan under Nimeiry has been actively pursued through the state appartus, and through the SSU. Channels of propaganda have been developed to win the allegiance and participation of a small minority. Thus, the emergence of an elitist, professional administration and the widening of the economic gap between the state and commercial bourgeoisie and the mass of the population have spurred the evolution of a well-defined context for power relations centering upon Nimeiry. The rest of the population has been either intimidated by occasional shows of force or mystified by grandiose visions of a prosperous future and of Nimeiry's presumed "enlightened leadership."

Presidential power has relied more on the valorization of authoritarian principles of rule than on the embodiment of impersonal principles in an institutionalized context for power relations. Communication, participation and responsiveness revolve around the office of the president, and are organized in a strict hierarchical pattern. Information is disseminated at increasing levels of generality as one descends the political and social ladders, and payoff varies proportionally to the quality of the individual's relationship with Nimeiry and his immediate entourage.

If political development and legitimization lie in the creation of institutions transcending the temporally limited authority of individual politicians, it seems that the justification of the regime's power does not rest on a solid consensual foundation. Even though Nimeiry's position is so central and dominant that it creates conformity in political behavior and continuity in the pattern of power relations, the political system he has developed cannot stand on its own, precisely because the hierarchy is not committed to the organization of its roles within a specific institutional framework, but to a single leader who can define to a large extent permissible forms of political interaction.

It is doubtful that the Sudan Socialist Union could routinely handle the problem of succession or even survive without the central bond provided by the presidential state. Similarly, competition for the economic benefits of political power would be unbridled were Nimeiry to disappear, since there are no solid national centers of economic policy-making. A new, greedy, corrupt and very mediocre group has emerged which, through the mediation of Nimeiry, has been able to amass wealth and to collaborate with foreign businessmen. Without Nimeiry's extravagant use of patronage and the coordination of parochial interests through his network of protégés and informers in the national, and supporters in the foreign spheres, his regime would probably have collapsed by now. So far, no amount of SSU propaganda regarding the need to develop a national consciousness or rational progress toward the achievement of collective goals can deeply touch the various levels of the political hierarchy when modes of participation related to these ideas have not been established. Indeed, the life of the SSU depends on that of Nimeiry, for the political system which he has built has constantly sacrificed institutionalization for the maintenance of control over real or imagined opponents who pose personal threats to the president.

A political system which is held together by the authority of an individual presidential monarch, and by the mechanisms of consent offered by payoff, will inevitably be vulnerable to variations in the nature of foreign financial support and international political and economic trends. This weakness has definitely been manifested in the Sudan partly as a result of Nimeiry's zigzag foreign policy, and partly as a result of foreign actors shifting and readjusting their interests and strategies accordingly. Major challenges to the regime bring with them an increased need for coercive security measures and further imperil the possibilities of developing a broad consensual context for the definition of "state coherence" and the exercise of power. Creative change must be continually postponed because of the need to reconsolidate power and authority. Henceforth, the relationship between the state and the governed remains one between domination or, alternatively, submission, acquiescence or rebellion.

Footnotes

The introductory section is adapted from my article on "Sources of Communal Conflicts and Secessionist Politics in Africa," Ethnic and Racial Studies, 1, 3 (July 1978).

1. Rupert Emerson, Self-Determination Revisited in the Era of Decolonization, Occasional Paper in International Affairs, Center for International Affairs, (Cambridge: Harvard University, 1964), pp. 25-26. For other views, see S.P. Panter-Brick, "The Right to Self-Determination: Its Application to Nigeria," International Affairs (London), 44, 2 (April 1968): 254-266; Jack Forbes, "Do Tribes Have Rights? The Question of Self-Determination for Small Nations," Journal of Human Relations (Wilberforce), 18, 1 (1970): 690-699; Onyconoro S. Kamanu, "Secession and the Right of Self-Determination: An O.A.U. Dilemma," Journal of Modern African Studies, 12, 3 (1974): 355-376; Charles R. Nixon, "Self-Determination: The Nigeria/Biafra Case," World Politics, 24, 4 (1972): 473-497.
2. Emerson, Self-Determination Revisited, p. 26.
3. Alfred Cobban, The Nation State and National Self-Determination (New York: Thomas Y. Crowell, 1966), pp. 105-106.
4. Ibid.
5. For a detailed analysis of the war and its temporary conclusion in the Sudan, see Dunstan M. Wai, The African-Arab Conflict in the Sudan (New York: Africana Publishing House, Holmes and Meier, 1978).
6. I am deeply indebted to Professor James S. Coleman for suggesting this definition of geoethnicity. For a detailed application of the concept of geoethnicity, see Howard Wolpe, Urban Politics in Nigeria: A Study of Port Harcourt (Berkeley and Los Angeles: University of California Press, 1974).

7. Pierre L. van den Berghe, "Ethnicity: The African Experience," International Social Science Journal, 23, 4 (April 1971): 507-518.
8. See Robert O. Collins, "The Sudan: Link to the North," in Fred G. Burke and Stanley Diamond, eds., The Transformation of East Africa (New York: Basic Books, Inc., 1966), pp. 359-406; and George W. Shepherd, Jr., "National Integration and the Southern Sudan," Journal of Modern African Studies, 4, 2 (October 1966): 193-212.
9. See Robert O. Collins, The Southern Sudan in Historical Perspective (Tel Aviv: University of Tel Aviv, The Israel Press, 1975); and Dunstan M. Wai, The African-Arab Conflict in the Sudan, especially chapters II and III.
10. Report of the Commission of Inquiry into the Disturbances in the Southern Sudan During August 1955 (Khartoum: McCorquedale and Co., Ltd., 1956).
11. For instance, see President Nimeiry's Policy Statement on the Southern Question, 9th June 1969 (Khartoum: Ministry of Information, 1969).
12. Of special relevance on the Nigerian-Biafran conflict, see Anthony Kirk-Greene, Crisis and Conflict in Nigeria, 2 volumes (London: Oxford University Press, 1971); John J. Stremlau, The International Politics of the Nigerian Civil War, 1967-1970 (Princeton: Princeton University Press, 1977); Victor A. Olorunsola, The Politics of Cultural Sub-Nationalism in Africa (Garden City, N.Y.: Doubleday, 1972); M.V. Akpan, The Struggle for Secession 1966-1970 (London: Frank Cass, 1971).
13. Crawford Young, The Politics of Cultural Pluralism (Madison: University of Wisconsin Press, 1976), p. 473.
14. Ibid., p. 503.
15. Herbert C. Kelman, "Patterns of Personal Involvement in the National System: A Social-Psychological Analysis of Political Legitimacy," in J.M. Rosenau, ed., International Politics and Foreign Policy (New York: Free Press, 1969), pp. 276-288.
16. Ibid.
17. Edward Shils, Center and Periphery: Essays in Macro Sociology (Chicago: University Press), pp. 164-182.
18. Dunstan M. Wai, "Revolution, Rhetoric and Reality in the Sudan," Journal of Modern African Studies, 17, 1 (1979): 71-93.
19. See Nimeiry's Policy Statement on the Southern Question, 9th June 1969.
20. See Dunstan M. Wai, ed., The Southern Sudan and the Problem of National Integration (London: Frank Cass, 1973), appendix VII.
21. Ibid., appendix VII, p. 225.
22. Wai, African-Arab Conflict in the Sudan, p. 159.
23. This section of the chapter was reproduced in my article on "Crisis in North-South Relations," Africa Report, 27, 2 (March-April, 1982): 20-26.
24. The advocates of decentralization were mainly from Western Sudan and felt that Khartoum, Northern and Blue Nile provinces were always favored by the ruling elites because most of them invariably come from these areas.

25. The North was divided into regions by amending Article 218 of the Permanent Constitution of Sudan by two-thirds majority of the People's National Assembly with the president of the Republic assenting to this decision. The recommendation and resolutions of the National Congress of the SSU were preliminary political measures to pave the way for the constitutionality and legality of this division.
26. The structure, power and functions of the regional governments in relation to the center have been aptly described:

> At the central level, the Presidential system of government would be retained with the President as head of the executive but also enjoying some legislative powers. Matters pertaining to the sovereignty of the state (such as defence, foreign policy, currency, customs, foreign trade, and national development schemes) would be exercised by the Central Government which would also reserve the right to veto regional legislation and to control the national purse. Regional administrations would consist of a Governor, a People's Regional Assembly, and a Regional Council of Ministers. The Governor would be appointed by the President from a list of three persons nominated by the Regional Assembly. The Council of Ministers would consist of seven members appointed by the President on recommendation of the Regional Governor. The elected People's Regional Assembly would have a membership of not less than 50 and not more than 70, depending on the density of the population in each region. At the provincial level, the Provincial Commissioner -- appointed by the President on the recommendation of the Regional Governor -- would enjoy Ministerial status and become the representative of the Regional Government, the chairman of the People's Province Executive Council (PPEC), and the head of the local government service. At the district level, members of the District Council who elect the PPEC members, would themselves be elected by rural and urban councils. Functions and powers would be delegated to them by the PPEC to which they would be directly responsible. The Base Councils, directly elected by the local people, would form the base of a pyramid of authorities, the apex of which would be the PPEC.

See Colin Legum, ed., _African Contemporary Record, 1979-1980_ (New York: Africana Publishing Co., 1980), p. B106.
27. Most of the critics of the Addis Ababa Agreement are members of the Islamic Front (Moslem fundamentalists) and remnants of the Umma and Democratic Unionist parties.
28. Nimeiry dismissed General Lagu in early 1980 on grounds that Lagu was unable to govern in the South with the cooperation of the regional assembly, and that he was corrupt. The move was unconstitutional and set a bad precedent.
29. This is the line of argument General Lagu developed in his celebrated document, _Decentralization: A Necessity for the Southern Provinces of the Sudan_ (Khartoum: Gamar P. Press, 1981).

30. *The Nile Mirror*, March 14, 1981.
31. President Nimeiry was, however, prevailed upon by his attorney general (a Moslem fundamentalist) and others, against the advice of legal advisers in his office.
32. The meeting took place between September 15-17, 1981.
33. See Solidarity Committee of the Southern Members of the 4th People's National Assembly, *The Redivision of the Southern Region, Why It Must Be Rejected* (Omdurman, 1981).

16
The Ogaadeen Question and Changes in Somali Identity

David Laitin

The problem of Somali nationalism within the Ethiopian state appears to be intractable. The Ethiopian state, armed with an ideology of the legitimacy of the current world order, and backed up by the principle in the OAU upholding the territorial integrity of member states, refuses to entertain demands from nationality groups within its boundaries for autonomy. The Somali nation, armed with an ideology of the legitimacy of national self-determination, and backed by Wilsonian principles, refuses to abandon the cause of the Western Somali Liberation Front. Given the intensity of feelings on this issue, it is not at all surprising that bargaining among parties has been notoriously unsuccessful. Nonetheless, the many attempts to bring about a negotiated settlement can be instructive to policy analysts who wish to understand the bases for failures in negotiation. Therefore, this chapter will first review a variety of attempts to foster a bargained outcome, and will then assess the reasons why such an outcome has as yet been unattainable.

Failures to bring about a negotiated settlement notwithstanding, the situation has not remained and will not remain static. States and nationalities are dynamic institutions, and as they change, so will the possibilities for a bargained solution. Part II of this chapter will examine (from a comparative perspective) the dynamics of multinational states to see what this means for changes within modern Ethiopia. Part III will speculate about changes in the identities of the national groups living in Africa's Horn. Parts II and III conjoined will provide a dialectical analysis of state coherence and national identity. It is suggested in the conclusion that to be successful, a bargaining strategy must be attuned to the constantly changing environment created by this dialectic.

Bargaining

To a considerable extent, the "problem" of the Ogaadeen was caused by diplomatic mismanagement. While this is not the place to give a history of treaty-making and treaty-breaking among the British, the Italians, the Ethiopians and the Somalis in the late nineteenth century,[1] suffice it to say that the great powers, all seeking to win favor in the Ethiopian emperor's court, were willing to grant him much of his territorial claims in Africa's Horn, even if they had no basis in reality. The British found themselves trapped by their treaty of 1897 in their future dealings with the Somalis because they had accepted Menelik's definition of his proper jurisdiction. Until Haile Selassie's return from exile during the Second World War, there was virtually no attempt to integrate the Somali areas into the imperial system. No land was alienated; nor were any infrastructural investments made. In fact, it was not until the end of the war that Somalis were fully cognizant of the British concessions to Menelik. They were outraged.

It was in this early postwar period that great power diplomacy missed a golden opportunity to "solve" the problem of crucial segments of the Somali nation living outside the boundaries of the future Somali state. The British were in control of virtually all the Somali lands (except only Djibouti), and Lord Bevin, the British foreign secretary, attempted to refashion the boundaries of Africa's Horn so that the Somali state would encompass the Somali nation. But the Ethiopians (who were in a morally impeccable position -- having been conquered by the Fascists) were aghast, the Russians suspicious (that the UK was attempting to expand her empire), and the Italians crushed (that they would lose status as a European power if they had no territory in Africa). The Bevin plan was therefore stillborn.

It was not until 1960 that there was a Somali state to fight for the interests of the Somali nation. This begins the era in which Somali diplomats worked in a variety of forums to rectify the damage they believe was caused by nineteenth- and mid-twentieth-century diplomacy. Somali diplomats worked in the context of the OAU, with the great powers, and directly with Ethiopia. What, then, is the diplomatic record since 1960, and what does that record suggest for a negotiated solution to the Ogaadeen problem? Let us first look at diplomacy within the OAU.

<u>The OAU</u>.[2] Although the Pan African Congress in 1945 strongly criticized the boundaries in Africa, the Afro-Asian People's Solidarity Conference in 1957 supported the Somali cause, once African colonies became African states, and clearly noticeable at the All-African People's Conference in Cairo in 1961, the view of the colonial boundaries began to change. The first Somali president, Adan Cabdille Cismaan, visited Kwame Nkrumah some months later, and they issued a joint communique advocating that African states should "restore...cultural...links arbitrarily destroyed by colonialism."[3] But in all their conferences from the 1958 conference in Accra to the 1963 Addis Ababa OAU founding conference, respect for sovereignty and territorial integrity of each state was reiterated. It was made one of the cardinal principles of the OAU.

While it is true that another cardinal principle of the OAU was the "absolute dedication to the total emancipation of the African territories which are still dependent" (Article III, paragraph 6), the OAU in 1963 agreed (and subsequently reaffirmed in its declaration on borders in 1964) that self-determination was applicable only to dependent territories and not to ethnic or national groups within each state. To have accepted a principle of self-determination for nationality groups within states would have been, in the view of nearly all OAU members, to open a Pandora's box of separatist claims. The fact that OAU principles were so much in favor of Ethiopian appeals is crucial to an understanding of OAU's unsuccessful role in peaceful settlement.

In 1963 at the inauguration conference of the OAU, and in two important OAU conferences in 1964 (the Emergency Council of Ministers held in Dar es Salaam and the Ordinary Council of Ministers meetings in Lagos), all the resolutions went against Somali interests. Article III, paragraph 3 (respect for territorial integrity), paragraph 4 (peaceful settlement) and Article VI (noninterference in internal affairs) were all pointed to in order to demonstrate to the Somalis that the _status quo ante_ was legitimate. Ethiopia's respected position in the OAU, the relative strength of Haile Selassie's army, and the fact that Somalia could get no diplomatic or military support anywhere in the world for its war aims all help explain why the Somalis had to accept an OAU ceasefire after border hostilities broke out in 1964. But the Somalis still would not give up hope. After the OAU unequivocally rejected Somali claims in its declaration on boundaries at the OAU summit in Cairo, the Somali National Assembly unanimously passed a resolution declaring that the Cairo resolution "should not bind the Somali Government."[4]

From 1965 through 1967 there were sporadic border conflicts, but no large-scale military confrontations. When in 1967 Ibraahiim Cigaal became prime minister, a subtle reorientation of Somali foreign policy ensued. Convinced by Western diplomacy that Somalia had no chance against the Ethiopian army, Cigaal charted a policy of detente. The period of detente outlasted the fall of Cigaal to the military regime headed by Maxamad Siyaad Barre in 1969. For the first few years of military rule, the Supreme Revolutionary Council of the Somali Democratic Republic emphasized the building of socialism and counseled patience and peace in regard to the Ogaadeen. All the while, of course, the Somalis began to build up a military establishment with Soviet aid which could rival any in Africa.

In 1973, in line with this commitment to "reunify" the Somali lands, President Siyaad and Foreign Minister Cumar Arteh took on a diplomatic offensive. In the May 1973 OAU Council of Ministers meeting, Somalia tried to place "its territorial dispute with Ethiopia" on the agenda. The Ethiopians objected, as the Somali position, they claimed, was "contrary to the OAU charter."[5] There was a deadlock as Somalia had the support of the North African states while Ethiopia had the support of all sub-Saharan Africa. To forestall a bitter confrontation, Nigerian Foreign Minister Arikpo engineered a five-man commission which he would chair, but when the commission was preparing to put the dispute before the Heads of State summit, the Somali foreign minister, reading the mood of the delegates, determined that his country did not plan to bring up the dispute again before the heads of state.

During the summit, an eight-nation "Good Offices Committee" was set up. The committee empowered President Nimeiry of Sudan and Gowon of Nigeria to explore new approaches to the problem. Gowon's plan, modeled on the Algeria-Morocco border pact, called for the joint development of the Ogaadeen in the context of Ethiopian sovereignty. This proposal could not satisfy the Somalis. In 1974, when Siyaad was chairman of the OAU Heads of State meeting, he tried again to convince his colleagues of the justice of the Somali cause. Selassie, whose position in Addis was becoming tenuous, at first remained in Ethiopia. Although Gowon was able to persuade Selassie to attend, he could report no progress on this peace proposal. He said that only the two countries would be able to find a satisfactory solution.[6]

Selassie's imperial rule was indeed in peril. A military group calling itself the Dergue soon began to emasculate the emperor's role, and ultimately deposed him in the name of socialist revolution. In the early years of the revolution, there were some signs that the Dergue would be open to the various cries for regional autonomy in the non-Amhara areas, but under President Mengistu Haile Mariam, the opponents of "imperial domination" were considered enemies of the state. Mengistu committed his country to the territorial integrity of the Ethiopian state at all costs. Although military conflict immediately escalated among the Eritreans and Oromos, the Somali border was surprisingly quiet. First, Siyaad was counseling restraint in the hope for a political solution and second, his Soviet advisers, who had considerable control over materiel, were strongly against Somali territorial goals.

But the army and forces within Somalia -- especially the Western Somali Liberation Front -- were already mobilizing. By 1977 regular Somali troops marched into the Ogaadeen to aid the WSLF. Early substantial victories by Somali forces in August forced Ethiopia to seek external aid from the OAU.[7] In response, President Omar Bongo of Gabon, then chairman of the OAU, heroically but unsuccessfully attempted to resuscitate the Good Offices Committee; this was something less than the Emergency Council of the OAU Ministers which Ethiopia requested to consider the "Somali invasion." The OAU failed because at that time Somali victories in the field led them to snub any offer for conciliation.

The military situation soon reversed itself. The Soviets refused to help the Somali war effort, and soon responded to Mengistu's pleas for external aid by putting the full weight of their military machine and Cuban soldiers behind the Ethiopians. Russian aid decisively altered the military situation, and the Somali troops, as well as countless civilians who lived in the Ogaadeen, retreated into the Somali Republic. In 1978, the OAU peace plan which provided for a six-mile-wide demilitarized zone along the border and an end to Ethiopian air attacks on Somalis was categorically rejected by the Ethiopians. They claimed that "it took no account of the guerilla war in the Ogaadeen waged by Somali regular troops dressed as nomads."[8] They also refused to allow the WSLF to participate in any formal discussions. Whatever their reason, the fact was that in a militarily comfortable position, they had no incentive to negotiate.

The war goes on, but now the WSLF stands by itself in the Ogaadeen without direct support from the Somali regular forces. The OAU has made many attempts to bring about a solution to this problem, without success, but the prospects are for continued war, the continued presence of Russian and Cuban forces in Ethiopia, and the gradual expansion of the American military establishment in Somalia. The prospects are dim for the fulfillment of OAU principles in the Horn. For one, the OAU's charter gives it a partiality toward the Ethiopian definition of the issue, and the Somalis do not see it able to perform its role as an emancipator of colonized peoples in regard to the Ogaadeen. Second, the OAU lacks coercive authority, making it unable to act authoritatively on either side. Finally, it lacks the organizational requisites for continuous attention to new opportunities for peaceful settlement.

The Great Powers. Somali diplomats have not, of course, relied only on the OAU to seek redress. They courted the great powers as well. While Somali diplomats toyed with their Soviet and American counterparts, the Somalis garnered a very successful foreign aid package from both superpowers.[9] And both superpowers saw the strategic possibilities of creating regional peace. In the early 1960s, even though the Soviets had been responsible for the growth and professionalization of the Somali army, the Americans were the dominant superpower in the region. They had a twenty-five year agreement with the Ethiopians signed in 1953, close relations with Kenya, and good relations with the civilian leaders in Somalia. The American diplomatic team promised the Somalis reasonably high economic aid in exchange for Somali support for regional detente. The Somalis could save face by arguing that all Somalis ultimately would be reunited in a wider East African federation. So the Somalis applied to join the East African Community, accepted foreign aid from the Americans, and withdrew support from Somali nationalists in the Ogaadeen and in Kenya's Northern Frontier District.[10] Most Somalis interpreted their government's concessions as betrayal, and in retrospect, this discontent represented the beginning of the end for the civilian government. Detente American style provided no real concessions to Somali nationalist aspirations.

The big power volte face on the Horn provided an opportunity for a pax Sovietica. The collapse of the civilian regime in Somalia led to the rise of the Soviet-supported military under Siyaad. Siyaad adopted a "scientifically socialist" ideology and developed close relations with the Soviet government. By 1974, he signed a Treaty of Friendship and Cooperation with the Soviets, the first African leader to have done so. Shortly thereafter, Selassie, America's ally in Ethiopia, was deposed by a creeping military coup, with socialist leaders. The United States already considered its communications facility at Kagnew to be defunct, and, with egregious human rights violations occurring in Ethiopia (a revolution, as they say, is no dinner party), America was reluctant to continue giving support to the Ethiopian Dergue. The Soviets found little trouble in supporting the Ethiopian revolution.

That of course presented a dilemma for the Soviets. They had equipped the Somali army sufficiently well such that it could make a reasonable effort at capturing the Ogaadeen. By 1975 Somalia had fifty-two combat aircraft and 250 tanks, while Ethiopia had thirty-seven aircraft and sixty-two tanks. Even the Ethiopian manpower advantage (44,500 to 23,000) was considered "illusory" because Ethiopian troops were already fighting in Eritrea.[11] So the Soviets began to support the Ethiopians, and in the spring of 1977, Colonel Mengistu, on an army shopping trip to Moscow, reached a number of agreements that ultimately led to a Treaty of Friendship and Cooperation in 1978.

The Soviets were not interested in becoming the patrons of both sides in a nasty war. They tried to create a "loose socialist federation under Soviet patronage, which would allow the Ogaadeen Somalis an unspecified measure of local autonomy."[12] Both Fidel Castro and President Nikolai Podgorny attempted to unify Somalia, Ethiopia, Djibouti and South Yemen. But the proposed pax Sovietica under an ideology of socialism had no more legitimacy in the eyes of the regional actors than the pax Americana under an ideology of a regional common market. Unlike the OAU, the superpowers have been able to supply vast amounts of weapons to the Horn; but like the OAU, they have proven themselves incapable of fashioning peace.

Other Parties. There have been other diplomatic initiatives, to be sure, besides those of the OAU and the great powers. Individual African and Asian states, and many private groups (from Yale University professors to Mennonite missionaries), have attempted to find the formula to solve the difficult problem of the Horn. Middle level powers, especially Italy and Great Britain, the two former metropoles, also have attempted to bring about a settlement. More recently the French, in collaboration with President Hassan Gouled of Djibouti (who is a Somali, and appears committed to the separate identity of Djibouti even though a majority of the people there are Somalis) have supported a general settlement under a rubric of a regional economic commission.[13] Whether any West European state, or a neutral African state, has the resources and the skills to pay off each of the parties in the late negotiation and early enforcement stages, as the great powers certainly are able, is an open question. Perhaps the most interesting diplomatic initiatives are those which have occurred directly between the Ethiopians and Somalis. Throughout the past twenty years there have been meetings at the highest levels between Somali and Ethiopian authorities. This is an as yet unstudied but vastly interesting political process. Nonetheless, like all other diplomatic initiatives, these meetings too have ended in failure.

To move from history to the dynamics of diplomacy, it seems that there is no discernible diplomatic solution, in part because no third party has sufficient power or authority to force concessions from both sides. The OAU suffers, at least from the Somali point of view, from "frontier fetishism,"[14] and it hasn't the diplomatic tradition to define a solution nor the military power to enforce one. The Soviet Union has lost its historical opportunity to create regional peace in large part because its diplomats have been insufficiently sensitive to the strong force of nationalism. The Soviets have been discredited in Somalia for at least a few years, and one could guess that their close relationship with the Ethiopians is also tenuous. They are therefore in no position to create a pax Sovietica.

As for the Americans, it is not impossible to think of a joint Anglo-U.S. negotiating position not dissimilar to the structure created for the Zimbabwe issue. My sense is that the time may be right in the Horn for such an effort, but that the present regime in Washington is more desirous of confronting the Soviet Union than of creating the conditions for regional peace. I do not see an incentive for the Americans to play such a role in the Horn today; they would prefer to arm the Somalis to stand up to the Russians.

Constructive bargaining between the Ethiopian state and the Somali nation is not, therefore, likely. But it would be politically naive to end analysis here, because both states and nations are subject to dynamic change. What might such change mean for the future of ethnic conflict in the Ogaadeen? This question will be explored in the two sections which follow.

Dynamics of Change in Multinational States

The Ethiopians recognize that theirs is a multinational state. They could point out, however, that in the early periods of state formation, so was "France" or "The United Kingdom." In both these nation-states, it would be an egregious anachronism to suggest that their boundaries were determined by the ethnic character of the population. What in fact happened in both France and the United Kingdom was that boundaries were based on military, dynastic and political realities, and monarchs fashioned a "nation" out of the congeries of people who lived within those boundaries.[15] Ethiopia need be no different; its emperors have acted in ways similar to French and English monarchs as they attempted to create states in the fifteenth and sixteenth centuries. The future of Ethiopia, from this point of view, is secure, and what looks like a multinational empire in the late twentieth century may, in the next century, be considered a "natural" nation-state.

To the Somalis, however, the Ethiopian empire of today is, like the Ottoman Empire in nineteenth century Europe, the "sick man" of Africa.[16] Just like the Serbs, the Greeks, the Rumanians, and the Bulgarians, whose national aspirations were fulfilled amid the demise of the Ottoman Empire, today the Oromos, the Tigres, the Eritreans, and the Somalis have only to wait for the "historically inevitable" collapse of the Ethiopian empire. Just like the Ottoman Empire in the early nineteenth century, Somali nationalists could claim, the Ethiopian empire survives only at the behest of the great powers.

Each of these analogies is, of course, one-sided and inadequate. Surely the Ethiopian ruler today has neither the legitimacy nor the relative power of the Tudor or Bourbon kings who fashioned nations in sixteenth- and seventeenth-century Europe. Surely, too, the Ethiopian state has greater integrity than the Ottoman Empire of the nineteenth century. The Ottoman Empire was so politically defunct, you might say it "shed" nation-states rather than "lost" them. Perhaps two better examples for purposes of historical comparison with contemporary Ethiopia are the Russian empire, which held together as the Soviet Union, and the Austro-Hungarian dual monarchy,[17] which dissipated into a group of self-determining nation-states. In these three empires some dominant minority, embued with a tradition of cultural superiority, ruled over a vast number of minority peoples. Is there a historical lesson for the Ethiopian empire in the experience of those two nineteenth-century empires?

The seeds of the destruction of the Habsburg monarchy are of many varieties. The dual monarchy with Hungary, for one, led to policies which counteracted each other. But the core reason for the collapse was the forces of self-determination of the subject peoples. The Hungarians at first attempted to repress the minorities, with little success. The Austrians attempted to grant the minorities a number of concessions to give them local autonomy. These were badly administered. But more important, they just led the subject peoples to articulate ever-increasing demands. Amid the Habsburg failures in the First World War, and with the Allies suporting the idea of self-determination of all minorities, the empire, already in disarray, dissipated.

The Russian monarchy collapsed for different reasons although the World War, like the Habsburg monarchy, acted as the catalyst. In Russia, not national self-determination but socialism constituted the dynamic force for change. While the civil war which was unleashed by the socialist revolution nurtured national feelings among the minorities, ideas of national self-determination did not themselves propel the Russian revolution. And in the vicissitudes of the revolution, the Allies found themselves often on the side of the Whites (the aristocracy), who were in most cases antagonistic to the minorities. In Ethiopia, as in Austria-Hungary, the minorities helped challenge the imperial system; but as in Russia, a socialist revolution dominated the political stage, and the great powers found it politically necessary to downplay the legitimacy of national self-determination.

What are the consequences of these differences for the minority nations of these two empires? For Austria-Hungary, the larger nationalities, despite their problems with the politics of peace, got more or less what they had agitated for. The result: a plethora of small Eastern European and Slav states, nominally independent but surviving only by the grace of the great powers. In 1939, Germany invaded Czechoslovakia and Poland. In 1956, the Soviet Union invaded Hungary. And in 1968, the Soviets invaded Czechoslovakia. As this is written, Poland is adjusting itself to a possible Soviet invasion. To be sure, Yugoslavia and perhaps Rumania have had considerable autonomy over the past thirty years -- with a security which has enabled them to enjoy many of the fruits of national autonomy. But the victory of the nationalities in the wake of the collapse of the Habsburg monarchy has not been without considerable compromise in autonomy.

For the Soviet Union, the answer is equally complex. While Lenin supported the claims of self-determination of all peoples during his battle with the Whites, he promised but had no intention of permitting the subject peoples to consider leaving the union. The call for self-determination was for him a political strategy to mobilize subject peoples around the world against the colonial system sustained by international capitalism. He put the problem of the nationalities in the Russian monarchy into this worldwide context. Again and again through the Russian civil war, Lenin made bargains with national leaders only to subvert them when the Bolsheviks had sufficient power to exert political control. When Armenia was fighting the Turks in 1920, to give but one example, ministers of Armenia and Soviet Russia signed a treaty in which "Armenia [was] proclaimed an independent socialist republic." Some nineteen days later, a "Revolutionary Committee of the Soviet Socialist Republic of Armenia," a group hand-picked by Stalin, arrived at the Armenian capital and decreed that all laws of the Soviet Russian government would be in force in Armenia.[18] All the Armenians got for their long efforts for national self-determination was starvation and near extermination. Make bargains to co-opt national leaders; then use Soviet Russian power to exert total control. Such was the Leninist strategy.

But there is a second side to the Soviet coin. Once the nationalities were under centralized control, the Soviet government often made efforts to support national languages and cultures. In the Ukraine, they tried (unsuccessfully) to get Russian-speaking administrators and factory managers to speak the language of the region to which they were sent. They tried (more successfully) to recruit peoples from peripheral areas to participate in party affairs and to speak Russian fluently enough to take on positions of responsibility. Some national languages and cultures within the Soviet Union flourish, even though political and economic control is centralized, and even though the Soviets were not beyond creating chasms between similar languages in order to divide and rule.[19] Without fear of Soviet military invasion, and participating in the gains of the rapid industrialization of the former empire since the 1930s, are these nationalities less well off than those which received their "autonomous" states?

There are no historical laws which can guide us to predict the future course of the Ethiopian state. But the analogy with the Russian and Austro-Hungarian experiences suggests that the collapse of the Ethiopian empire is not imminent.[20] The conditions favoring the breakup of multinational states do not seem to be met in the case of Ethiopia. For one, unlike Austria-Hungary, the Amharic center has the capacity and desire for ruthless and purposeful action. There will be no ambiguity about the place of the Eritreans, the Tigres, the Oromos within the Ethiopian state as there surely was about the Serbs, Croatians, and Rumanians in the dual monarchy. And second, like the Russian empire, the great powers have not directly defended the right of self-determination of the minorities amid the Ethiopian revolution. The Soviet Union and the United States are wary of supporting the claims of the minority peoples within Ethiopia, and without massive international support these movements have much less chance for success. From this comparative perspective, then, it seems that the Ogaadeen Somalis could well continue to be subject peoples to an increasingly coherent Ethiopian state, periodically garnering sufficient resources to mount bothersome but ineffective guerrilla attacks.

National Redefinition

If the prospects are for a more coherent Ethiopian state, what can we expect in terms of changes in ethnic identification in the Horn? Must these Ogaadeenis who consider themselves to be Somalis always do so? What about the Oromos? For centuries they distinguished themselves from the Somalis, but today in the refugee camps, however difficult it is for them to learn the Somali language, they are considered (and perhaps have begun to consider themselves) Somalis. Is it possible that Somalis, Oromos, Amhars, and Eritreans might all consider themselves to be of the "Hamitic" nation, and seek to rewrite their histories to show their common origins? We know a good deal about identity change in the Third World from the works of

Crawford Young, Abner Cohen and Paul Brass.[21] "Tribe" or "ethnic group" is not the primordial "given" that Geertz sometimes suggests it is,[22] but is a membership category which individuals and groups manipulate to advance political, economic, and social interests. The question to be asked in this section is whether the ethnic identities in the Horn are immutable. Perhaps changes in identity might occur autonomously, or be induced, which might have the effect of making the problem seem less intractable.

At first blush, these ideas in the context of the Somali situation appear to be preposterous. Almost all political analysts of the Horn, whether Saadia Touval or I.M. Lewis (or myself), have emphasized the depth of Somali nationalism. It has been traced to the integrating effects of the jihad of Ahmad Gran into the Ethiopian highlands in the sixteenth century and to the great wars of Maxamad Cabdille Xasan against the British, Italians and Ethiopians in the early twentieth century. Somali nationalism is based not only on a common identification with these historical events, but a common view of descent, common political institutions, a common religion, and a common language. Finally, because of the colonial experience, Somalis together defined an ideology of Somali nationalism in order to overcome colonialism. This ideology is, from all accounts, indelibly printed in the Somali consciousness. If national consciousness is ever immutable, it could be argued, the Somali case is surely one example of this.

Nevertheless, identity management and change are common occurrences in the Horn, and those scholars who read only the secondary literature may be unaware of this.[23] I shall give three examples -- that of Somalis seeing themselves as Arabs; that of Oromos seeing themselves alternately as Oromos, as Somalis, and as Amharas; and that of Somalis seeing themselves primarily in terms of clan affiliation. These examples of national redefinition are only intended to explode the myth that national identities are petrified in the Horn.

Somalis as Arabs. Are the Somalis Arabs? The standard answer in the national mythology and conventional anthropology is a qualified yes.[24] It is held that Arab shaykhs landed on the Red Sea coast from the tenth through twelfth centuries, mixed with the local population, and fathered large "clan-families" which grew in strength and level of internal organization. These clan-families, armed with the technological and social advances brought by their Arab forefathers, migrated southward and eventually conquered the Horn. Oromos and Bantus were pushed out of their lands by this "Somali" onslaught. Followers of this perspective point out that most Somalis can trace their ancestry back to an Arab forefather, and ultimately to the family of the Prophet.

Ironically, this conventional viewpoint coincides with a popular antipathy toward Arabs among Somalis. Arabs are considered to be exploitative shopkeepers, incompetent pastoralists, and culturally inferior. Somali popular folklore is therefore replete with stories mocking Arabs. The Somalis, then, consider themselves to be of Arab "stock," but sharply distinguish themselves from Arabs in their national self-conception.

Recently, there has been some important scholarly work which has contested the conventional wisdom. Linguistic evidence seems to suggest that the origins of the Somali nation are in the northern area of Kenya, close to where the Rendille people live today.[25] Some 2,000 years ago, it is surmised, there was migration to the northeast, and vibrant Somali clans had peopled the northern coasts of today's Somalia by the year 1000. It is about this time that Arabs came to the Somali coast, but never in large numbers. They came for trade, to proselytize for the Islamic religion and to escape the wars endemic to the Islamic world in its early centuries.

While this revisionist account acknowledges that Arab civilization brought to Somalia its religion, its contact with the outer world through access to trade, its material culture, and many of its "vices" (the words in Somali for prostitute, gambling, cards, drugs and alcohol are all of Arab derivation), Ali Abdirahman Hersi, who has formulated this thesis, argues that "beyond the noted influences in these...general areas there is nothing else in Somali life and culture, now or in the past, which could be attributed to the Arabs."[26] Hersi's argument is that while Arabs have had considerable influence in the development of Somali culture, the Somali people are Africans with an autonomous culture and history.

Ambiguity is, of course, a vacuum for politics. And Somalis have been quick to understand the opportunities available to them in their ambiguous Arab heritage for manipulating their self-conceptions to serve political or economic needs. The most interesting example of this occurred in the early twentieth century in Kenya, where the Isxaaq Somalis refused to be classified as Africans, and demanded to be counted as "Arabs." The British put this down to Somali "audacity" and "intelligence," but could not quite cope with the fact that the Isxaaq insisted on paying the higher tax rates associated with their perceived status. It is now clear that these Isxaaq knew that to be considered "African" was to be frozen out from the opportunity to engage in international trade, and to receive fewer public goods, so they were redefining themselves culturally in order to compete with migrants from the Indian subcontinent for the control of markets and for access to education.[27]

Whatever the cause, the Somalis bombarded the British with telegrams of this nature: "If the reason for the administration to classify us as Africans or natives [is] because of our country which is situated in Africa, this will not, however, coincide with the fact that many parts of Africa, e.g., Egypt, Tunisia, Algeria, Libya, Morocco, etc., though in Africa yet the inhabitants of these countries are not known as Africans but are classified as Arabs, thus natives of Asia."[28] This redefinition may be explained by utilitarian motives; but one effect of lobbying for it is that the lobbyists themselves promote behavior consistent with this redefinition. The Somalis began to adjust their behavior in other realms (e.g., to engage in more religious activity) to make their actions congruent with their claims. In acting like Arabs, they became more like Arabs.

A second example is more recent. Although the Somali military regime claimed to be "scientifically socialist" and "pan-Africanist," it proudly joined the Arab League in 1974. Cumar Arteh, then the Somali foreign minister, waxed brilliant on the cultural identity of the Arab and Somali people. "Somalia is ready," he declared, "to play its role fully...in the service of the great Arab cause."[29] The reasons why the Somalis changed their position on the Arab League (the civilian regime spurned it) are complex. Siyaad's government was more militantly internationalist than the civilians, and the Arab League provided one platform on which to stand up against Zionism. Also, with the fantastic increases in wealth of Arab states in the wake of the initial OPEC successes, Somalis correctly perceived that joining the Arab League would bring Somalia considerable foreign aid. From 1975-78, for example, Somalia received $361.1 million in OPEC development aid, which constituted about 14 percent of the total Africa (south of the Sahara) outlay. Only Mauritania and Sudan, both of which also emphasize their Arab identity, received more over this period.[30]

Again, whatever the causes, the effects are worth separate analysis. Membership in the Arab League has meant for the Somalis far more access to jobs in the Arabian Gulf. Numerous Somali businessmen in Oman, Qatar and Abu Dhabi have found it useful to partially assimilate with their Arab hosts. In the Somali Democratic Republic, much Arabic reading material is available to Somalis, and the ministry of education has committed itself to a fuller curriculum in the Arabic language. While it would be foolhardy to suggest that any large segment of the Somali population considers itself Arab, it would not be foolhardy to claim that over the long term, if military and economic rewards come to Somalis through redefining themselves as Arabs, a change in Somali national consciousness could occur.

<u>The Oromos</u>. The national identity of the Oromo peoples who straddle the Ogaadeen desert is often seen as the key to the future of Africa's Horn. Their population may be anywhere from 10 to 15 million people, and they are without a doubt the largest nationality group in modern Ethiopia. But who are they, and where do they stand in terms of the great conflicts in the Horn?[31]

To a considerable extent, they are culturally similar to the Somalis. The Oromo languages (most of which are mutually intelligible to native speakers) and Somali are both considered to be of the "lowland" variety of East Cushitic languages in the wider Cushitic family.[32] Again, like the Somalis, the Oromos have a nonhierarchical social structure, many differentiated clans, and very limited centralization of political authority. Many Oromos have been Islamicized. And, throughout a long period of cultural contact, many Oromos assimilated with Somalis of the Sab clan-family. In the early 1960s there was still a question of whether the Sab clan-family among the Somalis would consider themselves as "Somalis," but they did. It would not take too much greater a stretch of imagination to think of the Oromos as a part of the Somali nation.

But there are significant differences as well. First, certain Oromo groups were retainers to Somali families, and see themselves now as liberated from Somali overrule. Second, many Oromos are Christians, and some have not converted to any world religion. This is why the Oromos are popularly referred to as "Gallas" (with an implication that they are heathen) by the Somalis. Third, the Oromos have a sophisticated and well-developed age grade system -- a social structure alien to the Somalis. Fourth, however closely related the

Oromo languages are to Somali structurally, the two languages are not mutually intelligible. Since language is a key to nationality in most of Africa, this difference has important political ramifications.

Politically rather than culturally the Oromos found themselves tied to the Amharic nation to a greater extent that to the Somalis. In the seventeenth century, the Oromos, in the wake of the jihad of Ahmad Gran into the Ethiopian highlands, invaded and conquered much of what is today northern Ethiopia. The increasing strength of the Somalis to the east of the Oromos made western expansionism more profitable. Many Oromos became agriculturalists, and partially assimilated with the Amharas. The Amharas speak a Semitic language (of the same Afro-Asiatic superfamily as Cushitic, but a very distant relation), have a centralized and hierarchical social structure, and are mostly Christian, but they became part of a wide interrelated cultural and economic zone with the Oromos and other national groups. As Donald Levine concludes from his (controversial) book, Greater Ethiopia, "The differentiated peoples of Greater Ethiopia [including the Oromos] in time established various lines of communication and interdependence through trade, migration, and other processes. Through continued interaction and the coalescence of cultural traits, Greater Ethiopia became to some extent a cultural as well as an ecological community."[33]

With the escalation of the hostilities in the Ogaadeen, it has become clear that (should a common position develop among them) the Oromos have three plausible choices. First, they could emphasize their national distinctiveness. This is the path of the Oromo Liberation Front, whose leaders equate feudal landowners with foreigners (captured by the word naftaanya).[34] To the extent that hostilities increase and social mobilization of Oromo peoples continues, it is predictable that more Oromos will develop Oromo national consciousness.

But a Somali nationalism is not impossible. In the Kenyan Northern Frontier District, many Oromos, when questioned by a British team, claimed that they wished to put their fate with the Somali state.[35] In the 1960s, an Oromo leader, Wako Guto, developed links with the WSLF, and his forces were successful in 1977 in winning over territory from the Ethiopians in Bale province. To demonstrate the close affinity between Somali and Oromo freedom fighters, a Somali Abo Liberation Front (an offshoot of the WSLF) was formed. (Abo is the Oromo word to attract someone's attention.) It must be acknowledged that many Oromo activists have bridled at the idea of union with Somalia[36] and that the multitudes of Oromo refugees who now live in camps in Somalia feel like foreigners. But the Somali authorities have spared nothing to make the Oromos feel like brothers, and have cultivated the idea that the Oromos and Somalis are part of the same nation.

Finally, the Oromos can become "Ethiopians." While I believe that the Levine thesis is politically naive, there is more than a grain of truth to the idea that Oromos and Amharas have been interdependent. Amhara society has been open to cultural assimilation, and many Oromos have "become" Amhara. It is sometimes suggested that the late Emperor Haile Selassie was part Oromo. But more important, in Ethiopian cities, many Oromos have adopted Amharic cultural traits and have therefore become socially mobile within Ethiopian society. In a careful study of language identification, M.L. Bender and associates found, due no doubt to economic opportunities open to Amharas, "that substantial members of urban Gallas [i.e., Oromos] have adopted Amharic as mother tongue."[37] Given that the Oromos constitute a plurality among national groups in Ethiopia, it is to their political advantage to preserve the integrity of the Ethiopian state and to encourage its democratization. To capture the Ethiopian state would be a great achievement for the Oromo nation!

To be sure, it is entirely possible that the Oromos will never develop a common response to the problems and opportunities in political realignments in the Horn. But creative leaders can nonetheless reinterpret Oromo history, culture, and politics to accord with any of three sorts of identification.

The Somalis as Clan Nations. The emergence of a Somali nation out of a congeries of clan groups was not a historical necessity. The binding powers of Islam, the creative reaction of Maxamad Cabdille Xasan to the colonial threat at the turn of the century when he wove together a variety of clan groups with an ideology of a common Somali-ness, and the brilliant efforts of a few generations of poets who provided a cultural storehouse for the national language all worked toward the cogency of the Somali national idea.

But clan fission has been as normal as clan fusion in the Somali political experience. The unity of the (primarily nomadic) Samaale clan-family with the (primarily agriculturalist) Sab clan-family remains precarious. Here language differences are more than dialectical. The unity of (the former British) Somaliland in the North with (the former Italian) Somalia in the South is also tense. In an extraordinary political event, these two colonial territories merged in 1960, based on their sense of common nationality. But with the perceived predominance of Somalia, the Northern region voted against the independence constitution. Its capital, Hargeysa, lost out to Muqdisho as the national capital, and has become a political backwater. Resentment in the North remains.

Clan identifications based on ancestral identification -- Daarood, Isxaaq, and Hawiya -- despite the considerable efforts by political authorities to eliminate them as sources of political identification, have, if anything, strengthened in the past few years. And within the Daarood family, fissures beteen the Majeerteen, the Marreexaan and the Ogaadeen are widening. While there are hardly any cultural differences discernible here, each of these groups, in developing its position within Somali politics, points to characteristics unique to it.

Since Somali independence in 1960, almost no Somali has questioned the reality of the Somali nation. Clan differences were used in order to build coalitions to capture the Somali state; they were not used to develop a national ideology for purposes of withdrawing from the Somali state. But I am no longer sure that this pattern will continue to hold. Today in Somalia, the Majeerteen people have been mobilizing against the present regime without building a transclan coalition. Some Majeerteen are beginning to question the necessity of including the Ogaadeen clan in the Somali nation. Some are pointing to the fact that there is evidence that a Majeerteen nation existed before Islamic times and that it had developed its own (more centralized) political institutions distinct from the Somali.

The key clan is, of course, the Ogaadeen. While they are not the only Somali clan inside Ethiopia (the Isxaaq and Dir are both represented), they are the largest. Furthermore, many Ogaadeenis live in the Somali Democratic Republic. Part of the thrust of the WSLF Ogaadeenis is that the Ogaadeen clan is split by an international boundary. If united, the Ogaadeen people most probably would not live in a multiethnic Ethiopia and consider themselves "Ethiopians." But if Somali politics became increasingly fissiparous, such an outcome would not be impossible.

In this general discussion of identity management, I do not wish to suggest that nationality is not crucial in understanding the political alignments of the Horn; rather I wish to suggest that the current definitions of cultural identification are not, as is sometimes supposed, immutable.

As the guerrilla war continues in the Ogaadeen, as the Ethiopian revolution routinizes, and as the Somali state is forced to adjust to the realities of a refugee population constituting about 40 percent of its inhabitants, different groups in the Horn will continue the process of subtle redefinition of their identity. How will the Oromo groups define themselves in the next ten years? Will the Somalis, in their search for aid, find more and more reason to see themselves as Arabs? Will the pressures of the refugee situation create fissures within Somali society to erode its national solidarity? The sociology and politics of national redefinition will continue; and those who seek peace must be on the constant lookout for changing opportunities.

Conclusion

I share the commitment of the editors of this volume to the promotion of bargaining strategies among the parties to conflicts between states and ethnic minorities in Africa. Within the Ethiopian state, it is reasonable to ask the political authorities to give greater protection to the Somalis and to redistribute some resources to the Ogaadeen, so the Somali population there can develop some hope for a better life. Along the border areas, it should not be impossible to get boundary concessions by the Ethiopians so that the Somali state will be able to have some control over the nomadic cycle. For concessions of this sort, it is not beyond reason for the leaders of the WSLF to give up their demands for complete dissociation from the Ethiopian state, or for the Somali government to give up hope for extensive territorial readjustments.

But because there may be a package of rational suboptimal agreements for conflict regulation in the Ogaadeen does not imply, as the editors make clear, that there is an efficient means toward that end. As we have seen, bilateral negotiations between the Somali and Ethiopian states are too politically charged even to make public. The Ethiopian state will not negotiate with (and thereby give legitimacy to) the WSLF. Third parties have either been too biased, too weak, or too inept to be effective brokers. Reaching a conflict-reducing solution through negotiation has not yet, in this case at least, been politically feasible.

What this chapter suggests is that it might be useful to complement a theory of bargaining between the state and nationality groups in Africa with a theory which comprehends changes in the coherence of the state and in the definition of the nation, which are themselves products of the conflict.[38] If the coherence of the Ethiopian state grows, it should be less willing but better able to bargain with representatives of national minorities in the Ogaadeen. If the Oromo people develop a strong national consciousness, they could pose a threat to the coherence of the Ethiopian state far beyond what the Somalis (or even the Eritreans) have been able to muster. Alternatively the Oromos could develop a strategy to capture rather than break away from the Ethiopian state. If Somali national unity begins to dissolve, or if Somalis increasingly see themselves as part of the Arab world, their support for the WSLF could well dwindle. This chapter has provided little grasp on when would be the most propitious time for a successful bargaining strategy, but it has suggested that an understanding of the proper timing demands a better theory of the dialectics of state coherence and national identity.

Footnotes

1. For an overview, see John Drysdale, The Somali Dispute (London: Pall Mall Press, 1964); and Somali Republic, The Somali Peninsula: A New Light on Imperial Motives (Muqdisho: Somali Government Information Services, 1962). The best source on this period is A.M. Brockett, "The British Somaliland Protectorate to 1905" (doctoral dissertation, Oxford, 1969).
2. For this section, I am in debt to Olusola Ojo. His dissertation, written at the University of London, led to my rethinking the OAU role in peaceful settlement disputes. Together we collaborated on a paper, "The OAU and the Ogaadeen Question: Towards a Solution," which will be published in the Proceedings of the First Congress of the Somali Studies International Association, which was held in Muqdisho in July 1980. In the present paper, the comparison of the Russian, Austro-Hungarian and Ethiopian empires and the discussion of the OAU role are revised statements from that collaborative effort.
3. I.M. Lewis, A Modern History of Somalia, revised edition, (London: Longman, 1980), p. 197.
4. African Research Bulletin (Political, Social and Cultural Series), 7, 2 (February 1964): 146. Hereafter ARB.
5. ARB, 10, 3 (May 1973): 2845.
6. ARB, 11, 6 (June 1974): 3256.
7. In August, Ethiopia called for an emergency session of the OAU Council of Ministers to consider "Somalia's invasion" of the Ogaadeen, and later appealed to the UN to stop the fighting in the Horn. See ARB, 14, 8 (August 1977): 4525, 4529.

8. ARB, 15, 7 (July 1978): 4913.
9. O. Mehmet, "Effectiveness of Foreign Aid: The Case of Somalia," Journal of Modern African Studies, 9 (May 1971): 31-47.
10. R. Thurston, "Detente in the Horn," Africa Report, 14, 2 (February 1969): 6-13.
11. S. David, "Realignment in the Horn: The Soviet Advantage," International Security, 4, 2 (Fall 1979): 73.
12. Lewis, Modern History, p. 233.
13. Le Monde, June 12, 1981.
14. This phrase is from I.M. Lewis, Modern History, p. 250, in one of his rare moments of unrestrained Somaliphilia.
15. On the historical pattern of states creating nations, see E.B. Haas, Beyond the Nation-State (Stanford: Stanford University Press, 1964), pp. 464-475.
16. A good historical overview of the Ottoman Empire is in Marshall G.S. Hodgson, The Venture of Islam, vol. 3, The Gunpowder Empires and Modern Times (Chicago: University of Chicago Press, 1974).
17. The standard work on the fall of the Austro-Hungarian Empire is O. Jaszi, The Dissolution of the Habsburg Monarchy (Chicago: University of Chicago Press, 1929). For the national issue in the early period of Soviet rule, see Richard Pipes, The Formation of the Soviet Union (Cambridge, Mass.: Harvard University Press, 1957). On allied intervention into Soviet Russia, I relied on John Bradley, Allied Intervention in Russia (London: Weidenfeld and Nicolson, 1968).
18. Pipes, Formation of the Soviet Union, pp. 232-233.
19. The development of the Uzbek language by the Soviet state is analyzed by one authority as a strategy "to circumvent the development of pan-Turanism." See P. Rubel, "Ethnic Identity among the Soviet Nationalities," in E. Allworth, ed., Soviet Nationality Problems (New York: Columbia University Press, 1971), p. 234.
20. On the tenacity of the postindependence African states, see Crawford Young's contribution to this volume; and R. Jackson and C. Rosberg, "Why Africa's States Persist: The Emperical vs. the Juridical in Statehood." Paper presented to the 1981 American Political Science Association Annual Meeting. Although this paper does not address the Eritrean issue, it might be noted that while the Eritreans have less of a claim by the criteria of cultural unity than do the Somalis for self-determination, they have more of a claim by virtue of colonial boundaries. As the reality of these boundaries increases in salience, the prospects for Somali autonomy, but not necessarily Eritrean autonomy, dim.
21. Crawford Young, Politics in the Congo (Princeton: Princeton University Press, 1965), especially his discussion of the "Bangala," pp. 242-246; Abner Cohen, Custom and Politics in Urban Africa (Berkeley and Los Angeles: University of California Press, 1969); Paul Brass, Language, Religion and Politics in North India (Cambridge: Cambridge University Press, 1974).
22. C. Geertz, "The Integrative Revolution," in C. Geertz, ed., Old Societies and New States (New York: Free Press, 1963), pp. 105-157. Geertz qualifies his statement properly so that he does not claim that primordial attachments are immutable; but he

does suggest an intractability about them when he calls them "givens."

23. To be sure, there are hints of this in the secondary literature. I.M. Lewis, the doyen of Somali sociology, has written of the "long legacy of opportunist manipulation of identity and citizenship" among the Somalis. Modern History, p. 244. But these references are the exception.
24. See Ministry of Information and National Guidance, "Somalia and the Arab League: A Wider Role in Afro-Arab Affairs" (Muqdisho: June 1974), for a Somali statement of cultural similarity. For the standard anthropological view, see I.M. Lewis, "The Somali Conquest of the Horn of Africa," Journal of African History, 1, 2 (1960): 213-230.
25. Bernd Heine, "Linguistic Evidence on the Early History of the Somali People," in Hussein M. Adam, ed., Somalia and the World (Muqdisho, 1979).
26. Ali Abdirahman Hersi, "The Arab Factor in Somali History: The Origins and the Development of Arab Enterprise and Cultural Influences in the Somali Peninsula" (doctoral dissertation, Department of History, UCLA, 1977), p. 307. On the vices, see pp. 41-42.
27. E.R. Turton, "Somali Resistance to Colonial Rule and the Development of Somali Political Activity in Kenya 1893-1960," Journal of African History, 13, 1 (1972): 119-143.
28. Hersi, "The Arab Factor in Somali History," p. 285, quoting from A.B. Chilivumbo, "Tanganyika's Mono-Party Regime" (doctoral dissertation, Department of Sociology, UCLA, 1968), p. 42.
29. Ministry of Information and National Guidance, "Our Foreign Relations -- A Review of Our Revolutionary Policies" (Muqdisho: June 1974), p. 46.
30. See Organization for Economic Co-operation and Development, Development Co-operation, 1979 Review (Paris: OECD, 1979), Table G.9, "Total official flow of resources to Individual Developing Countries and Territories from OPEC Countries and Arab/OPEC Multilateral Institutions 1975 to 1978," pp. 278-279.
31. On the Oromo role in Ethiopia, see Donald Levine, Greater Ethiopia (Chicago: University of Chicago Press, 1974); and P.T.W. Baxter, "Ethiopia's Unacknowledged Problem: The Oromo," African Affairs, 77 (July 1978): 283-296. Population figures are from Baxter, p. 287.
32. M.L. Bender et al., Language in Ethiopia (London: Oxford University Press, 1976), p. 14.
33. Levine, Greater Ethiopia, p. 184.
34. Baxter, "Ethiopia's Unacknowledged Problem," p. 287. See also the journal STORM (Somali, Tigray, and Oromo Resistance Monitor) (London), which emphasizes the distinctiveness of each national movement inside Ethiopia.
35. Lewis, Modern History of Somalia, p. 184.
36. This was the message a number of Oromo activists gave to Somali President Siyaad in the course of their participation in the International Congress of Somali Studies held in Muqdisho, July 1980. The literature demonstrating the reality of the Oromo nation is beginning to grow. See Richard Greenfield and Mohammed Hasan, "Interpretation of Oromo Nationality," Horn of

Africa, 3, 3 (1981), and the supporting interviews and documents in that issue of the journal.

37. Robert Cooper and Ronald Horvath, "Language, Migration, and Urbanization," in Bender, et al., _Language in Ethiopia_, p. 194.

38. It seems to me crucial for an understanding of bargaining strategies, for example, that the coherence of the Nigerian state was strengthened in the face of the Biafran secession and that the coherence of the UDI regime in "Rhodesia" was weakened in the face of a guerrilla war. Or, on the other side of the equation, it seems crucial to an understanding of the future of the Eritrean struggle vis-à-vis Ethiopia whether nationality differences are becoming more or less socially relevant among the Eritreans.

Index

Acheampong, I.K., 187, 188
Amin, Idi, 128, 267
Angola, 59, 117-118

Bargaining models, 15, 18, 61-63, 130, 176, 233-234, 299, 345, 346
Barre, Maxamad Siyaad, 333, 335
Biafra. See Nigerian Civil War
Botha, P.W., 96

Cabral, Amilcar, 120-122
Chad, ethnicity in, 9, 49, 56-57, 59, 63; civil war, 29, 34
Colonialism, 5, 85, 106-107, 157; British, 9, 27-29, 31-33, 35, 45, 310-311; French, 9, 32-34, 37, 39-41, 45, 59; Portuguese, 40; internal, 92-93, 148, 308
Consociation, 138, 140-141, 149
Co-optation, 134, 135, 149, 150, 151, 229, 256, 259, 260, 273

Detribalization, 5, 72

Eritrea, 211-212, 215-219, 221, 223, 224, 225-226. See also Ethiopia
Ethiopia, Eritrea and Ogaadeen conflicts, 74-75, 108, 109, 275; Marxism in, 117; single party in, 262; religious and language policies, 262; land reform, 263; revolution in, 334, 335; empire compared to European empires, 337-338
Ethnic associations, 7-8, 80, 91, 158-159
Ethnic conflict, 50, 58, 85, 237, 272, 274, 275; bargaining in, 61-63; related to resource distribution, 79, 90, 160, 164; and modernization, 92-93, 94; in bureaucracies, 110; state responses to, 237-238. See also Ethnic self-awareness; Ethnicity; Nigeria; Sudan
Ethnic quotas, 100, 107, 259, 298
Ethnic self-awareness, 14, 16, 100, 101; creation of, 29, 106-107, 157, 161, 345; in South Africa, 55; in Zaire, 56; changes in, 104-105, 339-340; in Angola, 117-118; in Nigeria, 310, 311; in Sudan, 325; among Somalis, 339-340. See also Ethnic conflict; Ethnicity
Ethnicity, 50, 52, 53, 58-60, 85, 87, 94, 102, 153, 189-192, 281, 306, 325; state responses to, 10, 60-63, 78, 111-112, 114-119, 121-122, 165-166, 228, 237-238, 259, 309, 337; associated with urban elites, 55-56, 58, 72-73, 160-161, 162-163,

165, 306; and class, 57, 103-105, 119; coherence of state promoted by, 78, 309; in Europe, 86-91, 88; in armed forces, 107-108, 110. See also Ethnic conflict; Ethnic self-awareness

Ghana, school attendance in, 182; development, by region, 182; agriculture in, 184-185, 190-191; Northern region demands, 185-188, 190-191; 1972 coup, 187
Gowon, Yakubu, 109, 207, 285, 312
Guinea-Bissau, colonial, 54; as discussed by Amilcar Cabral, 120

Houphouet-Boigny, Félix, 32, 260

Internal colonialism. See Colonialism
Ivory Coast, recognition of Biafra, 209; "Dialogues," 272

Katanga, secession of, 201-204, 222, 224-225
Kaunda, Kenneth, 261, 271
Kenya, Somalis in, 12-13; war of liberation, 29; British-Kenyan relations, 34; peasantry in, 69; ethnicity in, 106, 178, 179, 259; Kenyatta regime, 116; medical services, 178; school enrollments in, 178; Harambee, 269, 271
Kenyatta, Jomo, 116, 259, 271

Liberia, 62, 172
Lumumba, Patrice, 201, 202, 203, 222

Mali, 30
Mauritania, 54
Mengistu Haile Mariam, 108, 109, 244, 334, 335
Miscegenation, 37-38
Mobutu, Sese Seko, 128, 261
Modernization, 55, 76-77, 80-82, 138, 152, 153, 156-157, 164
Moi, Daniel arap, 34, 61, 261
Mozambique, 118-119, 132
Mugabe, Robert, 259

Negritude, 30, 102, 271
Nigeria, federalism in, 4-5, 18, 62-63, 64, 80, 111-112, 165-166, 228, 238, 282-283, 285-292, 294, 298, 299-300; oil revenue, 63, 78-79, 263, 294; ethnic tensions in, 93-94, 155, 204-205, 205-206, 285, 288, 295-296, 298, 310, 311, 314; revenue allocation, 176, 288, 294, 297; First Republic, 204-205; 1964 elections, 206; January 1966 coup, 206-207, 285, 312; July 1966 coup, 207; Richards constitution, 283; colonial era, 283-284; demands for additional states, 287-288, 298, 300; federal civil service, 293-294; political parties, 295-296; university education in, 296. See also Nigerian Civil War
Nigerian Civil War, 28, 109, 204-211, 225; international reaction to, 208-210, 221, 222-223; compared to Sudan, 309-313; related to British colonial policies, 310-311
Nimeiry, Jaafar el, 75, 245, 317, 321, 322, 323, 324, 325, 326, 327
Nkomo, Joshua, 259
Nkrumah, Kwame, 186
Nyerere, Julius, 114, 270

Obote, A. Milton, 111, 266, 267
Ojukwu, Odumegwu, 207, 225, 312
Ogaadeen, Somalis in, 11, 345; history of, 332; Great Power involvement in, 332, 335-336; aid from Somalia to, 334; Oromo Liberation Front, 343. See also Somalia; Ethiopia

Organization of African Unity, 90, 199, 227, 228, 252, 332, 333, 334, 336

Peasant economies, 68, 69, 70-71, 75, 76, 77. See also Modernization
Policies, distributive, 255, 259-260, 273; regulatory, 256, 260-263, 274; redistributive, 256-257, 263-265, 273; reorganizational, 257, 266-268, 274; symbolic, 257-258, 268-272, 273-274
Power sharing, 175, 234, 236

Qaddafi, Muammar, 60, 129, 217

Rawlings, Jerry, 37
Resources, as an ethnic issue, 13, 14, 15, 18; scarcity of, 16, 176, 257, 275; redundancy of, 253, 273
Rural-urban relations, 2, 8, 56, 57, 71, 72, 159-160, 161-162
Rwanda, Hutu-Tutsi relations, 27, 50, 53-54, 222

Sadat, Anwar, 128
Secessionism, 9, 10-11, 91, 94, 173, 175-176, 199, 244, 253-254. See also Nigerian Civil War
Selassie, Haile, 212, 234, 343
Self-determination, 11-13, 200-201, 228, 304-305, 337-339; and existing states, 227, 251-252, 324, 332
Senegal, 113
Senghor, Léopold, 32, 271
Shagari, Shehu, 62, 112, 291-292
Somalia, socialism in, 116; joining Arab League, 341-342; creation of, 344. See also Somali nation; Somali question
Somali nation, 331, 340, 343, 344, 345, 346; origins of, 341, 344; Oromo ethnicity in, 12-13, 339-342, 344-345; clans in, 344-345
Somali question, Ogaadeen in, 11, 345; OAU in, 332, 333, 334, 336; Bevin plan, 332; negotiations between Ethiopia and Somali on, 336, 345. See also Ethiopia; Ogaadeen; Somalia
South Africa, "separate development," 28, 29, 36; Coloureds in, 37, 135, 136, 150, 151; apartheid, 39, 40, 95-96, 129, 175; economic development of, 51, 131-132; ethnicity in, 55; Nationalist Party, 93, 133, 148, 149, 151; prognoses for, 96, 132, 136, 142-144, 149, 151; African National Congress, 96, 143, 245; Afrikaner nationalism, 129, 133, 135, 148; nature of state in, 130, 137-138, 148; Western influence on, 131; Mozambique policy toward, 132; recent "deracialization," 133, 135, 140, 150; Ciskei independence, 134; co-optation in, 134, 135, 149, 150, 151; April 1981 election, 135-137; Black Consciousness Movement, 138, 139, 143; Inkatha, 138, 143
State, theories of, 2-5, 128-131, 253, 254-255, 272; survival of, 6, 252-253, 272, 309; "soft state," 6-9, 10, 20, 73-74, 75, 76, 81-82, 192; malleability of, 75, 346; legitimacy of, 137, 138, 139, 253, 254, 315-316, 326
Suboptimization, 15, 239, 242
Sudan, Addis Ababa Agreement, 18-19, 75, 237-238, 305, 309, 316, 317, 318, 319, 322, 325; redivision of South, 19, 322-324; civil war, 29, 75, 267, 308-309; oil in, 75; Anya-Nya, 218, 309, 318; South's role in independence talks, 307, 308, 315; North-South relations, 307-309, 313-315, 318-319, 320, 325; cultural

policies, 308, 309, 313, 319-320; Southern regional government, 309, 317-318, 320; legitimacy of state, 316, 326; 1969 coup, 316-317; decentralization in North, 320-321; Sudan Socialist Union, 322, 323, 324, 326; dissolution of assemblies, 323-324; patronage in, 326

Tanzania, peasantry in, 69; ethnicity in, 114; recognition of Biafra, 208-209; as one-party state, 261; Ujamaa Socialism, 269, 270-271
Tombalbaye, François, 63

Uganda, "soft state" in, 6-7, 20; 1966-67 coup, 28; Buganda, 106, 266-267; electoral procedures, 111; Railway African Union, 160; one-party state, 267
United Nations, 203, 204, 213-214, 225, 227

Westminster model, 45, 127, 281, 286

Zaire, 56, 102, 201-204
Zambia, precolonial era, 49; roads, 180-181; schools, 181; medical care, 181; recognition of Biafra, 209; political parties in, 261
Zimbabwe, war of liberation, 29; policy toward South Africa, 132; Mugabe's cabinet, 259; land policies in, 265

Contributors

Heribert Adam, professor of sociology at Simon Fraser University in Vancouver, Canada, was educated in Germany where he worked at the Frankfurt School of Critical Theory until 1965. His main research interests are comparative ethnic conflicts with a focus on the politics of Southern Africa and migrant labor in Western Europe. His books include *Modernizing Racial Domination* and, with H. Giliomee, *Ethnic Power Mobilized: Can South Africa Change?*

Robert H. Bates is a professor of political science at the California Institute of Technology. He currently is working on agricultural policy in Africa. His most recent book is *Essays on the Political Economy of Rural Africa*, to be published in 1983 by Cambridge University Press.

Henry Bienen is William Steward Tod Professor of Politics and International Affairs at Princeton University. He has been director of Princeton's Research Program in Development Studies.

J. Isawa Elaigwu is a senior lecturer in the Department of Political Science, University of Jos, Nigeria. His main areas of interest are state and nation building, federalism, civil-military relations, local governments, and higher education. His research experience includes East Africa, Nigeria and the United States and he has contributed articles to books and to international and national academic journals

Goran Hyden is currently representative of the Ford Foundation for Eastern and Southern Africa. He is a former professor of political science at the University of Dar es Salaam, and has also taught at universities in Kenya, Sweden, Uganda and the United States. He is the author of several books, most recently *Beyond Ujamaa in Tanzania*.

Edmond J. Keller is an associate professor in the Department of Political Science at Indiana University. He has written numerous articles on development policy in East Africa, and is currently completing a book entitled *Revolutionary Ethiopia*.

David Laitin is an associate professor of political science at the University of California, San Diego. He is the author of *Politics, Language and Thought: The Somali Experience* and is presently completing a manuscript entitled "Hegemony and Culture: The Politics of Religious Change Among the Yoruba."

René Lemarchand is a professor of political science at the University of Florida at Gainesville. His works include *Political Awakening in the Former Belgian Congo*, *Rwanda and Burundi*, and *Selective Genocide in Burundi*. He is coauthor and editor of *African Kingships in Perspective*, *American Policies in Southern Africa: The Stakes and the Stance*, and coeditor, with S.N. Eisenstadt, of *Political Clientelism, Patronage and Development*.

Ali A. Mazrui is a research professor at the University of Jos in Nigeria and a professor of political science and of Afroamerican and African Studies at the University of Michigan, Ann Arbor. His books include *Towards a Pax Africana*, *Violence and Thought*, *Cultural Engineering and Nation-Building*, *Soldiers and Kinsmen in Uganda*, *The Political Sociology of the English Language*, *A World Federation of Cultures: An African Perspective*, *Africa's International Relations*, and *Political Values and the Educated Class in Africa*.

Victor A. Olorunsola is the chairman of the Department of Political Science at Iowa State University. Among his previous publications are *The Politics of Cultural Sub-Nationalism in Africa*, *Soldiers and Power*, and *Societal Reconstruction in Two African States*.

Donald Rothchild is a professor of political science at the University of California, Davis. He has lectured at universities in Uganda, Kenya, Zambia and Ghana. His books include *Racial Bargaining in Independent Kenya* (with Robert L. Curry, Jr.), *Scarcity, Choice, and Public Policy in Middle Africa*, and (with Kenneth A. Oye and Robert J. Lieber) *Eagle Entangled* and *Eagle Defiant*.

John Stone is a reader in the Department of Social Science and Administration and the director of Graduate Studies at the University of London, Goldsmiths' College. Previously he taught social theory and race relations at St. Antony's College, Oxford, and at Columbia University. He is the chief editor of *Ethnic and Racial Studies* and author of *Colonist or Uitlander?*, *Race, Ethnicity and Social Change*, and *Tocqueville on Democracy, Revolution and Society*.

Pierre L. van den Berghe was born in Zaire of a French mother and a Belgian father. He is now a professor of sociology and anthropology at the University of Washington. His publications include a dozen books and many articles dealing mostly with social stratification and race and ethnic relations in Africa and Latin America, and, more recently, sociobiology.

Dunstan M. Wai was born in Kajo-Kaji, Southern Region, Sudan, and was educated in Oxford and Harvard universities. He has been a visiting research fellow in the International Relations Division of the Rockefeller Foundation, a fellow of the Woodrow Wilson International Center for Scholars, and a consultant for the United Nations. He is author of *The African-Arab Conflict in the Sudan*, and coauthor and editor of *The Southern Sudan and the Problem of National Integration* and *Interdependence in a World of Unequals: African-Arab-OECD Econmmic Cooperation for Development* (Westview).

Crawford Young is a professor of political science at the University of Wisconsin, Madison, where he has been associate dean of the graduate school, chairman of the department, and chairman of the African Studies Program. He has also been dean of the Faculty of Social Science, National University of Zaire (Lumumbashi); and visiting professor, Makerere University, Uganda. He is author of Ideology and Development in Africa, The Politics of Cultural Pluralism (winner of the Herskovits Award), Politics in the Congo, and coauthor of Cooperatives and Development and Issues of Political Development.

7